高职高专机电类规划教材

机械工业出版社精品教材

机电专业英语

(第3版)

English Course for Mechanical & Electrical Engineering

(Third Edition)

主　编　徐存善　周志宇

副主编　王　玉　尹丽芳

参　编　杨　阳　鲁宵映　杜　帅

机械工业出版社

本书共5章，包括机械与模具制造、计算机数字控制（CNC）、电子信息技术、应用技术和职场交际技能训练。前4章共24单元，每单元包括课文、生词（专业术语）、长难句解析、课后习题、翻译技巧（实用英语）、阅读材料等内容。在电子信息技术一章中，纳入了互联网、4G网络技术与物联网等内容；在应用技术一章中，新增了工业4.0、增材制造、大国工匠与世界技能大赛等内容。其中，在第4章各单元的实用英语部分，详细介绍了机电产品英文说明书，英文招聘广告，英文个人简历、求职信等应用文体的阅读和写作，以及英文面试过程中的常用技巧；第5章则汇编了8篇实际职场交际对话内容，实用性强。附录提供了各单元课文及部分阅读材料的参考译文、习题答案和生词表。

本书可作为高职高专院校机械制造与自动化、模具设计与制造、机电一体化技术、电子信息工程技术、数控技术和电气自动化技术等专业的英语教材，也适合继续教育学院、应用型本科院校机电类专业学生使用，并可供专业技术人员学习参考。

为方便教师教学，本书配有电子教案和电子课件，凡使用本书作为教材的教师均可登录机械工业出版社教育服务网（http：//www.cmpedu.com），注册后免费下载，咨询电话：010－88379375。

图书在版编目（CIP）数据

机电专业英语/徐存善，周志宇主编．—3版．—北京：机械工业出版社，2018.12（2023.6重印）
高职高专机电类规划教材　机械工业出版社精品教材

ISBN 978-7-111-61146-2

Ⅰ．①机…　Ⅱ．①徐…②周…　Ⅲ．①机电工程－英语－高等职业教育－教材　Ⅳ．①TH

中国版本图书馆CIP数据核字（2018）第236837号

机械工业出版社（北京市百万庄大街22号　邮政编码100037）
策划编辑：王　丹　责任编辑：陈　宾　王　丹　刘良超
责任校对：王　延　封面设计：张　静
责任印制：邓　博
北京盛通商印快线网络科技有限公司印刷
2023年6月第3版第7次印刷
184mm×260mm·17.25印张·417千字
标准书号：ISBN 978-7-111-61146-2
定价：49.00元

电话服务	网络服务
客服电话：010-88361066	机　工　官　网：www.cmpbook.com
010-88379833	机　工　官　博：weibo.com/cmp1952
010-68326294	金　书　网：www.golden-book.com
封底无防伪标均为盗版	机工教育服务网：www.cmpedu.com

前言

随着科技进步和社会的发展，我国对专业人才英语能力的要求越来越高。理工科学生除了应具备一定的听说能力以外，还应掌握一定的专业英文词汇，具有阅读本专业英文技术资料和进行专业交流的能力。机电类专业是当今发展最迅速、技术更新最活跃的领域之一，我国在该领域注重引进世界先进技术和设备，同时要发展和创造更多的外向型经济，因此该领域对具有专业英语能力的人才需求比以往任何时候都更加迫切。为了更好地培养学生的专业英语能力，培养更多具有国际竞争力的人才，编者根据高职高专机电类专业最新教学大纲，在积累多年教学经验的基础上，对《机电专业英语》第2版进行了修订。力求按专业培养的宽口径，使专业英语教材具有良好的通用性，并遵照高等职业教育的应用性特征，使机电专业英语具有较强的实用性和针对性，在内容上则力求通俗易懂、简明扼要、便于教学和自学。

全书共5章，分别为机械与模具制造、计算机数字控制（CNC）、电子信息技术、应用技术和职场交际技能训练。前4章共24单元，每单元包括课文、生词（专业术语）、长难句解析、课后习题、翻译技巧（实用英语）、阅读材料等内容。值得一提的是，在第2单元和第16单元选入了两篇创新名人传记的阅读材料，目的是为了激励和培养学生的创新能力。在第17单元中增添了“零件数据库网站介绍与英文网站注册”。同时，在电子信息技术一章中，纳入了互联网、4G网络技术与物联网等内容。在应用技术一章中，新增了工业4.0、增材制造、大国工匠与世界技能大赛等内容。在第4章各单元的实用英语部分，详细介绍了怎样阅读英文招聘广告，怎样写英文简历、英文求职信等应用文体，以及英文面试过程中的常用技巧，目的是使学生应聘外资或合资企业时有所参考。第5章汇编了8篇实际职场交际对话的内容，目的是培养学生在职业场景下用英语进行技术交流的能力。附录提供了各单元课文及部分阅读材料的参考译文、习题答案（为了培养学生的独立阅读能力，部分阅读材料的参考译文与答案仅在电子教案中给出）和生词表。

在经济、文化全球化大发展的现实形势下，学好专业英语显得尤为重要，拥有熟练的专业技术与优良的专业英语水平无疑就是高技能紧缺人才。为此，编者在编写过程中力求最大程度地减轻学生的学习困难，配备生词表、课后注释、翻译技巧等模块，方便学生自学。授课教师应以饱满的热情激励学生努力争取优异的专业英语成绩，从而增强学生在就业竞争中的优势。

本书可作为高职高专院校机械制造与自动化、模具设计与制造、机电一体化技术、电子信息工程技术、数控技术和电气自动化技术等专业的英语教材，也适合于继续教育学院、应用型本科院校机电类专业学生使用，并可供专业技术人员学习参考。教师可根据各专业的学生情况，不受教材编排顺序的限制，进行适当的删选。部分授课学时偏少的院校，可选学其中约15个单元的内容，每单元参考学时为2～3学时。对老师在授课中没有选入的单元，学生可根据自己的兴趣或需要自学该部分内容，以拓宽专业英语的知识面。

本书由河南工业职业技术学院徐存善和河北东方学院周志宇任主编。编写分工为：浙江省德清永德家用纺织品有限公司尹丽芳编写第1～4单元；河南工业职业技术学院杜帅编写第5～6单元，杨阳编写第7～8单元，鲁宵昳编写第9～10单元，徐存善编写第11～17单元和附录B，王玉编写第18~21单元；河北东方学院周志宇编写第22~24单元和第5章。附录A中的参考译文与习题答案由各单元相应的编者提供。

本书的编审工作得到了编者所在院校领导的高度重视和大力支持，他们为本书的编写提出了宝贵意见，在此表示衷心的感谢。

由于编者的水平和经验有限，书中难免有缺点和错误，恳请广大读者和同行批评指正。主编邮箱 jinfeng 20107396@163.com，欢迎来信交流。

编　者

Contents

前　言

Chapter Ⅰ　Machinery and Mould Manufacturing

Unit 1　Engineering Materials …… 1
Translating Skills：科技英语翻译的标准与方法 …… 4
Reading：Ferrous and Non-Ferrous Materials …… 5
Unit 2　Machine Elements …… 8
Translating Skills：词义的选择和确定 …… 11
Reading：Henry Ford …… 12
Unit 3　Machine Tools …… 15
Translating Skills：词义引申 …… 17
Reading：Jig Borers …… 18
Unit 4　Heat Treatment and Hot Working of Metals …… 20
Translating Skills：词性转译 …… 22
Reading：Soldering and Welding …… 23
Unit 5　Introduction to Mould …… 26
Translating Skills：增词译法 …… 29
Reading：Mould Materials …… 30
Unit 6　Mould Design …… 34
Translating Skills：省略译法 …… 37
Reading：Plastic Product Design …… 38
Unit 7　The Injection Moulding and Machines …… 40
Translating Skills：科技英语词汇的结构特征Ⅰ …… 44
Reading：Components of Injection Moulding Machines …… 45

Chapter Ⅱ　Computerized Numerical Control（CNC）

Unit 8　Computer Numerical Control Machine Tools …… 48
Translating Skills：科技英语词汇的结构特征Ⅱ …… 51
Reading：Advantages of NC …… 52
Unit 9　Elements of CNC Machine Tools …… 55
Translating Skills：被动语态的译法 …… 59
Reading：Safety Precautions of CNC Machines …… 60
Unit 10　CNC Programming …… 63
Translating Skills：非谓语动词 V-ing 的用法 …… 65

Reading: An Example of Program Showing ······ 67
Unit 11 CAD ······ 69
Translating Skills: 非谓语动词 V-ed 和 to V 的用法 ······ 72
Reading: Computer – Aided Manufacturing ······ 73
Unit 12 Flexible Machining Systems ······ 75
Translating Skills: 句子成分的转换 ······ 79
Reading: Group Technology ······ 80

Chapter Ⅲ Electronics and Information Technology

Unit 13 Alternating Current ······ 82
Translating Skills: 定语从句的译法 ······ 86
Reading: Three-Phase Circuits ······ 87
Unit 14 Electronic Components ······ 90
Translating Skills: And 引导的句型的译法 ······ 94
Reading: Testing and Measuring Instruments ······ 96
Unit 15 Transistor Voltage Amplifiers ······ 99
Translating Skills: 科技术语的翻译 ······ 102
Reading: Integrated Circuits ······ 103
Unit 16 Introduction to Internet ······ 106
Translating Skills: 反译法 ······ 110
Reading: Bill Gates ······ 111
Unit 17 4G Network Technology ······ 114
Useful Information: 零件数据库网站介绍与英文网站注册 ······ 117
Reading: Internet Protocol Television (IPTV) ······ 119
Unit 18 Internet of Things ······ 123
Translating Skills: 长难句的翻译 ······ 127
Reading: How Connected Cars Might Actually Make Driving Better ······ 128

Chapter Ⅳ Application Technology

Unit 19 Programmable Logic Controllers (PLC) ······ 131
Practical English: 如何阅读机电产品的英文说明书 ······ 134
Reading: PLC Programming ······ 136
Unit 20 Automatic Control Systems ······ 138
Practical English: 机电产品说明书范例 (中英文对照) ······ 141
Reading: Control System Components ······ 145
Unit 21 Basic Robots ······ 147
Practical English: 如何阅读英文招聘广告 ······ 150
Reading: The Robot Applications ······ 154
Unit 22 3D Printing ······ 157

Practical English：如何写英文个人简历 …… 161
Reading：American Scientists Work on Printing of Living Tissue Replacements …… 164
Unit 23 Industry 4.0 Introduction …… 167
Practical English：如何写英文求职信 …… 171
Reading：Made in China 2025 and Industrie 4.0 Cooperative Opportunities …… 172
Unit 24 National Craftsmen …… 176
Practical English：英文面试技巧 …… 182
Reading：WorldSkills Competition …… 184

Chapter V The Communication Skills Training for Careers

Dialogue 1 Pick Up New Customers in Airport …… 188
Dialogue 2 Email Writing …… 188
Dialogue 3 AM Machines …… 189
Dialogue 4 New Job Orientation …… 190
Dialogue 5 New Buzzwords Online …… 191
Dialogue 6 Finding a Job Online …… 192
Dialogue 7 Living in China …… 194
Dialogue 8 Introduction of New Products …… 195

Appendixes

Appendix A 参考译文与习题答案 …… 197
Appendix B Glossary（总词汇表） …… 243

References （参考文献） …… 265

Chapter I Machinery and Mould Manufacturing

Unit 1 Engineering Materials

Text

In the design and manufacture of a product, it is essential that the material and the process be understood. [1] Materials differ widely in physical properties, *machinability characteristics*, methods of forming, and possible service life. The designer should consider these facts in selecting an economical material and a process that is best suited to the product. [2]

Engineering materials are of two basic types: *metallic* or *nonmetallic*. Nonmetallic materials are further classified as *organic* or inorganic substances. Since there is an *infinite* number of nonmetallic materials as well as pure and *alloyed* metals, considerable study is necessary to choose the *appropriate* one. [3]

Few *commercial* materials exist as elements in nature. For example, the natural *compounds* of metals, such as *oxides*, *sulfides*, or *carbonates*, must *undergo* a separating or refining operation before they can be further processed. Once separated, they must have an atomic structure that is stable at ordinary temperatures over a *prolonged* period. In metal working, *iron* is the most important natural element. Iron has little commercial use in its pure state, but when combined with other elements into various alloys it becomes the leading engineering metal. The nonferrous metals, including copper, tin, zinc, nickel, magnesium, aluminum, lead, and others all play an important part in our economy; each has specific properties and uses.

Manufacturing requires tools and machines that can produce economically and accurately. Economy depends on the proper selection of the machine or process that will give a satisfactory finished product, its *optimum* operation, and maximum performance of labor and support facilities. [4] The selection is influenced by the quantity of items to be produced. Usually there is one machine best suited for a certain output. In small-lot or job shop manufacturing, general-purpose machines such as the lathe, drill press, and milling machine may prove to be the best because they are adaptable, have lower initial cost, require less maintenance, and possess the *flexibility* to meet changing conditions. [5] However, a special-purpose machine should be considered for large quantities of a standardized product. A machine built for one type of work or operation, such as the grinding of a piston or the surfacing of a cylinder head, will do the job well, quickly, and at low cost with a semi-skilled operator.

Many special-purpose machines or tools differ from the standard type in that they have built into them some of the skill of the operator. A simple bolt may be produced on either a lathe or an automatic screw machine. The lathe operator must know not only how to make the bolt but must also be sufficiently skilled to operate the machine. On the automatic machine the sequence of operations and movements of tools are controlled by *cams* and stops, and each item produced is identical with the previous one. This "transfer of skill" into the machine, or automation, allows less skillful operators but does require greater skill in *supervision* and maintenance. Often it is uneconomical to make a machine completely automatic, because the cost may become prohibitive.

The selection of the best machine or process for a given product requires knowledge of production methods. Factors that must be considered are volume of production, quality of the finished product, and the advantages and limitations of the equipment capable of doing the work. Most parts can be produced by several methods, but usually there is one way that is most economical.

New Words and Phrases

machinability [məˌʃiːnəˈbiləti] n. 可加工性，可切削性，机械加工性

characteristic [ˌkærəktəˈristik] adj. 典型的，特有的 n. 特征，特性

metallic [məˈtælik] adj. 金属的，含金属的 n. 金属纤维

nonmetallic [ˌnɔnməˈtælik] adj. 非金属的 n. 非金属物质

organic [ɔːˈgænik] adj. 有机的，组织的，器官的

infinite [ˈinfinət] adj. 无穷的，极大的 n. 无限，无穷大

alloyed [ˈælɔid] adj. 合金的；合铸的；熔合的 vt. 将……铸成合金

appropriate [əˈprəupriət, əˈprɔpriət] adj. 适当的，合适的 vt. 占用，拨出

commercial [kəˈməːʃl] adj. 商业上的，商业的 n. 商业广告

compound [ˈkɔmpaund] n. 化合物 vt. 合成 vi. 和解 adj. 复合的

oxide [ˈɔksaid] n. 氧化物，氧化层，氧化合物

sulfide [ˈsʌlfaid] n. 硫化物，含硫系列，硫醚

carbonate [ˈkɑːbəneit] n. 碳酸盐 vt. 使充满二氧化碳，使变成碳酸盐

undergo [ˌʌndəˈgəu] vt. 经历，经受；忍受

prolonged [prəˈlɔŋd] adj. 延长的，拖延的，持续很久的

iron [ˈaiən] n. 铁，熨斗；烙铁 adj. 铁的；刚强的

optimum [ˈɔptiməm] n. 最佳效果，最适合条件 adj. 最适宜的

flexibility [ˌfleksəˈbiləti] n. 机动性，灵活性，适应性

cam [kæm] n. 凸轮，偏心轮 vt. 给（机器）配置偏心轮

supervision [ˌsjuːpəˈviʒn] n. 监督，管理，指导

Technical Terms

physical property 物理性质

nonferrous metal 非铁金属

small-lot 小批量

general-purpose 普通用途
special-purpose 专用的，特殊用途的
standardized product （工程设计）标准化产品

Notes

1. In the design and manufacture of a product, it is essential that the material and the process be understood.

译文：在产品的设计和加工过程中，对材料和加工工序进行充分了解是非常必要的。

说明：句中 it 是形式主语，真正的主语是 that the material and the process be understood。

2. The designer should consider these facts in selecting an economical material and a process that is best suited to the product.

译文：设计师应该在选择最适合产品的经济性材料和加工工序时考虑到上述因素。

说明：句中 that is best suited to the product 作定语从句，修饰 an economical material and a process。

3. Since there is an infinite number of nonmetallic materials as well as pure and alloyed metals, considerable study is necessary to choose the appropriate one.

译文：因为非金属材料与纯金属、合金的种类一样繁多，所以在选择合适的材料时需要做大量的研究工作。

说明：句中 as well as 意为“和……一样”，since there is an infinite number of... 作原因状语从句，to choose the appropriate one 作目的状语。

4. Economy depends on the proper selection of the machine or process that will give a satisfactory finished product, its optimum operation, and maximum performance of labor and support facilities.

译文：要达到经济性要求，就要选择合适的设备和工序，这样可以得到理想的产品，还要有最佳的操作、高效的人工执行和设备支持。

说明：句中 depend on 意为“取决于，依赖于”；that will give a satisfactory... 为定语从句，修饰 the machine or process；its optimum operation, and maximum... 作 depends on 的并列宾语。

5. In small-lot or job shop manufacturing, general-purpose machines such as the lathe, drill press, and milling machine may prove to be the best because they are adaptable, have lower initial cost, require less maintenance, and possess the flexibility to meet changing conditions.

译文：对于小批量生产，通用机器如车床、钻床和铣床是最好的选择，因为它们适应性强，生产费用低，维护简便，能灵活适应条件的改变。

说明：句中 such as 意为“例如”；because 引导原因状语从句；they are adaptable... 是并列成分，即 they are adaptable, they have..., they require..., they possess...。

Exercises

I. Complete the following sentences with the proper form of the words given below.

metallic, alloy, special-purpose, pure, fact, general-purpose, result, nonmetallic, standard

1. The designer should consider these ________ in selecting an economical material and a

process that is best suited to the product.

2. ________ materials are further classified as organic or inorganic substances.

3. Iron has little commercial use in its ________ state, but when combined with other elements into various ________ it becomes the leading engineering metal.

4. Many ________ machines or tools differ from the ________ type in that they have built into them some of the skill of the operator.

Ⅱ. Translate the following sentences into English.

1. 材料在物理性能、机加工性能、成形方法以及使用寿命等方面有很大不同。

2. 天然金属化合物如氧化物、硫化物和碳酸盐在能够进行深加工前必须经过分离或提纯。

3. 在自动机床上，工具的操作顺序和运动是由凸轮和制动器来控制的，并且每一零件的生产过程都与前一个过程相同。

Translating Skills：

科技英语翻译的标准与方法

翻译是一种再创造语言活动，即译者根据原作者的思想，用另一种语言将其所表达的内容再次表达出来。这就要求译者必须准确理解和掌握原作的内容与含意，在此基础上，熟练运用译文语言把原文内涵通顺流畅地再现给读者。

一、翻译的标准

科技英语的翻译标准可概括为“忠实、通顺”四个字。

忠实，首先指忠实于原文内容，译者必须把原作的内容完整而准确地表达出来，不得任意发挥或增删；忠实还指忠实于原作风格，尽量表现其本来特色。

通顺，即指译文语言必须通俗易懂，符合规范。

忠实与通顺是相辅相成的，缺一不可。忠实而不通顺，读者会看不懂；通顺而不忠实，脱离原作的内容与风格，通顺就失去了意义。例如：

1. The electric resistance is measured in ohms.

误译：电的反抗是用欧姆测量的。

正译：电阻的测量单位是欧姆。

2. All metals do not conduct electricity equally well.

误译：全部金属不导电得相等好。

正译：并非所有的金属都具有同样好的导电性。

3. The moment the circuit is completed, a current will start flowing the coil.

正译：电路一旦接通，电流即开始流向线圈。

4. Some special alloy steels should be used for such parts because the alloying elements make them tougher, stronger, or harder than carbon steels.

正译：对这类零件可采用某些特殊的合金钢，因为合金元素能提高钢的韧性、强度、硬度。

从以上例句可以清楚地看到，翻译不能任意删改内容，更不是逐词死译；汉语译文规范化，并非是离开原文随意发挥。此外，还应注意通用术语的译法，比如例 1 中的“electric resistance”译作“电阻”已成为固定词组，不能用别的译法。

二、翻译的方法

翻译的方法一般来说有直译（literal translation）和意译（free translation）。直译，即指既忠实于原文内容，又忠实于原文形式的翻译。意译，就是指翻译忠实于原文的内容，但不拘泥于原文的形式。

翻译时，我们应灵活运用上述两种方法，能够直译的就直译，需要意译的就意译。因为对同一个句子来说，有时并非只能用一种方法，所以我们可以交替使用或同时使用以上两种方法。

请看下面的句子：

1. Milky Way　银河（意译）（不可直译为：牛奶路）

2. bull's eye　靶心（意译）（不可直译为：牛眼睛）

3. New uses have been found for old metals, and new alloys have been made to satisfy new demands. 老的金属有了新用途，新的合金被冶炼出来，以满足新的需要（本句前半部分用了意译法，后半部分用了直译法）。

4. The ability to program these devices will make a student an invaluable asset to the growing electronic industry. 对这些器件进行编程的能力将使学生成为不断发展的电子工业领域的无价人才（asset 原意为资产，这里我们根据上下文意译成“人才”）。

三、翻译的专业性特点

科学技术本身的性质要求科技英语的翻译与专业内容相一致，这就决定了专业英语与普通英语有很大的差异。专业英语以其独特的语体，明确表达作者在其专业方面的见解，表达方式直截了当，用词简练。即使同一个词，在不同学科的专业英语中，其含义也有可能是不同的。例如：

1. The computer took over an immense range of tasks from workers muscles and brains.

误译：计算机代替了工人大量的肌肉和大脑。

正译：计算机取代了工人大量的体力和脑力劳动（muscles and brains 引申为“体力和脑力劳动”）。

2. In any case work doesn't include time, but power does.

正译：在任何情况下，功都与时间无关，但功率与时间有关（work, power 在物理专业中分别译为“功”“功率”）。

3. Like charges repel each other while opposite charges attracted.

正译：同性电荷相互排斥，异性电荷相互吸引（charge 含义有“负载、充电、充气、电荷”，按专业知识译为“电荷”）。

从以上例句可知，专业英语专业性强，逻辑性强，翻译要力求准确、精练、正式。这不仅要求我们能熟练掌握汉语表达规范，还要求具有较高的专业水平。

Reading: Ferrous and Non-Ferrous Materials

Metals are found everywhere in our life. They are classified into two *categories*, “*ferrous*” and "non-ferrous" metals (Table1-1 and Table 1-2). Ferrous means relating to, or containing iron and non-ferrous means having no iron. These two kinds of metals can be used to manufacture an equally large range of items.

Table 1-1 Some ferrous metals and properties

Name	Alloy of	Properties	Uses
Mild steel	Carbon 0.1% ~0.3% Iron 99.9% ~99.7%	Tough. High tensile strength. Can be *case-hardened*. Rusts very easily	Most common metal used in school workshops Used in general metal products and engineering
High carbon steel	Carbon 0.6% ~1.4% Iron 99.4% ~98.6%	Tough. Can be hardened and *tempered*	Cutting tools such as drills
Stainless steel	Iron, nickel and *chromium*	Tough, resistant to rusting and *staining*	*Cutlery*, medical instruments
Cast iron	Carbon 2% ~6% Iron 98% ~94%	Strong but *brittle*. Comprehensive strength very high	*Castings*, *manhole* covers, engines
Wrought iron	Almost 100% iron	*Fibrous*, tough, *ductile*, resistant to rusting	Ornamental gates and railings, not in much use today

Table 1-2 Some non-ferrous metals and properties

Name	Colour	Alloy of	Properties	Uses
Copper	Reddish brown	Not an alloy	Ductile, can be beaten into shape. Conducts electricity and heat	Electrical wiring, *tubing*, kettles, bowls, pipes
Brass	Yellow	Mixture of copper and *zinc*. Zinc 35%-65% (most common ratio)	Hard. Casts and machines well. Surface *tarnishes*. Conducts electricity	Parts for electrical fittings, *ornaments*
Aluminum	Light grey	Aluminum 95% Copper 4% *Manganese* 1%	Ductile, soft, *malleable*, machines well. Very light	Window frames, aircraft, kitchen ware
Silver	*Whitish* grey	Mainly silver but alloyed with copper to give *sterling* silver	Ductile, malleable, *solders*, resists corrosion	Jewelry, solders, ornaments
Lead	*Bluish* grey	Not an alloy	Soft, heavy, ductile, loses its shape under pressure	Solders, Pipes, batteries, roofing

New Words and Phrases

ferrous [ˈferəs] adj. [化] 亚铁的；铁的，含铁的

category [ˈkætigəri] n. 种类，分类；范畴

case-hardened adj. 表面硬化的；定型的；无情的

temper [ˈtempə] n.（钢等）回火；脾气，性情；倾向 vt. 调和，使回火

stain [stein] vt. 玷污，给……着色 vi. 污染，玷污 n. 污点，瑕疵
cutlery [ˈkʌtləri] n. 刀剑制造业；餐具，刀叉
chromium [ˈkrəumiəm] n. [化] 铬（24 号元素，符号 Cr）
brittle [ˈbritl] adj. 易碎的，脆弱的；易生气的
casting [ˈkɑːstiŋ] n. 投掷；铸造；铸件 v. 投掷，投向，扔掉；铸造
manhole [ˈmænhəul] n. 检修孔，检查井
wrought [rɔːt] adj. 锻造的；加工的；精细的
tubing [ˈtjuːbiŋ] n. 管子；装管 vi. 把……装管；使成管状
zinc [ziŋk] vt. 镀锌于……；用锌处理 n. 锌
ornament [ˈɔːnəmənt] n. 装饰；装饰物；教堂用品 vt. 装饰，修饰
fibrous [ˈfaibrəs] adj. 纤维的，纤维性的，纤维状的
ductile [ˈdʌktail, ˈdʌktil] adj. 易教导的；易延展的，柔软
malleable [ˈmæliəbl] adj. 可锻的，有延展性的，易适应的，可塑的
aluminium [ˌælјuˈminiəm] adj. 铝的 n. 铝
manganese [ˈmæŋgəniːs] n. [化] 锰
tarnish [ˈtɑːniʃ] n. 污点；无光泽 vt. & vi. 玷污；使……失去光泽
whitish [ˈwaitiʃ] adj. 带白色的；发白的
sterling [ˈstəːliŋ] adj. 英币的；纯正的；纯银制的 n. 英国货币；标准纯银
solder [ˈsɔldə] vi. 焊接 vt. 焊接；使连接在一起 n. 焊料；接合物
bluish [ˈbluːiʃ] adj. 带蓝色的，有点蓝的

Exercises

Translate the following phrases into Chinese.

1. a ferrous metal used to make cutting tools
2. a non-ferrous metal used to make parts for electrical fittings
3. a ferrous metal used to make cutlery and medical instruments
4. a non-ferrous metal used to make electrical wiring
5. a metal used to make manhole covers
6. a non-ferrous metal used to make aircraft

Unit 2 Machine Elements

Text

However simple, any machine is a *combination* of *individual components* generally referred to as machine elements or parts. Thus, if a machine is completely *dismantled*, a collection of simple parts remains such as *nuts*, *bolts*, springs, *gears*, cams and *shafts*—the building block of all *machinery*. [1] A machine element is, therefore, a single unit designed to perform a specific function and capable of combining with other elements. Sometimes certain elements are *associated* in pairs, [2] such as nuts and bolts or *keys* and shafts. In other instance, a group of elements is combined to form a *subassembly*, such as bearings, *couplings*, and *clutches*.

The most common example of a machine element is a gear, which, *fundamentally*, is a combination of the wheel and the *lever* to form a toothed wheel. [3] The rotation of this gear on a *hub* or shaft drives other gears that may *rotate* faster or slower, depending upon the number of teeth on the basic wheels.

Other fundamental machine elements have *evolved* from wheel and lever. A wheel must have a shaft on which it may rotate. The wheel is fastened to the shafts with couplings. The shaft must rest in bearings, may be turned by a *pulley* with a belt or a chain connecting it to a pulley on a second shaft. The supporting structure may be *assembled* with bolts or *rivets* or by *welding*. [4] Proper application of these machine elements depends upon knowledge of the force on the structure and the strength of the materials employed.

The individual *reliability* of machine elements becomes the basis for estimating the overall life expectancy of a complete machine.

Many machine elements are thoroughly *standardized*. Testing and practical experience have *established* the most suitable *dimensions* for common structural and mechanical parts. Through standardization, *uniformity* of practice and resulting economics are obtained. Not all machine parts in use are standardized, however. In the *automotive* industry only fasteners, bearings, *bushings*, chains, and belts are standardized. *Crankshafts* and connecting rods are not standardized.

New Words and Phrases

combination [ˌkɔmbiˈneiʃən] n. 组合，结合
individual [ˌindiˈvidjuəl] adj. 单独的，各个的，个别的，特殊的
component [kəmˈpəunənt] n. 元件，构件，部件
dismantle [disˈmæntl] vt. 分解（机器），拆开，拆卸
nut [nʌt] n. 螺母；难对付的人；难解的问题；坚果　vi. 采坚果
bolt [bəult] n. 螺栓；（门、窗等的）插销

gear [ɡiə] n. 仪器，装置；传动装置，齿轮
shaft [ʃɑːft] n. 柱身；连杆；传动轴，旋转轴；轴
machinery [məˈʃiːnəri] n. 机器（总称）
associate [əˈsəuʃieit] v. 联合，结合，参加，连带
key [kiː] n. 键，电键，开关；楔，销；钥匙
subassembly [ˌsʌbəˈsembli] n. 组合件，部件，机组
coupling [ˈkʌpliŋ] n. 联轴节，联轴器；联结器，联合器
clutch [klʌtʃ] n. 离合器，联轴器；夹紧装置
fundamentally [ˌfʌndəˈmentli] adv. （从）根本上
lever [ˈliːvə] n. 杠杆，控制杆，操作杆
hub [hʌb] n. （轮）毂；中心；木片
rotate [rəuˈteit] vt. 使旋转或转动
evolve [iˈvɔlv] vi. 进化，演变；开展，发展，展开
pulley [ˈpuli] n. 滑轮（组）；滑车；带轮
assemble [əˈsembl] v. 安装，装配，组合；集合，集中 n. 组件
rivet [ˈrivit] n. 铆钉 vt. 铆接，铆
weld [weld] vt. 焊接，熔焊 vi. 焊牢 n. 焊接点
reliability [ˌrilaiəˈbiliti] n. 可靠性，安全性，准确性
standardize [ˈstændədaiz] vt. 标准化，统一标准；标定，校准
establish [iˈstæbliʃ] vt. 确定，制定；建立，创办，产生；使固定
dimension [diˈmenʃən] n. 尺寸，尺度；范围，方面
uniformity [ˌjuːniˈfɔːmiti] n. 均匀性，一致性
automotive [ˌɔːtəˈməutiv] adj. 汽车的，自动车的
bushing [ˈbuʃiŋ] n. 衬套；轴衬；轴瓦；[电]（绝缘）套管
crankshaft [ˈkræŋkʃɑːft] n. 曲轴
refer to as 称为，叫作；当作；参考作为；所说的；提到的

Notes

1. Thus, if a machine is completely dismantled, a collection of simple parts remains such as nuts, bolts, springs, gears, cams and shafts—the building block of all machinery.

译文：因此，如果把机床完全拆开，就可以得到像螺母、螺栓、弹簧、齿轮、凸轮及轴等简单零件——所有机器的基础元件的集合。

2. Sometimes certain elements are associated in pairs.

译文：某些特定的零件必须成对地工作。

说明：be associated with 意为“与……有关；涉及”。例如：

I associated with him in business. 我与他合伙经商。

These concerns may be associated with strong feelings such as anger or shame. 这些忧虑很可

能与恼火、羞愧之类的强烈情绪有关联。

3. The most common example of a machine element is a gear, which, fundamentally, is a combination of the wheel and the lever to form a toothed wheel.

译文：机械零件中最常用的是齿轮，它实际上是由轮子和杆组成的带有齿的轮子。

说明：which 引导非限定性定语从句，修饰 gear，动词不定式 to form a toothed wheel 作结果状语。

4. The supporting structure may be assembled with bolts or rivets or by welding.

译文：支撑结构可由螺栓、铆钉连接，或通过焊接固定在一起。

说明：with bolts or rivets or by welding 作方式状语，be assembled with 意为“用……安装”。例如：

The bookcase can easily be assembled with a screwdriver. 这书柜用一把螺钉旋具就可以很容易地安装起来。

Exercises

Ⅰ. Answer the following questions according to the passage.

1. What is referred to as a machine element? Which is the most common machine element?
2. What are other fundamental machine elements?
3. Which machine elements are standardized in the automotive industry?

Ⅱ. Decide whether the following statements are True (T) or False (F) according to the passage.

1. Gears are used as building blocks for the construction of most devices, apparatus, and machinery.
2. A pair of gears with different numbers of teeth will rotate at different speeds.
3. Spur gears and bevel gears are different in their applications because their shapes are different.
4. The most common machine element is the gear from which other fundamental machine elements have developed.
5. A coupling is a machine element, which joins two shafts.
6. All machine parts have been developed into standardized designs.

Ⅲ. Translate the following sentences into Chinese.

1. The most common machine element is the gear, which combines the feature of the wheel and the lever to form a toothed wheel.
2. The hardness of a gear determines its ability to resist wear.
3. Manufacturing engineers have centered their efforts on the development of standardized elements.
4. These parts are produced in large quantities with a high degree of perfection at reduced cost.

Translating Skills:

词义的选择和确定

英、汉两种语言都有一词多类、一词多义的现象。一词多类是指一个词往往属于多个词类，具有多种不同的词性：如 display 一词，既可用作名词“显示（器）”，又可用作形容词“展览的、陈列用的”，还可用作动词“显示、表现”等。一词多义是说一个词在同一个词类中，往往有多个不同的词义：如 power 一词，用作名词时，其汉语意思为“电力、功率、次方”等。在英汉翻译过程中，我们在弄清句子结构后就要善于选择和确定句子中关键词的意义。选择和确定词义通常从以下几个方面入手。

一、根据词类选择和确定词义

选择某个词的词义时，首先要确定这个词在句中应属于哪一种词类，然后再进一步确定其词义。下面以 display 为例：

Here, you have the option of defining your own **display** variants.

这里，你有权定义你自己的**显示**变式。（display 为名词）

Often, it is best to **display** materials on an information table.

通常，最好是把资料放在提供各类资讯的桌子上**展示**。（display 为动词）

The reverse side of a control panel, **display** panel, or the like; the side with the interconnecting wiring.

控制面板、**显示**面板或类似的面板的反面，即带有互连接线的那一面。（display 为形容词）

二、同一词类不同词义的选择和确定

英语中同一个词，同一类词，在不同的场合，往往具有不同的含义。此时，必须根据上下文的联系及整个句子的意思加以判断和翻译。下面以 as 为例：

Wear resistance improves **as** cutting tool hardness increase.

当切削刀具的硬度提高**时**，其耐磨性也越好。（as 引导时间状语从句）

As heat makes things move, it is a form of energy.

因为热能使物体运动，**所以**热是一种能量。（as 引导原因状语从句）

三、根据单词搭配情况选择和确定词义

英译汉时，不仅必须根据上下文的联系理解词义，还需要根据词的搭配情况来理解词义。尤其在科技文献中，由于学科及专业不同，同一个词在不同的专业中具有不同的意义。例如：

The fifth **power** of two is thirty-two.

二的五**次方**是三十二。（数学）

With the development of electrical engineering, **power** can be transmitted over long distances.

随着电气工程的发展，**电力**能输送得非常远。（电学）

Friction can cause a loss of **power** in every machine.

摩擦能引起每一台机器**功率**的损耗。（物理学）

四、根据名词的单复数选择和确定词义

英语中有些名词的单数和复数表达的词义可能完全不同。例如：

名词	单数词义	复数词义
facility	简易，灵巧	设施，工具
charge	负荷，电荷	费用
spirit	精神	酒精

Although they lost, the team played with tremendous **spirit**.

尽管他们输了，但这支队伍表现了极其顽强的**精神**。

Whisky, brandy, gin and rum are all **spirits**.

威士忌、白兰地、杜松子酒和朗姆酒都是烈酒。

Reading: Henry Ford

Henry Ford was born in 1863, on a farm near Detroit. He loved putting machines together as well as taking them *apart*. [1] He was always dreaming of machines to make his work easier.

In 1896, the first Ford car was completed. It was like a box placed on four big bicycle *wheels*. It had an engine with two *cylinders* and no *brakes*. If you wanted to go backwards, you had to get out and push. He drove his *quadricycle*, with more noise than speed, through the empty streets. [2] It worked! Indeed, it ran well for several years. But Henry soon lost interest. Already he was planning a bigger and better car.

In 1898 his new car was finished. The motor car was starting to *attract* attention. Years later, Henry opened a car factory himself, with his own staff. He sold his first car in 1903. By 1904 the company was doing very well indeed with its Model A Ford.

Henry Ford dreamed about a car for the ordinary man. What he wanted to do was to produce cars in large *quantities*, all exactly the same, so that a part from one car would fit all the others. That would keep down the cost. As Henry Ford saw it, cheaper cars would lead to more people buying cars, which would lead to better roads. This, in turn, would lead to more people buying cars, which would lead to cheaper cars. In 1908 the first Model T Ford *rolled* off the production line. It was an ugly little car, but so simple that a child could drive it.

Fifteen million Model T Fords were sold all over the world. *Meanwhile*, the Model T was selling all over the world. By 1919, the Ford factories were turning out 86, 000 cars a month. By November 1922, the *figure* was 240, 000. And he started to buy other businesses too.

In 1927 the new Model A came on the scene. Crowds rushed to see it. Then, two years later, at the age of sixty-eight, when most men are enjoying a well-earned rest, Henry Ford did it again. He built the big V8 Ford. [3]

Some people say he was the richest man in history. Some say he was the richest man in the world then. But Henry Ford made it possible for ordinary people to go from place to place easily and cheaply.

New Words and Phrases

apart [ə'pɑːt] adj. 分离的；与众不同的 adv. 与众不同地；分离着
wheel [wiːl] n. 车轮；方向盘 vt. 转动；给……装轮子 vi. 旋转
cylinder ['silində(r)] n. 圆筒；气缸；[数] 柱面；圆柱状物
brake [breik] n. 闸，制动器
quadricycle ['kwɒdrisaikl] adj. 四轮的 n. 脚踏四轮车
attract [ə'trækt] vt. 吸引；引起 vi. 吸引；有吸引力
quantity ['kwɔntəti] n. 量，数量；大量；总量
roll [rəul] v. 滚动；卷 n. 滚；卷
meanwhile ['miːnwail] adv. 同时，其间 n. 其间，其时
figure ['fɪgə(r)] n. 图形；数字；人物；体型
dream of/about 梦见；渴望
attract attention 引起注意
in large quantities 大量地；批量地
keep down 抑制；控制
come on the scene 问世；来到
well-earned rest 应得的休息

Proper Names

Henry Ford 亨利·福特（福特汽车公司的建立者，他也是世界上第一位使用流水线大批量生产汽车的人）
Detroit 底特律
Model A Ford 福特 A 型汽车

Notes

1. He loved putting machines together as well as taking them apart.

译文：他喜欢拆卸机器，也喜欢把各种零件组装起来。

说明：这是一个简单句。loved 是及物动词，后面跟随并列的动名词短语作宾语。take … apart 意为“拆开，拆卸；分辨，区分”。例如：

One day, she waited for her parents to leave the house so she could take apart the television and reassemble it before they returned. 有一天，她就等她父母离开家，以便能拆开电视机，并赶在他们回来之前把它装好。

2. He drove his quadricycle, with more noise than speed, through the empty streets.

译文：他驾驶着那辆速度有限、噪声很大的四轮车，穿行在空荡荡的街道上。

说明：这也是一个简单句。with more noise than speed 是介词短语作后置定语，修饰 quadricycle；through the empty streets 是介词短语作状语，修饰 drove。

3. Then, two years later, at the age of sixty-eight, when most men are enjoying a well-earned rest, Henry Ford did it again. He built the big V8 Ford.

译文：两年之后，亨利·福特在他68岁的年纪——这个年纪的大多数人正安享辛勤工作换来的安逸之时，福特又一次一展身手。他造出了大车身的V8福特车。

说明：这是一个复合句。when引导时间状语从句，后面是主句；well-earned rest意为“应得的休息”。例如：

Instead, it's a good chance for anyone who has spent a long time preparing for the celebrations to stay at home and take a well-earned rest. 反倒是为了准备过年辛苦忙了好一阵子的人，可以趁此机会好好地在家休息一下。

Exercises

Ⅰ. Decide whether the following statements are True (T) or False (F) according to the passage.

1. The first Ford car had an engine with four cylinders and no breaks.
2. Ford was always dreaming of making cars for the rich man.
3. The method that Ford used to make cars would keep down the cost.
4. Crowds rushed to see the new Modal A in 1927.

Ⅱ. Translate the following sentences into English by using the words in brackets.

1. 他既种菜也种花。(as well as)
2. 我已经安排好派一个职员到飞机场接你。(staff)
3. 你算出假期得花多少钱了吗？(figure out)
4. 我希望他成为钢琴家的梦想可以成真。(dream of)

Unit 3 Machine Tools

Text

Most of the mechanical operations are commonly performed on five basic machine tools:

The *drill* press;

The *lathe*;

The *shaper* or *planer*;

The *milling* machine;

The *grinder*.

Drilling

Drilling is performed with a rotating tool called a drill. Most drilling in metal is done with a *twist* drill. The machine used for drilling is called a drill press. Operations, such as *reaming* and *tapping*, are also classified as drilling. Reaming consists of removing a small amount of metal from a hole already drilled. Tapping is the process of cutting a thread inside a hole so that a cap screw or bolt may be threaded into it. [1]

Turning and *Boring*

The lathe is commonly called the father of the entire machine tool family. For turning operations, the lathe uses a single-point cutting tool, which removes metal as it travels past the *revolving* workpiece. [2] Turning operations are required to make many different *cylindrical* shapes, such as *axes*, gear blanks, pulleys, and threaded shafts. Boring operations are performed to enlarge, finish, and accurately locate holes.

Milling

Milling removes metal with a revolving, *multiple* cutting edge tools called milling cutter. Milling cutters are made in many styles and sizes. Some have as few as two cutting edges and others have 30 or more. Milling can produce flat or angled surfaces, *grooves*, slots, gear teeth, and other *profiles*, depending on the shape of the cutters being used. [3]

Shaping and Planing

Shaping and *planing* produce flat surfaces with a single-point cutting tool. In shaping, the cutting tool on a shaper *reciprocates* or moves back and forth while the work is fed automatically towards the tool. In planing, the workpiece is attached to a worktable that reciprocates past the cutting tool. The cutting tool is automatically fed into the workpiece a small amount on each stroke.

Grinding

Grinding makes use of *abrasive* particles to do the cutting. Grinding operations may be classified as precision or nonprecision, depending on the purpose. Precision grinding is concerned with grinding to close *tolerances* and very smooth finish. [4] Nonprecision grinding involves the removal of metal where accuracy is not important.

New Words and Phrases

drill [dril] vi. 钻孔 n. 钻床，钻头；训练
lathe [leið] n. 车床 vt. 用车床加工
shaper [ˈʃeipə] n. 牛头刨床；造型者
planer [ˈpleinə] n. 龙门刨床，刨床；刨工
milling [ˈmiliŋ] n. 磨，制粉；[机] 铣削 v. 碾磨，磨成粉
grinder [ˈgraində] n. 磨床，研磨机
twist [twist] n. 扭曲 vt. 捻，拧，扭伤 vi. 扭动，弯曲
ream [riːm] vt. 铰孔 n. 令（纸张的计数单位）；大量
tap [tæp] vt. 轻敲，轻拍；攻丝 vi. 轻拍，轻击 n. 水龙头
boring [ˈbɔːriŋ] n. 钻孔，镗削 vt. 钻（孔），[机] 镗（孔） adj. 无聊的
revolve [riˈvɔlv] vi. & n. 旋转；循环出现 vt. 使……旋转；使……循环
cylindrical [siˈlindrikl] adj. 圆柱形的；圆柱体的
axes [ˈæksiːz] n. 轴线；轴心；坐标轴；斧头（axe 的复数）
multiple [ˈmʌltipl] adj. 多样的；许多的；多重的 n. 并联；倍数
groove [gruːv] n. 槽 vt. 开槽于 vi. 形成沟槽
profile [ˈprəufail] n. 侧面，轮廓，外形，剖面
plane [plein] n. 平面；飞机 vi. 刨 vt. 刨平，用刨子刨；掠过水面
reciprocate [riˈsiprəkeit] vt. 互换；报答 vi. 互给；酬答；往复运动
abrasive [əˈbreisiv] adj. 有研磨作用的；粗糙的 n. 研磨料
tolerance [ˈtɔlərəns] n. 公差，容忍

Technical Terms

twist drill 麻花钻
gear blank 齿轮毛坯
threaded shaft 丝杠
back and forth 反复地，来回地
single-point cutting tool 单刃刀具
abrasive particle 磨料颗粒
precision grinding 精磨

Notes

1. Tapping is the process of cutting a thread inside a hole so that a cap screw or bolt may be threaded into it.

译文：攻螺纹是在孔里加工出螺纹的过程，以便螺钉或螺栓能旋合进孔里。

说明：so that a cap screw or bolt may be threaded into it 作目的状语。

2. For turning operations, the lathe uses a single-point cutting tool, which removes metal as it

travels past the revolving workpiece.

译文：在车削操作中，车床用一个单刃刀具在旋转的工件上进行金属切削。

说明：which removes metal as it travels past the revolving workpiece 是非限制性定语从句，修饰 a single-point cutting tool。

3. Milling can produce flat or angled surfaces, grooves, slots, gear teeth, and other profiles, depending on the shape of the cutters being used.

译文：根据所用的铣刀形状，铣削可加工平面或斜面、槽口、缝、齿轮的齿和其他的轮廓。

说明：depending on the shape of the cutters being used 作伴随状语。

4. Precision grinding is concerned with grinding to close tolerances and very smooth finish.

译文：精磨应用于精密公差和表面粗糙度要求非常高的磨削中。

说明：close tolerances 意为"精密公差、严格公差"；smooth finish 意为"表面粗糙度"；is concerned with... 意为"与……联系/有关"。例如：

He is concerned with the real estate business. 他从事房地产事务。

Exercises

Ⅰ. Answer the following questions according to the passage.

1. What parts can be processed with lathes?
2. Which machine tools which we have learned can produce flat surfaces?
3. What is known as the tapping?

Ⅱ. Decide whether the following statements are True (T) or False (F) according to the passage.

1. For turning operations, the lathe uses multiple cutting edge tools to process.
2. The milling machine is commonly called the father of the entire machine tool family.
3. In shaping, the workpiece is attached to a worktable that reciprocates past the cutting tool.
4. Grinding makes use of the metal material tools to do the cutting.

Ⅲ. Translate the following phrases or sentences into Chinese.

1. twist drill 2. cutting edge 3. smooth finish

4. Milling operations are employed in producing flat or angled surfaces, grooves, slots, gear teeth, and other profiles, depending on the shape of the cutters being used.

5. Turning operations are required to make many different cylindrical shapes, such as axes, gear blanks, pulleys, and threaded shafts.

Translating Skills:

词义引申

英、汉两种语言在表达上有很大差别。翻译时，有些词或词组不能直接搬用词典中的释义，若生搬硬套，会使译文生硬晦涩，难以理解，甚至造成误解。所以，要在弄清原文内涵的基础上，根据上下文的逻辑关系和汉语的搭配习惯，对词义加以引申。若遇到有关专业方面的内容，必须选用专业的术语。引申后的词义应能更确切地表达原文意义。例如：

1. However, colors can give more **force** to the form of the product.
欠佳译法：然而，色彩能给予产品更多的**力量**。
引申译法：然而，色彩能使产品外形**增添美感**。

2. Power plugs are **male electrical connectors that fit into female electrical sockets.**
欠佳译法：电源插头是**雄性插接器适配雌性插接器**。
引申译法：电源**插头可以插入电源插座**。

3. There is **a wide area of performance duplication** between numerical control and automatics.
引申译法：数控和自动化机床有**很多相同的性能**。

4. High-speed grinding does not **know** this disadvantage.
引申译法：高速磨床**不存在**这个缺点。

5. The charge current **depends upon** the technology and capacity of the battery being charged.
引申译法：充电电流的**大小根据**充电技术和电池容量的不同而不同。

Reading: Jig Borers

The *jig borer* resembles both the drill press and the *vertical* milling machine. The head of this unit is similar to a drill press, and the table is very much like that of a vertical milling machine. The jig borer is used in the manufacture of jigs, fixtures, and dies which require extremely accurate dimensions. Since the work can be *clamped* securely to the table that moves back and forth and in and out and since the drilling head is made with great precision, holes may be located and drilled more accurately on this machine than on any other machine tool.

The *coordinate* measuring system of a jig borer is designed to provide *longitudinal* and *transverse* table movements. One of the simplest systems is based on two lead screws. Another system, which is used on many jig borers, is based on precision end measuring rods. Some jig borers are provided with electrical or *optical* systems.

New Words and Phrases

jig [dʒig] n. 夹具　vi. 抖动；用夹具或钻模等加工
borer [ˈbɔːrə] n. [机] 镗床，镗孔刀具
vertical [ˈvəːtikəl] adj. 垂直的，直立的，竖式的　n. 垂直线
clamp [klæmp] n. 夹钳，夹子　vt.（用夹钳）夹住，夹紧
longitudinal [ˌlɔndʒiˈtjuːdinəl] adj. 长度的，纵向的
transverse [ˈtrænzvəːs] adj. 横向的，横断的
optical [ˈɔptikəl] adj. 光学的，视觉的，有助视力的
coordinate [kəuˈɔːdineit] n. 坐标　adj. 并列的，坐标的
vertical milling machine　立式铣床

Exercises

Ⅰ. Answer the following questions according to the passage.

1. Why is it said that the jig borer resembles both the drill press and the vertical milling machine?

2. Why the holes may be located and drilled more accurately on the jig borer than on any other machine tool?

3. Which coordinate measuring system is used on many jig borers?

Ⅱ. Translate the following phrases into English.

1. 纵向和横向工作台的运动 2. 导螺杆 3. 测量杆 4. 光学系统

Unit 4 Heat Treatment and Hot Working of Metals

Text

We can alter the characteristics of steel in various ways. In the first place, steel which contains very little carbon, will be milder than steel, which contains a higher percentage of carbon, up to the limit of about 1.5%. Secondly, we can heat the steel above a certain critical temperature, and then allow it to cool at different rates. At this critical temperature, changes begin to take place in the *molecular* structure of the metal. In the process known as *annealing* we heat the steel above the critical temperature and permit it to cool very slowly. This causes the metal to become softer than before, and much easier to machine. Annealing has a second advantage. It helps to *relieve* any internal stresses which exist in the metal. These stresses are *liable* to occur through hammering or working the metal, or through rapid cooling. Metal which we cause to cool rapidly *contracts* more rapidly on the outside than on the inside. This produces unequal contractions, which may give *rise* to *distortion* or *cracking*. [1] Metal which cools slowly is less liable to have these internal stresses than metal, which cools quickly.

On the other hand, we can make steel harder by rapid cooling. We heat it up beyond the critical temperature, and then *quench* it in water or some other liquid. The rapid temperature drop *fixes* the structural change in the steel, which occurred at the critical temperature, and makes it very hard. But a bar of this hardened steel is more liable to fracture than normal steel. We therefore heat it again to a temperature below the critical temperature, and cool it slowly. This treatment is called *tempering*. It helps to relieve the internal stresses, and makes the steel less brittle than before. The properties of tempered steel enable us to use it in the manufacture of tools, which need a fairly hard steel. High carbon steel is harder than tempered steel, but it is much more difficult to work.

These heat treatments take place during the various *shaping* operations. We can obtain bars and sheets of steel by rolling the metal through huge *rolls* in a rolling mill. The roll *pressures* must be much greater for cold rolling than for hot rolling, but cold rolling enables the operators to produce rolls of great accuracy and uniformity, and with a better surface finish. Other shaping operations include *drawing* into wire, casting in moulds, and forging.

The *mechanical* working of metal is the shaping of metal in either a cold or a hot state by some mechanical means. This does not include the shaping of metals by *machining* or *grinding*, in which processes metal is actually *machined* off, nor does it include the casting of molten metal into some form by use of molds. [2] In mechanical working processes, the metal is shaped by pressure-acutely forging, *bending*, *squeezing*, drawing, or *shearing* it to its final shape. In these processes the metal may be either cold or hot worked. Although normal room temperatures are ordinarily used for cold working of steel, temperatures up to the *recrystallization* range are sometimes used. Hot working of metals takes place above the recrystallization or work-*hardening* range. For steel, recrystallization starts around 650 ~ 700℃, although most hot work on steel is done at temperatures considerably above this range. There is no tendency for hardening by mechanical work until the lower limit of the re-*crystalline* range is reached. Some metals, such as lead and tin, have a low re-crystalline range and can be hot worked at room temperature, but most commercial metals require some heating. Alloy composition has a great *influence* upon the proper working range, the usual result being to raise the re-crystalline range temperature. This range may also be increased by *prior* cold working.

New Words and Phrases

molecular [mə'lekjulə] adj. 由分子组成的，分子的
annealing [ə'niːliŋ] n. （低温）退火，焖火，磨炼
relieve [ri'liːv] v. 解除，减轻，使不单调乏味
liable ['laiəbl] adj. 有责任的，有义务的，应受罚的，有……倾向的
contract [kən'trækt] vt. 收缩，缩紧；感染；订约
rise [raiz] vi. & n. 上升，增强，起立，高耸
distortion [dis'tɔːʃən] n. 变形，扭曲，曲解，失真
crack [kræk] vt. 使破裂，使裂纹，使打开
quench [kwentʃ] vt. 熄灭，[机] 淬火
fix [fiks] vt.& n. 安装，修理，使固定，准备
tempering ['tempəriŋ] n. 回火
shaping [ʃeipiŋ] n. 成形，造型，塑造
rolls [rəuls] n. 薄板卷，泛指钢材
pressure ['preʃə] n. 压力，压强
draw [drɔː] vt. 拉，画，吸引
mechanical [mi'kænikəl] adj. 机械的，呆板的，力学的
machining [mə'ʃiːniŋ] n. 切削，制造，机械加工
grinding ['graindiŋ] n. 磨削加工；研磨 adj. 刺耳的；令人难以忍受的
machined [mə'ʃiːnd] adj. 机械加工的
bend [bend] vt. 使弯曲，使屈服，使致力，使朝向
squeeze [skwiːz] vt. 挤（压），压榨，勒索，紧握
shear [ʃiə] vt. 剪断，切断，修剪，剥夺 n. 切变；修剪，大剪刀
recrystallization [riːˌkristəlai'zeiʃən] n. 再结晶
hardening ['hɑːdəniŋ] n. 硬化，淬火，锻炼
crystalline ['kristəlain] adj. 结晶的 n. 结晶体
influence ['influəns] n. 影响，作用
prior ['praiə] adj. 在先的，在前的，优先的
mechanical means 机械设备（工具）
be known as... 称为……
work-hardening range 淬火温度区

Notes

1. This produces unequal contractions, which may give rise to distortion or cracking.

译文：这将引起不均匀收缩，增加其出现畸变和破裂的可能。

说明：此句为复合句，which 引导的非限制性定语从句是对 unequal contraction 进一步解释说明。

2. This does not include the shaping of metals by machining or grinding, in which processes metal is actually machined off, nor does it include the casting of molten metal into some form by use of molds.

译文：这既不包括去除金属材料的切削或磨削加工成形，也不包括需借助模具的金属铸造成形。

说明：in which processes metal is actually machined off 为非限制性定语从句，对前面的主句起进一步解释说明的作用；nor does it include the casting of molten metal into some form by use of molds 为倒装句，由于 nor 提前，第二个分句采用了部分倒装。

Exercises

Ⅰ. Answer the following questions according to the passage.

1. Explain the ways that can alter the characteristics of steel.
2. What are the advantages of annealing?
3. What is the meaning of tempering?
4. What is the temperature at which steel starts to re-crystallize?

Ⅱ. Translate the following phrases or sentences into English.

1. 金属的分子结构　　2. 不均匀压缩　　3. 在临界温度
4. 各种成形工序　　5. 金属的机械加工　　6. 金属的冷热加工
7. 我们可以通过快速冷却提高金属的硬度。
8. 金属通过锻造、弯曲、挤压、拉拔或剪切达到其最终形状。

Translating Skills:

词 性 转 译

在翻译过程中，由于英、汉两种语言的表达方式不同，不能逐词对译。原文中有些词在译文中需要转换词性，才能使译文通顺流畅。词性转译的几种典型情况如下：

一、英语动词、形容词、副词译成汉语名词

1. Telecommunications **means** so much in modern life that without it our modern life would be impossible.

电信在现代生活中**意义**重大，没有它就不可能有我们现在的生活。

2. Gases **differ** from solids in that the former has greater compressibility than the latter.

气体和固体的**区别**在于气体的可压缩性比固体大。

3. The instrument is **characterized** by its compactness and portability.

这个仪器的**特点**是结构紧凑，携带方便。

二、英语名词、介词、形容词、副词译成汉语动词

1. **Substitution** of manual finishing is one example of HSM application.

替代手工精加工是高速加工应用的一个例子。

2. Scientists are **confident** that all matter is indestructible.

科学家们**深信**一切物质都是不灭的。

三、英语名词、副词、动词译成汉语形容词

1. This wave guide tube is **chiefly** characterized by its simplicity of structure.

这种波导管的**主要**特点是结构简单。

2. It is a fact that no structural material is **perfectly** elastic.

事实上没有一种结构材料是**十全十美**的弹性体。

四、英语形容词、名词译成汉语副词

1. Quasi-stars were discovered in 1963 as a result of an **effort** to overcome the shortcomings of radio telescopes.

类星体是 1963 年发现的，是人们**努力**克服射电望远镜的缺点所取得的一项成果。

2. A **continuous** increase in the temperature of a gas confined in a container will lead to a **continuous** increase in the internal pressure within the gas.

不断提高密封容器内气体的温度，会使气体的内压力**不断**增大。

Reading: Soldering and Welding

There are a number of methods of joining metal articles together, depending on the type of metal and the strength of the joint that is required.

1. Soldering and Brazing

Soldering gives a satisfactory joint for light articles of steel, *copper* or *brass*, but the strength of soldered joint is rather less than a joint which is *brazed*, *riveted* or welded. These methods of joining metal are normally adopted for strong permanent joint. Soldering is the process of joining two metals by a third metal to be applied in the *molten* state. Solder consists of tin and lead, while *bismuth* and *cadmium* are often included to lower the melting point. One of the most important operations in soldering is that of cleaning surfaces to be joined, this may be done by some *acid* cleaner. Although the *oxides* are removed by the cleaning operation, a new oxide coating forms immediately after cleaning, thus is preventing the solder to unite with the surface of the metal. *Flux* is used to remove and prevent *oxidation* of the metal surface to be soldered, allowing the solder to flow freely and unite with the metal. Zinc *chloride* is the best flux to use for soldering most ferrous and non-ferrous metals, for soldering aluminum *stearin* acid or *vaseline* is to be used as fluxes. The soldering copper is a piece of copper attached to a steel rod having a handle. Soldering coppers are made in different lengths, forms and weights. The quality of soldering depends to a great degree on the form and size of the soldering copper. Two parts may be perfectly soldered only when the surfaces to be joined have absorbed enough heat to keep solder melted for some time.

In some cases it may be necessary to connect metal surfaces by means of hard zinc solder which *fuses* at high temperature. This kind of soldering is called brazing.

2. Welding

Welding is the joining of two metal pieces by *softening* with heat and then pressing, hammering, or fusing them together. Many parts of machines, automobiles, airplanes, ships, bridges, and buildings are welded.

Oxyacetylene welding is the heating of two pieces of metal with a flame that burns a mixture of oxygen and *acetylene* gas. The oxygen and acetylene gas are kept in two separate steel *tanks* from which they flow to a *torch*; there the two gases mix and then pass into the flame. It is the hottest flame known for ordinary use; its temperature is about 6,300 degrees Fahrenheit. The oxyacetylene flame may be used to cut iron and steel. Electric or arc welding is the heating of two pieces of metal to be welded by electricity. This heat is the hottest that can be obtained for engineering purposes; it is about 7,232 degrees Fahrenheit. The ends are thus melted together, making a welded joint. Spot welding is welding two pieces of metal in spots with electricity and is done with a machine called spot welder. A forged weld is made by softening the ends of two metal pieces in a *furnace* and then hammering them together.

New Words and Phrases

brass [brɑːs] n. 黄铜，黄铜制品
braze [breiz] vt. 用铜锌合金焊接
copper [ˈkɔpə] n. 铜，铜币；铜制物 adj. 铜的
rivet [ˈrivit] n. 铆钉 vt. 铆，铆接
molten [ˈməultən] adj. 融化的，熔融的
bismuth [ˈbizməθ] n. 铋
cadmium [ˈkædmiəm] n. 镉
acid [ˈæsid] n. 酸 adj. 酸的，酸味的
oxide [ˈɔksaid] n. 氧化物
flux [flʌks] n. 焊剂，助焊剂
oxidation [ˌɔksiˈdeiʃən] n. 氧化
chloride [ˈklɔːraid] n. 氯化物
stearin [ˈstiːərin] n. [化] 硬脂精；硬脂
vaseline [ˈvæsiliːn] n. [化] 石油冻；矿脂；凡士林（商品名）
fuse [fjuːz] vi. 熔（化），熔合 n. 熔丝
soften [ˈsɔfn, ˈsɔːfn] vt. 使软化，使柔和，变软弱
oxyacetylene [ˌɔksiəˈsetiliːn] adj. [化] 氧乙炔的
acetylene [əˈsetiliːn] n. [化] 乙炔
tank [tæŋk] n. （盛液体、气体的）大容器，槽，罐；坦克
torch [tɔːtʃ] n. 火炬，火把；[机] 气炬；吹管

furnace [ˈfəːnis] n. 炉子，熔炉；[冶] 鼓风炉，高炉

two separate steel tanks 两个独立的钢制储气瓶

soldering iron 焊铁，烙铁

Exercises

Answer the following questions according to the passage.

1. How many methods are there in joining metal articles together?
2. What does the soldering mean?
3. What's the brazing?
4. Explain the method of welding.

Unit 5 Introduction to Mould

Text

Mould is a *fundamental* technological *device* for industrial production. Industrially produced goods are formed in moulds which are designed and built specially for them. The mould is the core part of *manufacturing* process because its *cavity* gives its shape. There are many kinds of moulds, such as casting and forging *dies*, drawing dies, injection moulds, extruding moulds, compressing moulds, etc.

The following is the introduction of some of the moulding processes and the *corresponding* moulds used.

Compression Moulding

Compression moulding is the least *complex* of the moulding processes and is the ideal for large parts of low-quantity production. For low-quantity requirements it is more economical to build a compression mould than an injection mould. Compression moulds are often used for proto-typing, where samples are needed for testing fit and forming into assemblies. [1] This allows for further design modification before building an injection mould for high-quantity production. [2] Compression moulding is best suited for designs where tight tolerance is not required.

Injection Moulding

Injection moulding is the most complex of the moulding processes. Due to the more complex design of the injection mould, it is more expensive to *purchase* than a cast or compression mould. Although tooling costs can be high, cycle time is much faster than other processes and the part cost can be low, particularly when the process is *automated*. [3] Injection moulding is well suited for moulding *delicately* shaped parts because high pressure (as much as 29000 psi) is maintained on the material to push it into every corner of the mould cavity.

Casting Moulding

There are two types of casting, open casting and pressure casting. With open casting, the liquid *mixture* is poured into the open cavity in the mould and allowed to cure. With pressure casting, the liquid mixture is poured into the open cavity, the cap is put in place and the cavity is *pressurized*. Pressure casting is used for more complex parts and when moulding foam materials.

In principle, pressure casting is identical to injection moulding with a different class of materials. Cast moulding can, in fact, produce parts that have identical *geometries* to injection-moulded ones. In many cases, injection moulding has been a *substitute* for casting moulding due to decreased parts cost. However, for *structural* parts, particularly those parts with thick-walls, cast moulding can often be the better selection.

Extrusion Moulding

Although extrusion moulds are quite simple, the extrusion moulding requires great care in the setting up and manufacture and final processing to ensure *consistency* of product. Pressure is forced through the die plate that has the correct profile cut into it. *Variations* in feed rate, temperature and pressure need to be controlled.

Most extrusion moulds are simply one round piece of steel with the profile of the intended extrusion wire cut into them. *Allowances* are made for the *shrinkage*, expansion of the intended compound. Extrusion dies are the least complex of the moulds.

New Words and Phrases

fundamental [ˌfʌndəˈmentl] adj. 基本的，根本的，重要的

device [dɪˈvaɪs] n. 装置，设备；策略；设计

manufacture [ˌmænjuˈfæktʃə(r)] v. 制造；捏造

cavity [ˈkævəti] n. 洞，空穴，型腔，蛀洞

die [daɪ] vt. & vi. 死亡，熄灭；凋零，枯萎；n. 模具；冲模；压模

corresponding [ˌkɒrəˈspɒndɪŋ] adj. 相当的；相应的；一致的

complex [ˈkɒmpleks] adj. 复杂的；合成的；复合的

purchase [ˈpɜːtʃəs] vt. 购买

automated [ˈɔːtəʊmeɪtɪd] adj. 自动化的

delicately [ˈdelɪkətli] adv. 优美地；微妙地；精致地；谨慎地；巧妙地

mixture [ˈmɪkstʃə(r)] n. 混合；混合物

pressurized [ˈpreʃəraɪzd] adj. 加压的，受压的

geometry [dʒiˈɒmətri] n. 几何，几何学；几何形状，几何图形，几何结构

substitute [ˈsʌbstɪtjuːt] vt. & vi. 代替，替换，代用 n. 代替者；替补

structural [ˈstrʌktʃərəl] adj. 结构的，构造的；建筑的，建筑用的

consistency [kənˈsɪstənsɪ] n. 连贯，一致性；强度；硬度；浓稠度

variations [ˌveəriːˈeɪʃənz] n. 变更；变化

allowance [əˈlaʊəns] n. 津贴；定量，余量；允许

shrinkage [ˈʃrɪŋkɪdʒ] n. 收缩；减少；损失，损耗

casting and forging dies 铸造模具与锻造模具

drawing dies 拉丝模具；拉伸模具

injection moulds 注塑模具

extruding moulds 挤出模具

compressing moulds/compression moulds 压缩模具

cycle time 工作周期

in principle 原则上

be identical to 与……相同

feed rate 进给速率

Notes

1. Compression moulds are often used for proto-typing, where samples are needed for testing fit

and forming into assemblies.

译文：压缩模具常用于产品原型制作，所制作的试样用于组合件间的配合测试并最终成型为组件的一部分。

说明：该句为复合句。主句是 compression moulds are often used for proto-typing，由关系副词 where 引导的定语从句修饰先行词 proto-typing，此处的 where 相当于 at/in/on + which。例如：

They reached there yesterday, where a negotiation of sale will be held.

他们昨天抵达那里，有一个关于销售的谈判在那儿举行。

2. This allows for further design modification before building an injection mould for high-quantity production.

译文：这样一来，在制作适用于大批量生产的注塑模具之前，可在设计上做进一步的变更和完善。

说明：该句为简单句。介词 before 意为“在……之前”；但也可以根据具体语意翻译成“在……之后，再……”。例如：

Look before you leap.　　三思而后行。

3. Although tooling costs can be high, cycle time is much faster than other processes and the part cost can be low, particularly when the process is automated.

译文：尽管模具的费用很高，但产品的生产周期比其他工艺要短，且产品的生产成本低，尤其是在成型过程实现自动化的条件下。

说明：该句为并列复合句，而且两个分句都是复合句。在第一个分句中，although 引导的让步状语从句置于句首，表示“尽管，虽然”；第二个分句中主句在前，when 引导的时间状语从句在后，与第一个分句之间用 and 连接，存在因果关系。第二个分句中的副词 particularly 在翻译时还可以放到句尾，译为“……尤其如此”，使汉语句子读起来更加通顺、完整。例如：

Particularly in Alaska, we understand the inherent link between energy and prosperity, energy and opportunity and energy and security. 我们理解能源和繁荣、机遇、安全的内在关系，在阿拉斯加尤其如此。

Exercises

Ⅰ. Answer the following questions according to the passage.

1. How many types of mould are mentioned? What are they?
2. Which mould is the most complex of the moulding processes?
3. Why is pressure casting identical to injection moulding with a different class of materials?
4. What does “allowances” mean in the last paragraph?

Ⅱ. Translate the following phrases into Chinese or English.

1. injection moulds	2. extruding moulds	3. compressing moulds
4. feed rate	5. 周期	6. 原则上

Ⅲ. Complete the following sentences with the proper form of the words in brackets.

1. ______ (automate) Teller Machine (ATM) is pervasive（无处不在的）nowadays.

2. Life is a ______ (mix) of joy and sorrow.

3. Mary performed her dance ________ (delicate) in the Spring Festival Gala.

4. Kunming, known as The City of Spring, offers people obvious ______ (vary) throughout the year.

Ⅳ. Translate the following sentences into English by using the words in brackets.

1. 现在他们在制造大规模杀伤性武器方面领先。(manufacture)

2. 我们在原则上接受了这个条款。(in principle)

3. 什么都替代不了努力工作和奉献。(substitute)

4. 这个方案涉及许多复杂的技术问题。(complex)

Translating Skills:

增 词 译 法

增词译法，是翻译的基本方法之一，就是在译句中增加或补充句子中原本没有或省略掉了的词语，以便更完善、更清楚地表达原意。

一、增加表示时态概念的词

英语中通常用动词形式的变化或加上助动词表示时态，翻译时要将其用汉语明确表达出来。例如：

The mechanical ignition system **was** mechanical and electrical and **used no** electronics before 1975.

1975 年之前，机械式点火系统**只采用**机械装置和电气装置，**并未采用**电子装置。

二、增加表示数的概念的量词

翻译数词时，常常需要加上量词，如“个”“只”“量”“台”等。此外，还要体现其复数含义。例如：

1. The **moving parts** of a machine are often oiled that friction may be greatly reduced.

机器的**各个可动部件**经常加油润滑，可大大减少摩擦。

2. **The first** electronic computers used vacuum tubes and others components and this made equipment very large and bulky.

第一代电子计算机使用真空管和其他元件，这使设备又大又笨。

三、增加被省略的句子成分

英译汉时，有时所有的词汇都相应地译出后，汉语句子的结构仍显得不完整。这时，应增补缺少的句子成分。例如：

1. An exact solution **demands** much calculation.

一个精确的解答**需要进行**很多次计算。

2. Fuses are cheap, other **equipment** is much more expensive.

熔丝很便宜，但其他**保险设备**却昂贵得多。

四、增加被动语态或动名词中没有具体指出的动作执行者或暗含的逻辑主语

To explore the moon's surface, rockets **were launched** again and again.

为了勘探月球的表面，**人们一次又一次地发射火箭**。

五、在形容词前加名词

1. According to Newton's Third Law of Motion action and reaction are **equal and opposite**.

根据牛顿运动第三定律，作用力和反作用力是**大小相等方向相反的**。

2. The washing machine of this type is indeed **cheap and fine**.

这种类型的洗衣机真是**物美价廉**。

六、增加表示数量意义的概括性的词，起修饰润色作用

1. **The frequency, wavelength and speed** of sound are closely related.

声音的**频率、波长与速度三者**密切相关。

2. A designer must have a good foundation in **statics, kinematics, dynamics and strength of materials.**

一个设计人员必须在**静力学、运动学、动力学和材料力学这四个方面**有很好的基础。

七、增加使译文语气连贯的词

1. Manganese is a **hard, brittle**, gray-white metal.

锰是一种灰白的、**又硬又脆的**金属。

2. In general, all the metals are good conductors, **with silver the best and copper the second**.

一般来说，金属都是良导体，**其中银最佳，铜次之**。

Reading: Mould Materials

Depending on the processing parameters for the various processing methods as well as the length of the production run, i.e., the number of finished products to be produced, moulds must satisfy a great variety of requirements. It is therefore not surprising that, moulds can be made from a very broad *spectrum* of materials, including such *exotic* materials as paper match and plaster. However, because most processes require high pressures, and often combined with high temperatures, metals still represent by far the most important material group, with steel being the *predominant* metal. It is interesting in this regard that, in many cases, the selection of the mould material is not only a question of material properties and an *optimum* price-to-performance ratio but also that the methods used to produce the mould, and thus the entire design can be influenced.

A typical example can be seen in the choice between cast metal moulds, with their very different cooling systems, compared to machined moulds. In addition, the production technique can also have an effect. For instance, it is often reported that, for the sake of simplicity, a prototype mould is frequently machined from solid stock with the aid of the latest technology such as computer-aided design (CAD) and computer-integrated manufacturing (CIM). In contrast to the previously used methods based on the use of patterns, the use of CAD and CIM often represents the more economical solution today, not only because this production capability is available in house but also because with any other technique an order would have to be placed with an outside supplier.

Overall, although high-grade materials are often used, as a rule, standard materials are used in mould making. New, high-performance materials, such as ceramics, for instance, are almost completely absent. This may be related to the fact that their desirable characteristics, such as constant properties up to very high temperature, are not required in moulds. Whereas their negative characteristics, e. g., low tensile strength and poor thermal conductivity, having a clearly related to ceramics, such as sintered material, is found in mould making only to a limited degree. This refers less to the modern materials and components produced by powder *metallurgy*, and possibly by hot *isostatic* pressing, than to sintered metals in the sense of *porous*, air-permeable materials.

Removal of air from the cavity of a mould is a necessary with many different processing methods, and it has been proposed many times that this can be accomplished using porous metallic materials. [1] The advantages over specially fabricated *venting* devices, particularly in areas where melt flow fronts meet, i. e., at weld lines, are as obvious as the potential problem areas: on one hand, preventing the texture of such surfaces from becoming visible on the finished product; and on the other hand, preventing the *micropores* from quickly becoming *clogged* with residues. It is also interesting in this case that completely new possibilities with regard to mould design and processing technique result from the use of such materials. The process steps of venting, cooling, and ejecting in relation to the use of sintered metals can be best illustrated with the aid of the sketches shown in Fig. 5-1.

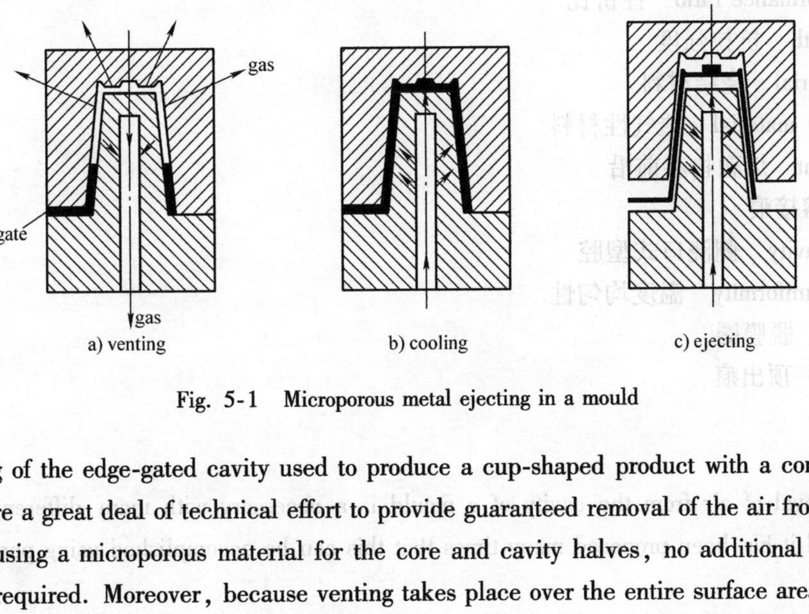

Fig. 5-1 Microporous metal ejecting in a mould

Venting of the edge-gated cavity used to produce a cup-shaped product with a complex bottom would require a great deal of technical effort to provide guaranteed removal of the air from the bottom region. By using a microporous material for the core and cavity halves, no additional measures for venting are required. Moreover, because venting takes place over the entire surface area of the cavity, filling of the cavity can occur faster, and there is, in principle, freedom in selecting the location of the gate.

It is particularly difficult to remove the necessary amount of heat in regions with long, narrow cores. In this case, it is possible to distribute cold gas via the system of micropores in the core and in this way intensify the cooling, with the possible result of achieving a shorter cycle time. Improved temperature uniformity over the mould surface can be another beneficial side effect.

It is also possible, in combination with other means (*ejectors*, stripper rings) or even without them, to eject the product by introducing gas at high pressure into the core half of the mould. In this way, the risk of ejector marks is reduced or eliminated, and there is no need to overcome a *vacuum* during ejection.

New Words and Phrases

spectrum [ˈspektrəm] n. 范围；光谱
exotic [ɪɡˈzɒtɪk] adj. 奇异的，外来的
predominant [prɪˈdɒmɪnənt] adj. 占优势的，主要的
optimum [ˈɒptɪməm] adj. 最优化的
metallurgy [məˈtælədʒi] n. 冶金学，冶金术
isostatic [ˌaɪsəˈstætɪk] adj. 均衡的
porous [ˈpɔːrəs] adj. 多孔的
micropore [ˈmaɪkrəpɔː(r)] n. 微孔
clog [klɒɡ] v. 堵塞，障碍
venting [ˈventɪŋ] n. 通风，排气
ejector [ɪˈdʒektə] n. 顶出器
vacuum [ˈvækjuəm] n. 真空
price-to-performance ratio 性价比
tensile strength 抗拉强度
sintered material 烧结材料
air-permeable material 透气性材料
melt flow front 熔融物料前沿
weld line 熔接痕
edge-gated cavity 侧浇口式型腔
temperature uniformity 温度均匀性
stripper ring 脱膜圈
ejector mark 顶出痕

Notes

1. Removal of air from the cavity of a mould is a necessary with many different processing methods, and it has been proposed many times that this can be accomplished using porous metallic materials.

译文：在许多不同的成型方法中，人们都必须将型腔中的气体排出，使用多孔金属材料来解决上述问题的呼声很高。

说明：该句为用 and 连接的并列句。在英语中，“无灵”名词作主语指的是由没有生命的事物充当主语，如物品、抽象概念、情感、时间、地点等。第一个分句中 removal 是“无灵”名词作主语，译文添加了“有灵”主语“人们”，更加符合汉语的表达习惯。此外，科技英语具有客观的文体特征，第二个分句用 it 作形式主语，真正主语是其后的 that 从句。例如：

It must be admitted that only when we get a taste of its abuse will we feel the full meaning of cloning. 必须承认，只有当我们尝到克隆被滥用的恶果时，我们才会彻底明白它意味着什么。

Exercises

Complete the following sentences with the proper form of the words in brackets.

1. The country is famous for its ______ (exotic) atmosphere.
2. Translation occupies an ______ (predominate) role in English studies.
3. Your illness must be closely ______ (relate) to the irregular sleep habit.
4. This type of robot is featured by high ______ (perform).

Unit 6 Mould Design

Text

Die-casting is a technique for mass-producing metal products and components. Mould design is one of the most important steps in the process because the shape and *attributes* of the mould will directly affect the final product. The die casting procedure forces *molten* metal into moulds using high pressure and it requires a mould with exact *specifications* to achieve the task. [1]

The Importance of Mould Design

Mould design affects the shape, *configuration*, quality, and uniformity of a product created through the die casting procedure. Improper specifications can result in tool or material *corrosion*, as well as *inferior* product quality, while an effective design can improve efficiency and production time. [2]

Factors *Contributing* to Quality Mould Design

There are a number of mould design factors to consider when deciding on the *appropriate* specifications for a project. Some of these factors include: ①die draft; ②*fillets*; ③parting lines; ④*bosses*; ⑤*ribs*; ⑥holes and windows; ⑦ symbols; ⑧ wall thickness.

Die draft

Die draft is the degree to which a mould core can be *tapered*. A precise draft is needed to smoothly *eject* the casting from the die, but since draft is not constant and varies according to the angle of the wall, features such as the type of molten alloy used, shape of the wall, and depth of the mould can affect the process. Mould geometry can also influence draft. In general, untapped holes require tapering, due to the risk of *shrinkage*. Likewise, inner walls can also shrink, and therefore require more drafting than outer walls.

Fillets

A fillet is a *concave junction* used to smooth an angled surface. Sharp corners can *hinder* the casting process, so many moulds have fillets to create rounded edges and reduce the risk of production errors. With the exception of the parting line, fillets can be added nearly anywhere on a mould.

Parting Line

The parting line, or parting surface, connects different sections of the mould together. If the parting line is imprecisely positioned or becomes *deformed* from work strain, material may *seep* through the gap between the mould pieces, leading to non-uniform moulding and excessive *seaming*. [3]

Bosses

Bosses are die cast *knobs* that serve as mounting points or stand-offs in mould design. Manufacturers often add a hole to the interior structure of the boss to ensure uniform wall thickness in a moulded product. Metal tends to have difficulty filling deep bosses, so filleting and ribbing may be necessary to *alleviate* this problem.

Ribs

Die cast ribs can be used to improve material strength in products lacking the wall thickness required for certain applications. Selective rib placement can reduce the chance of stress cracking and non-uniform thickness. It is also beneficial for decreasing product weight and improving fill capabilities.

Holes and Windows

Including holes or windows in a die cast mould directly affects the ease of ejecting a completed moulding and enables the creation of *substantial* drafts. Additional features, such as *overflows*, *flashovers*, and cross feeders may be necessary to prevent unwanted casting within the holes or poor material flow around the holes.

Symbols

Manufacturers often include brand names or product logos in the mould design of die-cast products. While symbols do not typically complicate the die casting process, their use can affect production costs. In particular, a raised logo or symbol requires additional molten metal volume for each manufactured part. Conversely, a recessed symbol requires less raw material and can reduce expenses.

New Words and Phrases

die-casting [ˈdaɪkɑːstɪŋ] n. 压模法；铸造法

attribute [ˈætrɪbjuːt] vt. 把……归于 n. 属性，标志，象征，特征

molten [ˈməʊltən] adj. 熔化的；炽热的；铸造的

specification [ˌspesɪfɪˈkeɪʃn] n. 规格；详述

configuration [kənˌfɪɡəˈreɪʃn] n. 结构，布局，形态

corrosion [kəˈrəʊʒn] n. 侵蚀；腐蚀；锈蚀

inferior [ɪnˈfɪəriə(r)] adj. 较低的，次等的，不如的 n. 下级，属下；[印] 下标符号

contribute [kənˈtrɪbjuːt] vt. 捐助；投稿 vi. 投稿；贡献；是原因之一

appropriate [əˈprəʊpriət] adj. 适当的；相称的 vt. 占用；拨出（款项）

fillet [ˈfɪlɪt] n. 圆角 vt. 倒圆角

boss [bɒs] n. 老板；首领；凸起，凸台

rib [rɪb] n. 肋骨，排骨，肋状物 vt. 戏弄；装肋于

taper [ˈteɪpə(r)] n. 蜡芯；尖锥形；渐弱 v. 逐渐变小，逐渐消失

eject [iˈdʒekt] v. 喷射；放逐，驱逐

shrink [ʃrɪŋk] vi. 收缩，退缩，萎缩；畏惧，害怕 vt. 使收缩

concave [kɒnˈkeɪv] adj. 凹的 n. 凹面，凹线，凹形

junction [ˈdʒʌŋkʃn] n. 连接，会合处，交叉点

hinder [ˈhɪndə(r)] v. 阻碍；打扰 adj. 后面的

deformed [dɪˈfɔːmd] adj. 变形的，变丑的，破相了的，畸形的

seep [siːp] v. 渗出，渗漏 n. 渗漏；小泉；水坑
seaming [siːmɪŋ] n. 接缝缝合（卷边接合）
knob [nɒb] n. 疙瘩；球形把手；小块；旋钮
alleviate [ə'liːvieɪt] vt. 减轻；使……缓和
substantial [səb'stænʃl] adj. 大量的；坚固的；实质的；可观的 n. 本质；重要部分
overflow [ˌəʊvə'fləʊ] v. 泛滥，溢出，充满，洋溢 n. 泛滥，溢值，剩出
flashover ['flæʃˌəʊvə] n. 飞弧；击穿；闪络；跳火
die casting mould 压铸型
die draft 起模斜度
parting line/ parting surface 分型线/分型面
with the exception of 除……之外
mounting point 安装位置点
stand-off 托脚

Notes

1. The die casting procedure forces molten metal into moulds using high pressure and it requires a mould with exact specifications to achieve the task.

译文：压铸工艺使用高压将熔化了的金属注入模腔，该过程的完成要求模腔的规格必须精准。

说明：这是一个并列句，两个分句都是简单句。第一个分句中的 using high pressure 是现在分词短语作方式状语；第二个分句中的 to achieve the task 是动词不定式短语作目的状语。此外，specification 意为"规格、详述"，常出现于科技英语中。例如：

You can increase model precision through the specification and assemblage of model constraints. 您可以通过模型约束的规范和集合来提高模型的精确度。

2. Improper specifications can result in tool or material corrosion, as well as inferior product quality, while an effective design can improve efficiency and production time.

译文：不合适的规格会导致工具或原材料受到腐蚀，还会造成产品质量低劣，而有效的设计会提高效率，节省生产时间。

说明：这是一个并列句，两个分句之间是转折关系。句中的 while 表示前后两种情况的对比，原作者重在利用该词的功能来说明有效的设计能够提高效率，进一步彰显设计规格精准的重要性，从而使前后文形成一定反差。例如：

The walls are green, while the ceiling is white. 墙是绿色的，而天花板是白色的。

3. If the parting line is imprecisely positioned or becomes deformed from work strain, material may seep through the gap between the mould pieces, leading to non-uniform moulding and excessive seaming.

译文：如果分型线的位置不准确，或由于工作压力而变形，原材料可能会从模具工件之间的间隙渗出，会引起成型不均匀及接缝溢出。

说明：这是一个包含有条件状语从句的复合句，主句在后，leading to... 是现在分词短语作结果状语。

Exercises

Ⅰ. Answer the following questions according to the passage.

1. Why mould design is important?

2. What are the factors contributing to quality mould design?

3. What is a fillet?

4. What is the use of ribs?

Ⅱ. Choose the best answer(s) to complete the following statements.

1. Woman is inferior ________ man in strength.

A. to　　B. with　　C. for　　D. of

2. Honesty and hard work __________ to success and happiness.

A. attribute　　B. contribute　　C. tribute　　D. pay

3. The government has ____________ some funds for education.

A. appropriated　　B. prospered　　C. recovered　　D. discovered

4. The storm is beginning to ________ now.

A. taper off　　B. take off　　C. put off　　D. get off

5. He could only __________ the sorrow by drinking.

A. repeat　　B. alleviate　　C. repay　　D. return

6. The reforms have made ________ headway.

A. substantial　　B. substance　　C. subway　　D. suburban

Ⅲ. Translate the following sentences into English by using the words in brackets.

1. 我相信我们每一个人都能为世界的未来做出贡献。(contribute to)

2. 如今，很多方法都可以用来减轻背部疼痛。(alleviate)

3. 文件需要做大量修改。(substantial)

4. 我喜欢所有的水果，就是不喜欢葡萄。(with the exception of)

Translating Skills:

省 略 译 法

英译汉时，英文中有些词语不必译出，而句子意思仍然清楚完整，这种译法称为省略译法。省略译法用于如下情况：

一、省略代词

1. By the word "alloy" **we** mean "mixture of metal".

用"合金"一词表示"金属混合物"。

2. **You** can think of CNC offsets as memories on an electronic calculator.

可以将 CNC 偏置想象成电子计算机的存储器。

二、省略冠词

1. You can think of CNC offsets as memories on **an** electronic calculator.

可以将 CNC 偏置想象成电子计算机的存储器。

2. 定冠词 the 与某些形容词连用，使形容词名词化，代表一类人或物，the 不译。如：the positive 正极，the rich 富人。

三、省略介词

1. Many water power stations have been built **in** our country.

我国已建成许多水电站。

2. The barometer is a good instrument **for** measuring air pressure.

气压计是测量气压的好仪器。

四、省略连词

1. The consequences are higher costs, a decrease in output **and** a lower reliability.

结果是成本提高、产量减少、可靠性降低。

2. **If** there were no heat-treatment, metals could not be made so hard.

没有热处理，金属就不会变得如此硬。

五、省略逻辑上或修辞上不需要的词

1. As we know, electrons revolve about the nucleus, **or center of an atom.**

正如我们所知，电子围绕着原子核旋转。

2. A generator can not produce energy, **what it does** is to convert mechanical energy into electrical energy.

发电机不能产生能量，只能将机械能转为电能。

Reading: Plastic Product Design

It is well known that plastics possess many valuable characteristics such as low weight, unlimited color ranges, *esthetic* values, low costs, and excellent mechanical, electrical and chemical properties, to name a few. [1] The task of designing work is to take the right combination of all these characteristics and *embody* them in the product to be moulded.

First of all, the designer must consider what the articles are to accordingly determine the material, shape and process to be taken. Generally speaking, the following points should be noticed particularly in designing products:

1. Keep wall thickness uniform. If change in wall thickness is necessary, avoid abrupt change; *slope* gradually from one thickness to next.

2. Include draft in all walls, ribs and bosses.

3. Internal corners must be *radiused* at all times. External corners, if design allows, should also be radiused.

4. Rib thickness should be 60% to 70% of *adjoining* wall thickness.

5. Allow sufficient *tolerance* to compensate for variables in material, tool construction and process.

6. Avoid running threads to end of part. Allow *clearance* at both ends.

7. Provide inserts with proper *anchorage* to prevent *pullout* or rotation.

8. Maintain adequate wall thickness over, around, and between inserts.

9. Avoid locating holes too close to an edge.

10. Holes to be tapped should be counter- sink.

11. Use *raised lettering* in place of depressed lettering where possible.

12. Allow additional draft for *textured* parts.

13. *Undercuts* are costly to produce and should be avoid if possible.

New Words and Phrases

esthetic(al) [iːs'θetik] adj. 美（学）的；审美的；雅致的
embody [im'bɔdi] vt. 表现，使具体化；包括，包含
slope [sləup] n. 斜坡，坡度；山坡 vi. 倾斜；悄悄地走，溜
radius ['reidjəs] n. 半径（距离），半径范围；桡骨
adjoining [ə'dʒɔɪnɪŋ] adj. 毗邻的，邻近的
tolerance ['tɒlərəns] n. 宽容，容忍；忍受性；（配合）公差，容限
clearance ['kliərəns] n. 净空，余隙，间隙；清除，排除
anchorage ['æŋkərɪdʒ] n. 抛锚，停泊；抛锚处
pullout [pulaut] n. 拔；拉；撤退
counter-sink vt. 钻（孔），穿（孔），使（钻头等）插入 n. 埋头钻；凹陷
raised [reizd] adj. 凸起的，阳文的，浮雕的；有凸起的花纹（或图案）
lettering ['letərɪŋ] n. 刻字 v. 用字母写；用印刷体写（letter 的 ing 形式）
textured ['tekstʃəd] adj. 起纹理的，构造成的；使具有某种结构的
undercut [ˌʌndə'kʌt] vt. 从下部切开；n. 底切；砍口；切球凸雕；浮雕

Notes

1. It is well known that plastics possess many valuable characteristics such as low weight, unlimited color ranges, esthetic values, low costs, and excellent mechanical, electrical and chemical properties, to name a few.

译文：众所周知，塑料具有许多宝贵的特性，例如质量轻、无限的染色范围、审美价值、低成本，优良的机械性能、电性能和化学性能等。

说明：这是 it is + 过去分词 + that 的特殊句式。句中 it 作形式主语，真正主语是 that 引导的主语从句。such as 表示列举，译为"例如"；而句尾的 to name a few 具有异曲同工之妙，在翻译时可以采用省略译法。

Exercises

Answer the following questions according to the passage.

1. What are the advantages of plastics?

2. What' s the task of plastic product designing work?

3. How to design the plastic products?

4. What' s the specification of thickness ?

Unit 7 The Injection Moulding and Machines

Text

Injection moulding is the most commonly used manufacturing process for the fabrication of plastic parts. A wide variety of products are manufactured using injection moulding, which vary greatly in their size, complexity, and application. The injection moulding process requires the use of an injection moulding machine (Fig. 7-1), raw plastic material, and a mould. The plastic is melted in the injection moulding machine and then injected into the mould, where it cools and *solidifies* into the final part. [1]

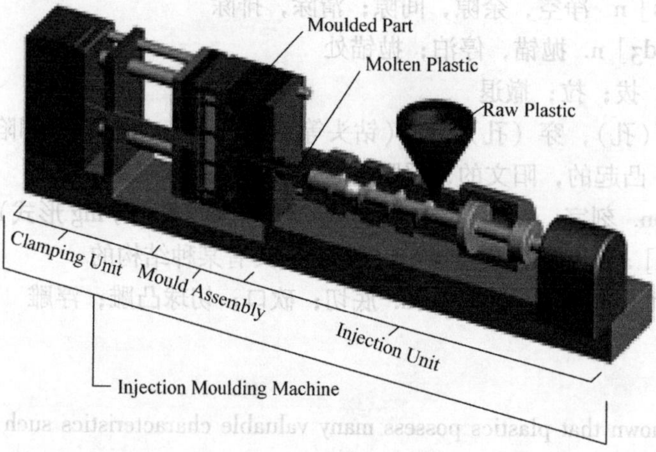

Fig. 7-1 Injection moulding Overview

Injection moulding is used to produce thin-walled plastic parts for a wide variety of applications, one of the most common being plastic housings. [2] Plastic housing is a thin-walled *enclosure*, often requiring many ribs and bosses on the interior. These housings are used in a variety of products including household appliances, consumer electronics, power tools, and as automotive dashboards. Other common thin-walled products include different types of open containers, such as buckets. Injection moulding is also used to produce several everyday items such as toothbrushes or small plastic toys.

Process Cycle

The process cycle for injection moulding is very short, typically between 2 seconds and 2 minutes, and consists of the following four stages:

1) *Clamping*: Prior to the injection of the material into the mould, the two halves of the mould must first be securely closed by the clamping unit. Each half of the mould is attached to the injection

moulding machine and one half is allowed to slide. The *hydraulically* powered clamping unit pushes the mould halves together and *exerts* sufficient force to keep the mould securely closed while the material is injected. The time required to close and clamp the mould is dependent upon the machine — larger machines (those with greater clamping forces) will require more time. This time can be *estimated* from the dry cycle time of the machine.

2) Injection: The raw plastic material, usually in the form of *pellets*, is fed into the injection moulding machine, and advanced towards the mould by the injection unit. During this process, the material is melted by heat and pressure. The molten plastic is then injected into the mould very quickly and the buildup of pressure packs and holds the material. The amount of material that is injected is referred to as the shot. The injection time is difficult to calculate accurately due to the complex and changing flow of the molten plastic into the mould. However, the injection time can be estimated by the shot volume, injection pressure, and injection power.

3) Cooling: The molten plastic that is inside the mould begins to cool as soon as it makes contact with the interior mould surfaces. As the plastic cools, it will solidify into the shape of the desired part. However, during cooling some shrinkage of the part may occur. The packing of material in the injection stage allows additional material to flow into the mould and reduce the amount of visible shrinkage. The mould can not be opened until the required cooling time has *elapsed*. The cooling time can be estimated from several *thermodynamic* properties of the plastic and the maximum wall thickness of the part.

4) Ejection: After sufficient time has passed, the cooled part may be ejected from the mould by the ejection system, which is attached to the rear half of the mould.[3] When the mould is opened, a mechanism is used to push the part out of the mould. Force must be applied to eject the part because during cooling the part shrinks and adheres to the mould. In order to *facilitate* the ejection of the part, a mould release agent can be *sprayed* onto the surfaces of the mould cavity prior to injection of the material. The time that is required to open the mould and eject the part can be estimated from the dry cycle time of the machine and should include time for the part to fall free of the mould. Once the part is ejected, the mould can be clamped shut for the next shot to be injected.

After the injection moulding cycle, some post processing is typically required. During cooling, the material in the channels of the mould will solidify attached to the part (Fig. 7-2). This excess material, along with any flash that has occurred, must be *trimmed* from the part, typically by using cutters. For some types of material, such as *thermoplastics*, the scrap material that results from this trimming can be recycled by being placed into a plastic grinder, also called regrind machine or *granulator*, which regrinds the scrap material into pellets.[4] Due to some degradation of the material properties, the regrind must be mixed with raw material in the proper regrind ratio to be reused in the injection moulding process.[5]

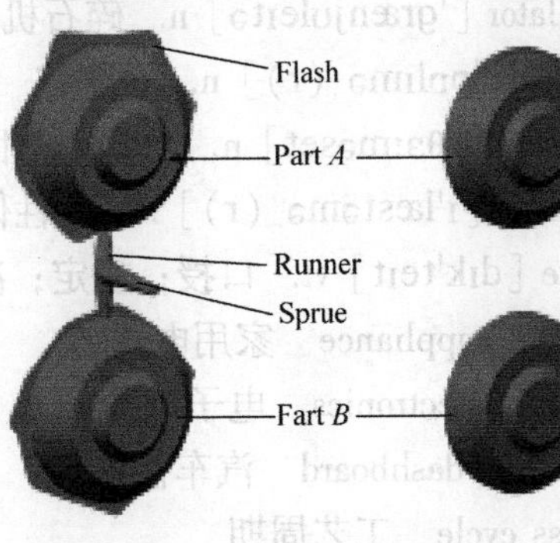

Fig. 7-2 Injection Moulded Part

Equipment

Injection moulding machines have many components and are available in different configurations, including a horizontal configuration and a vertical configuration. However, regardless of their design, all injection moulding machines utilize a power source, injection unit, mould assembly, and clamping unit to perform the four stages of the process cycle.

Materials

There are many types of materials that may be used in the injection moulding process. Most *polymers* may be used, including all thermoplastics, some *thermosets*, and some *elastomers*. When these materials are used in the injection moulding process, their raw form is usually small pellets or a fine powder. Also, colorants may be added in the process to control the color of the final part. The selection of a material for creating injection moulded parts is not solely based upon the desired characteristics of the final part. While each material has different properties that will affect the strength and function of the final part, these properties also *dictate* the parameters used in processing these materials.

New Words and Phrases

solidify [səˈlɪdɪfaɪ] v. 变固体，凝固；变坚定；使团结
enclosure [ɪnˈkləʊʒə(r)] n. 附件；围墙；围绕
clamp [klæmp] n. 夹子；螺钉钳 vt. 夹住；强加；压制
hydraulic [haɪˈdrɔːlɪk] adj. 水力的，水压的，液压的
exert [ɪɡˈzɜːt] vt. 运用；施加
estimate [ˈestɪmət] n. 估价；估计 v. 估计；估价；评价
pellet [ˈpelɪt] n. 小球，小子弹
elapse [ɪˈlæps] v. 逝去，过去 n.（光阴）逝去
thermodynamic [ˌθɜːməʊdaɪˈnæmɪk] adj. 热力的；热力学的
facilitate [fəˈsɪlɪteɪt] vt. 促进，帮助，使……容易
spray [spreɪ] v. 喷雾，喷射，扫射 n. 喷雾，喷雾器，水沫
trim [trɪm] v. 修剪，削减；装饰；调整帆以适应风向
thermoplastic [ˌθɜːməʊˈplæstɪk] adj. 热塑性的 n.［塑料］热塑性塑料
granulator [ˈɡrænjʊleɪtə] n. 碎石机（成粒器）
polymer [ˈpɒlɪmə (r)] n. 聚合体
thermoset [ˈθɜːməset] n. 热凝，热固性；热变定法；热熔塑料
elastomer [ɪˈlæstəmə (r)] n. 弹性体；合成橡胶
dictate [dɪkˈteɪt] vt. 口授；规定；决定，影响 vi. 听写
household appliance 家用电器
consumer electronics 电子消费品
automotive dashboard 汽车仪表板
process cycle 工艺周期
mould release agent 脱模剂

plastic grinder 塑料磨粉机

mould assembly 模具组件

Notes

1. The plastic is melted in the injection moulding machine and then injected into the mould, **where** it cools and solidifies into the final part.

译文：塑料在注塑机里熔化后被注入模腔，并在模腔里冷却、凝固成最终的产品。

说明：这是一个复合句。主句是 the plastic is melted in the injection moulding machine and then injected into the mould；从句是由关系副词 where 引导的定语从句，其先行词为 mould，在定语从句中作地点状语，此时的 where 相当于 in which。例如：

The police railed the spot where the accident happened. 警察将事故发生的地点用栏杆围了起来。

2. Injection moulding is used to produce thin-walled plastic parts for a wide variety of applications, one of the most common being plastic housings.

译文：注射成型可用于生产各种用途的薄壁塑件，其中最常见的是塑料外壳。

说明：这是一个简单句。one of the most common being plastic housings 是独立主格结构作状语，其中 one of the most common 为名词短语，与后面的 being plastic housings 具有逻辑上的主谓关系。例如：

Almost all metals are good conductors, silver being the best of all. 几乎所有的金属都是良导体，而银是最好的导体。

3. After sufficient time has passed, the cooled part may be ejected from the mould by the ejection system, **which** is attached to the rear half of the mould.

译文：经过足够的时间，冷却了的工件会通过弹出系统从模腔脱出，脱出系统附着在模腔的后半部。

说明：这是一个复合句。which 引导的非限制性定语从句，在翻译时可以采用重复法，即重复先行词，使读者能够准确通晓原文的意义。该句的翻译便采用了这种译法，重复提及了“脱出系统”一词。例如：

He never switched off his TV, which always stayed on. 他从来不关电视机，他的电视机始终开着。

4. For some types of material, such as thermoplastics, the scrap material that results from this trimming can be recycled by being placed into a plastic grinder, also called regrind machine or granulator, **which** regrinds the scrap material into pellets.

译文：对于有些种类的材料，如热塑材料，因修剪而产生的废料可放进塑料磨粉机中再循环利用。塑料磨粉机也叫研磨机或粉碎机，可把废料研磨成小颗粒。

说明：一个句子若混合使用 that 引导的定语从句和 which 引导的非限制性定语从句，一般采取分译法，即将长句进行有效切分，译为两句。该句的翻译便采用了这种译法。除此之外，第二个分句的翻译还采用了重复译法，重复了先行词。

5. Due to some degradation of the material properties, the regrind must be mixed with raw material in the proper regrind ratio to be reused in the injection moulding process.

译文：由于重磨后的材料存在一定程度的降解，重磨后的材料必须以适当的重磨比例与原材料混合，才能在注射成型加工过程中实现再利用。

说明：这是一个简单句。由于英语在句法上具有“先原因后结果”的特点，如本句将due to引导的原因状语放在前边，而后边才是结果部分。此外，in the proper ratio to... 意为以适当的比例，等同in the proper proportion to...。例如：

The labor operations of all must be in the proper proportion to one another. 从事这一工作，各方面的劳动者之间必须有适当的比例。

Exercises

Ⅰ. Decide whether the following statements are True（T）or False（F） according to the passage.

1. The process cycle for injection moulding is very short, and is composed of 5 stages.
2. After the injection moulding cycle, some post processing is typically required.
3. Injection moulding machines have many components and are available in different configurations, including a horizontal configuration only.
4. The selection of a material for creating injection moulded parts is solely based upon the desired characteristics of the final part.

Ⅱ. Translate the following phrases into Chinese or English.

1. household appliances	2. consumer electronics
3. process cycle	4. 汽车仪表板
5. 脱模剂	6. 模具组件

Ⅲ. Complete the following sentences with the proper form of the words in brackets.

1. With the ________（elapse）of time, the steel begins to corrode.
2. The medicare（医疗）cost is ________（estimate）to be one billion dollars.
3. The new underground railway will ________（facility）the journey to all parts of the city.
4. The mixture is cooled and then ________（solidify）into toffee（太妃糖）。

Ⅳ. Translate the following sentences into English by using the words in brackets.

1. 对自己要有正确的评价。(estimate)
2. 我们发挥的影响力是很强大的。(exert)
3. 这座岛屿拥有丰富多样的景致。(a wide variety of)
4. 香槟酒从瓶子里喷洒而出。(spray)

Translating Skills:

科技英语词汇的结构特征 I

科技英语（English for Science and Technology，简称EST）可以泛指一切论及或谈及科学和技术的书面语和口语。科技英语词汇概括起来主要有5个特点：①科技英语词汇多源于希腊语和拉丁语；②词义专一；③前后缀出现频率高；④专业词汇出现的频率低；⑤广泛使用缩略词。科技英语词汇可以分为三类：科技词汇、半科技词汇和非科技词汇。

1. 科技词汇（Technical Terms）

科技英语有许多专门的术语，只在某一特定的科技领域来用。许多情况下，虽然同为科学家，但某领域的科学家可能并不了解另一领域的专业词汇，比如物理学家就可能不明白生物学领域的专业术语。且随着科学技术的发展，科技词汇（术语）也日益增多。我们在阅读或翻译一些科技文章时，一定要了解该领域的专业术语，否则可能会翻译不出原文的本意。

例如，化学领域中的 isotope（同位素）、物理领域的 photon（光子）以及生物领域的 chromosome（染色体）等，都是科技词汇。

2. 半科技词汇（Semi-technical Terms）

半科技词汇指的是那些不同于生活中的语言词汇，而且在不同领域又有不同含义的词汇，也称通用科技术语。

例如，orbit 在医学领域指的是“眼眶”，而在物理学领域则表示“轨道”。很普通的词如 web（网）、net（网）、site（站点）在计算机科学中可组成与原词仍然关联的专业词汇，即 website（网址、网站）、internet（互联网）等；mouse（老鼠）、memory（记忆）、read（阅读）在计算机科学中则具有完全不同的含义，其含义分别为“鼠标”“内存”和“读取”。

3. 非科技词汇（Non-technical Terms）

非科技词汇指经常用于科学技术领域，但在日常生活当中不常用的词汇，在修辞学上指的是那些正式用语（learned and formal words）。因为科技英语是一种描述各种自然现象、客观事实的发生过程和特性的语言，所以科技英语的文章所涉及的日常生活中的常用词汇（common words）并不多。

例如，日常生活中我们用 give、have 表达“拥有”的意思，而科技英语中使用 possess；日常生活中用 refer to 表达“涉及，意指”的意思，而科技英语中使用 allude to；日常生活中用 mad 表达“疯狂”的意思，而科技英语中使用 insane 等。

Reading: Components of Injection Moulding Machines

For thermoplastics, the injection moulding machine converts granular or pelleted raw plastic into final moulded parts via a melt, inject, pack, and cool cycle. A typical injection moulding machine consists of the injection system, the mould system, the hydraulic system, the control system, and the clamping system.

The Injection System

The injection system consists of a hopper, a reciprocating screw and barrel assembly, and an injection *nozzle*. This system confines and transports the plastic as it progresses through the feeding, compressing, degassing, melting, injection, and packing stages.

(1) The hopper The thermoplastic material is supplied to moulders in the form of small pellets. The hopper on the injection moulding machine holds these pellets. The pellets are gravity-fed from the hopper through the hopper throat into the barrel and screw assembly.

（2） The barrel The barrel of the injection moulding machine supports the reciprocating plasticizing screw. It is heated by the electric heater bands.

（3） The reciprocating screw The reciprocating screw is used to compress, melt, and convey the material. The reciprocating screw consists of three zones: the feeding zone, the compressing (or transition) zone and the metering zone.

While the outside diameter of the screw remains constant, the depth of the *flights* on the reciprocating screw decreases from the feed zone to the beginning of the metering zone. [1] These flights compress the material against the inside diameter of the barrel, which creates *viscous* (shear) heat.

（4） The nozzle The nozzle connects the barrel to the *sprue bushing* of the mould and forms a seal between the barrel and the mould. The temperature of the nozzle should be set to the material's melt temperature or just below it, depending on the recommendation of the material supplier.

The Mould System

The mould system consists of tie bars, stationary and moving platens, as well as moulding plates (bases) that house the cavity, sprue and runner systems, ejector pins, and cooling channels. The mould is essentially a heat exchanger in which the molten thermoplastic solidifies to the desired shape and dimensional details defined by the cavity.

The Hydraulic System

The hydraulic system on the injection moulding machine provides the power to open and close the mould, build and hold the clamping *tonnage*, turn the reciprocating screw, drive the reciprocating screw, and *energize* ejector pins and moving mould cores. A number of hydraulic components are required to provide this power, which includes pumps, valves, hydraulic motors, hydraulic fittings, hydraulic tubing, and hydraulic *reservoirs*.

The Control System

The control system provides consistency and repeatability in machine operation. It monitors and controls the processing parameters, including the temperature, pressure, injection speed, screw speed and position, and hydraulic position. The process control has a direct impact on the final part quality and the economics of the process. Process control systems can range from a simple relay on/off control to an extremely sophisticated microprocessor-based, closed-loop control.

The Clamping System

The clamping system opens and closes the mould, supports and carries the *constituent* parts of the mould, and generates sufficient force to prevent the mould from opening. Clamping force can be generated by a mechanical (*toggle*) lock, hydraulic lock, or a combination of the two basic types.

New Words and Phrases

nozzle [ˈnɔzl] n. 喷嘴；管口；鼻

viscous [ˈviskəs] adj. 黏性的

sprue [spruː] n. 主流道

bushing [ˈbuʃɪŋ] n. 套管；轴衬

tonnage [ˈtʌnɪdʒ] n. 连接，肘节
reservoir [ˈrezəvwɑː(r)] n. 水库，蓄水池；液压油箱
toggle [ˈtɒg(ə)l] vt. 拴牢，系紧 n. 开关，触发器；拴扣；[船] 套索钉
flight [flaɪt] n. 飞行，班机；几节楼梯；几圈螺纹 vt. 射击；使惊飞 vi. 迁徙
energize [ˈenədʒaiz] vt. 激励；使活跃；供给……能量 vi. 活动；用力
constituent [kənˈstɪtjʊənt] n. 成分；选民；委托人 adj. 构成的；选举的
clamping tonnage 锁模力
microprocessor-based 基于微处理器的

Notes

1. While the outside diameter of the screw remains constant, the depth of the flights on the reciprocating screw decreases from the feed zone to the beginning of the metering zone.

译文：当螺杆外径保持恒定时，自加料段到计量段开始处，往复式螺杆螺纹区段的槽深逐渐变浅。

说明：这是一个复合句。while 引导时间状语从句，后面是主句。while 引导时间状语从句时与 when 稍有不同，主要体现在：while 只表示在一段时间内，而 when 可以表示在一段时间内或者在某个时间点。例如：

I can sit your children while you go out. 在你外出时，我可以临时照看一下你的小孩。

I had only danced with Mary about three minutes when someone cut in. 我和玛丽刚刚跳了大约三分钟的舞，就有人来截舞。

Exercises

Translate the following sentences into Chinese.

1. Children have to learn to communicate effectively.

2. This is in addition to having a sound knowledge of the theoretical and practical aspects of the various manufacturing methods.

3. Many products, such as those made from colored plastics or other special materials, are saleable because of appearance.

Chapter Ⅱ Computerized Numerical Control (CNC)

Unit 8 Computer Numerical Control Machine Tools

Text

Early machine tools were operated by craftsmen who decided many variables such as speeds, feeds, and depth of cut, etc. [1] With the development of science and technology, a new term, *numerical* control (NC) appeared. Controlling a machine tool using a *punched* tape or stored program is known as numerical control (NC). NC has been *defined* by the *Electronic* Industries Association (EIA) as "A system in which actions are controlled by the direct insertion of numerical data at some points. [2] The system must *automatically interpret* at least some *portion* of this data."

In the past, machine tools were kept as simple as possible in order to keep their costs down. Because of the ever-rising cost of labor, better machine tools, complete with electronic controls, were developed so that industry could produce more and better goods at prices that were competitive with those *offshore* industries. [3]

NC is being used on all types of machine tools, from the simplest to the most complex. The most common machine tools are the *single-spindle drilling* machine, lathe, *milling* machine, *turning* center, and machining center.

1. Single-Spindle Drilling Machine

One of the simplest numerically controlled machine tools is the single-spindle drilling machine. Most drilling machines are programmed on three axes:

1) The *X-axis* controls the *table* movement to the right and left.

2) The *Y*-axis controls the table movement toward or away from the *column*.

3) The *Z*-axis controls the up or down movement of the spindle to drill holes to depth.

2. Lathe

The engine lathe, one of the most productive machine tools, has always been a very efficient means of producing round parts (Fig. 8-1). Most lathes are programmed on two axes:

1) The *X*-axis controls the cross motion (in or out) of the cutting tool.

2) The *Z*-axis controls the *carriage* travel toward or away from the *headstock*.

3. Milling Machine

The milling machine has always been one of the most versatile machine tools used in industry (Fig. 8-2). Operations such as milling, *contouring*, gear cutting, drilling, boring, and *reaming* are only a few of the many operations that can be performed on a milling machine. The milling machine can be programmed on three axes:

1) The X-axis controls the table movement left or right.

2) The Y-axis controls the table movement toward or away from the column.

3) The Z-axis controls the vertical (up and down) movement of the *knee* or spindle.

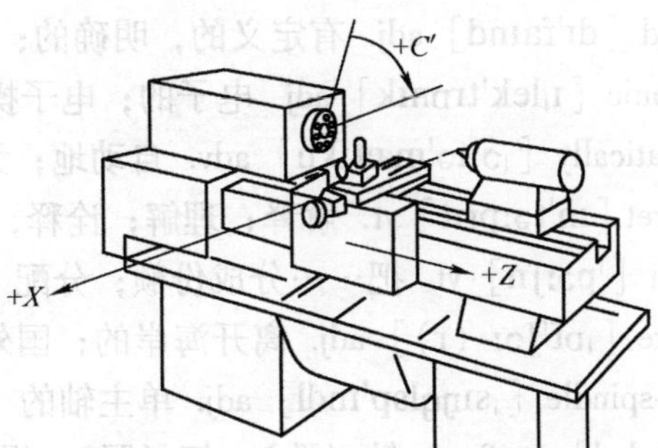

Fig. 8-1 The engine lathe-cutting tool moves only on the X and Z axes

4. Turning Center

Turning centers were developed in the mid-1960s after studies showed that about 40 percent of all metal cutting operations were performed on lathes. These numerically controlled machines are capable of greater *accuracy* and higher production rates than were possible on the engine lathe. The basic turning center operates on only two axes:

1) The X-axis controls the cross motion of the turret head.

2) The Z-axis controls the lengthwise travel (toward or away from the headstock) of the turret head.

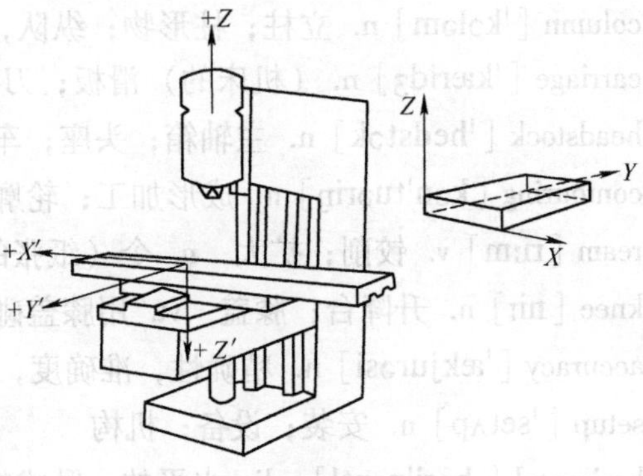

Fig. 8-2 The vertical knee and column milling machine

5. Machining Center

Machining centers were developed in the 1960s so that a part did not have to be moved from machine to machine in order to perform various operations. These machines greatly increased production rates because more operations could be performed on a work-piece in one *setup*. There are two main types of machining centers, the *horizontal* and the *vertical* spindle types.

(1) The horizontal spindle-machining center operates on three axes:

1) The X-axis controls the table movement left or right.

2) The Y-axis controls the vertical movement (up and down) of the spindle.

3) The Z-axis controls the horizontal movement (in or out) of the spindle.

(2) The vertical spindle-machining center operates on three axes:

1) The X-axis controls the table movement left or right.

2) The Y-axis controls the table movement toward or away from the column.

3) The Z-axis controls the vertical movement (up and down) of the spindle.

New Words and Phrases

numerical [njuːˈmerɪkl] adj. 数字的，用数字表示的，数值的

punch [pʌntʃ] v. 打孔；用拳猛击

defined [dɪˈfaɪnd] adj. 有定义的，明确的；轮廓分明的 v. 给……下定义，解释
electronic [ɪˌlekˈtrɒnɪk] adj. 电子的；电子操纵的；用电子设备生产的
automatically [ˌɔːtəˈmætɪklɪ] adv. 自动地；无意识地；不自觉地；机械地
interpret [ɪnˈtəːprɪt] vt. 解释；理解；诠释，体现；口译 vi. 做解释；做口译
portion [ˈpɔːʃn] vt. 把……分成份额；分配 n. 一部分；一份遗产
offshore [ˌɒfˈʃɔː（r）] adj. 离开海岸的；国外的 adv.（指风）向海地，离岸地
single-spindle [ˌsɪŋglspˈɪndl] adj. 单主轴的
drill [drɪl] vt. & vi. 钻（孔）；打（眼）；操练；训练 n. 操练；钻头；军事训练
milling [ˈmɪlɪŋ] n. 铣削 v. 碾磨；磨成粉；滚（碾轧）金属
turning [ˈtəːniŋ] n. 车削；旋转；转向；转弯处
axis [ˈæksɪs] n. 轴，轴线；[政] 轴心（复数 axes [ˈæksiːz]）
table [ˈteibl] n. 工作台，台面 vt. 嵌合；搁置；制表
column [ˈkɔləm] n. 立柱；柱形物；纵队，列；圆柱；专栏
carriage [ˈkæridʒ] n.（机床的）滑板；刀架；拖板，客车；运费
headstock [ˈhedstɔk] n. 主轴箱；头座；车头
contouring [kɔnˈtuəriŋ] n. 成形加工；轮廓线；做等值线；外形修整
ream [riːm] v. 铰削；扩大 n. 令（纸张的计数单位）；大量
knee [niː] n. 升降台；膝盖 vt. 用膝盖碰
accuracy [ˈækjurəsi] n. 精确性，准确度，精度
setup [ˈsetʌp] n. 安装；设备；机构
horizontal [ˌhɔriˈzɔntl] adj. 水平的，卧式的；地平线的 n. 水平线，水平面
vertical [ˈvəːtikəl] adj. 垂直的，直立的；顶点的 n. 垂直线
a punched tape 穿孔纸带
the Electronic Industries Association（EIA） （美国）电子工业协会
gear cutting 齿轮加工
turning center 切削中心
machining center 加工中心
turret head 转塔头
cross motion 横向运动
lengthwise travel 纵向运动

Notes

1. Early machine tools were operated by craftsmen who decided many variables such as speeds, feeds and depth of cut, etc.

译文：早期机床通常由工人操作并由他们决定机床速度、进给量、切削深度等变量。

说明：who decided many variables…是定语从句修饰 craftsmen。例如：

The engineers who we met yesterday have designed a new automatic device. 我们昨天碰到的那些工程师设计了一种新的自动化装置。

2. NC has been defined… as "A system in which actions are controlled by the direct insertion of numerical data at some points. . . ".

译文：数控被定义为"通过在点位直接插入数字数据来控制系统动作……"。

说明：句中 as 的宾语是 a system，in which 引导的定语从句修饰 system。

3. Because of the ever-rising cost of labor, better machine tools, complete with electronic controls, were developed so that industry could produce more and better goods at prices that were competitive with those of offshore industries.

译文：随着劳动成本日益增长，人们研制出性能更好，并配有电控设备的机床，这样企业可以生产更多、更好、具有价格优势的产品，以便和海外企业的产品竞争。

说明：句中 better machine tools 是句子的主语，were developed 是句子的谓语，so that 引导结果状语从句，that 引导的定语从句修饰 goods。

Exercises

Ⅰ. Answer the following questions according to the passage.

1. What does the term NC stand for?
2. What are the most common machine tools?
3. Has the engine lathe always been a very efficient means of producing round parts?
4. Which machine has always been one of the most versatile machine tools used in industry?
5. How many main types of machining center are there? What are they?

Ⅱ. Translate the following terms into English.

1. 单轴钻床 2. 普通车床 3. 加工中心 4. 生产率
5. 齿轮加工 6. 垂直运动 7. 穿孔纸带 8. 电子工业协会

Ⅲ. Translate the following sentences into English.

1. 最简单到最复杂的机床都会用到数控技术。
2. 普通车床是生产效率最高的机床之一。
3. 曾有研究表明，所有金属材料的切削操作大约 40% 是在车床上进行的。
4. 卧式加工中心在三个坐标轴上工作。

Translating Skills:

科技英语词汇的结构特征Ⅱ

科技英语词汇的构词主要有派生、缩略和合成三种方法。下面介绍这三种构词法：

1. 派生（Derivation）

科技英语词汇主要是通过缀合法，也称派生法构成的，即通过在原有的单词前面或后面加词缀来构成新词，这种词就叫派生词（derivative）。根据词根所加词缀的位置，派生法有：前缀（prefix）和后缀（suffix）。

例如：前缀 photo-表示光，照相，相片，加在其他词前即构成包含相关含义的词，如 photoammeter（光电电流表）、photocopy（影印）、photocell（光电池）、photoscope（透视镜）、photosynthesis（光合作用）等。

例如：后缀-logy 表示某种学科，是一个十分活跃的词缀，由它构成的新词如 futurology（未来学）、planetology（行星学）、paleontology（古生物学）等。

2．缩略（Abbreviation）

缩略词可分为三种：字母缩略词（acronyms），截短词（clippings）和混合词（blends）。

（1）字母缩略词（acronyms）　科技英语词汇中字母缩略词有两种构成方式：

一种是首字母缩略词：指取用各单词的第一个字母组成新的科技词汇或者机构名称。

例如，不明飞行物：UFO（unidentified flying object）；艾滋病：AIDS（acquired immune deficiency syndrome）；激光：laser（lightwave amplification by stimulated emission of radiation）；声呐：sonar（sound navigation and ranging）；重症急性呼吸系统综合症，俗称“非典”：SARS（severe acute respiratory syndrome）。

另一种是取单词的第一个字母和单词中的另外一个字母或者部分字母形成新的科技词汇。

例如，肺结核：TB（tuberculosis）；电视：TV（television）；脱氧核糖核酸：DNA（deoxyribonucleic acid）。

（2）截短词（clippings）　科技英语词汇中截短词的形成也和普通英语词汇一样，用单词的一部分来代替整个单词。代替单词的可以是整个单词的第一个音节（auto：automobile；flu：influenza；diam：diameter），也可以是最后一个音节（copter：helicopter；plane：aeroplane），还可以是单词中间的部分（fridge：refrigerator；morph：metamorphosis）。

（3）混合词（blends）　科技英语中混合词的构成是把两个单词以某种规则结合在一起形成一个单词。例如，telex（电传）：teleprinter + exchange；digicam（数码相机）：digital + camera；comsat（通信卫星）：communication + satellite；bit（比特）：binary + digit 等。

3．合成（Compounding）

合成法是把两个或两个以上的单词按照一定次序排列成新词的方法。用这种方法构成的新词叫作复合词（compound）。科技英语词汇中名词、动词、形容词、副词、介词都存在着合成词的形式，其中以合成名词为最常见。例如，greenhouse（温室）：green（绿色）+ house（房子）；mousepointer（鼠标指针）：mouse（老鼠）+ pointer（指针）；meteorogram（气象图）：meteor（气象）+ gram（图）；barothermograph（气压温度计）：baro（气压）+ thermo（热）+ graph（记录器）。

Reading: Advantages of NC

Recent studies show that of the amount of time an average part spent in a shop, only a *fraction* of that time was actually spent in the machining process. Let us assume that a part spent 50 hours from the time it arrived at a plant as a rough casting or bar stock to the time it was a finished product. During this time, it would be on the machine for only 2.5 hours and be cut for only 0.75 hour. The rest of the time would be spent on waiting, moving, setting up, loading, unloading, inspecting the part, setting speeds and feeds, and changing cutting tools.

NC reduces the amount of non-*chip*-producing time by selecting speeds and feeds, making rapid moves between surfaces to be cut, using automatic *fixtures*, automatic tool changing, controlling the *coolant*, in-process *gauging*, and loading and unloading the part. These factors, plus the fact that it is no longer necessary to train machine operators, have resulted in considerable savings throughout the entire manufacturing process and caused tremendous growth in the use of NC. Some of the major advantages of NC are as follows:

1) There is automatic or *semiautomatic* operation of machine tools. The degree of *automation* can be selected as required.

2) Flexible manufacturing of parts is much easier. Only the tape needs changing to produce another part.

3) Storage space is reduced. Simple work-holding fixtures are generally used, reducing the number of *jigs* or fixtures that must be stored.

4) Small part lots can be run economically. Often a single part can be produced more quickly and better by NC.

5) Nonproductive time is reduced. More time is spent on machining the part, and less time is spent on moving and waiting.

6) Tooling costs are reduced. In most cases complex jigs and fixtures are not required.

7) Inspection and *assembly* costs are lower. The quality of the product is improved, reducing the need for inspection and ensuring that parts fit as required.

8) Machine utilization time is increased. There is less time that a machine tool is idle because work piece and tool changes are rapid and automatic.

9) Complex forms can easily be machined. The new control unit features and programming capabilities make the machining of *contours* and complex forms very easy.

10) Parts inventory is reduced. Parts can be made as required from the information on the punched tape.

Since the first industrial revolution, about 200 years ago, NC has had a significant effect on the industrial world. The developments in the computer and NC have extended a person's mind and muscle. The NC unit takes symbolic input and changes it to useful output, expanding a person's concepts into creative and productive results. NC technology has made such rapid advances that it is being used in almost every area of manufacturing, such as machining, welding, press-working, and assembly.

If industry's planning and logic are good, the second industrial revolution will have as much or more effect on society as the first industrial revolution had. As we progress through the various stages of NC, it is the entire manufacturing process that must be kept in mind.

Computer-assisted manufacturing (CAM) and computer-*integrated* machining (CIM) are certainly where the future of manufacturing lies, and considering the developments of the past, it will not be too far in the future before the automated factory is a reality.

Developed originally for use in aerospace industries, NC is enjoying widespread acceptance in manufacturing. The use of CNC machines continues to increase, becoming visible in most *metalworking* and manufacturing industries. Aerospace, defense contract, automotive, electronic, appliance, and tooling industries all employ numerical control machinery. Advances in microelectronics have lowered the cost of acquiring CNC equipment. It is not unusual to find CNC machinery in contract tool, die, and mould-making shops. With the *advent* of low cost OEM (original equipment manufacturer) and *retrofit* CNC vertical milling machines, even shops specializing in one-of-a-kind prototype work are using CNCs.

Although numerical control machines traditionally have been machine tools, bending, forming, *stamping*, and inspection machines have also been produced as numerical control systems.

New Words and Phrases

fraction [ˈfrækʃən] n. 小部分，部分；稍微；［数］分数
chip [tʃip] n. 碎片，碎屑 v. 切成碎片，削，凿
fixture [ˈfikstʃə] n. 夹具，固定装置，设备
coolant [ˈkuːlənt] n. 切削液，冷却剂
gauge [ɡeidʒ] v. 检验，校准（同 gage） vt. 估计；测量；给……定规格 n. 测量的标准；
semiautomatic [ˌsemiɔːtəˈmætik] adj. 半自动的
automation [ˌɔːtəˈmeiʃən] n. 自动化；自动操作
jig [dʒiɡ] n. 夹具；带锤子的钓钩 v. 抖动；用夹具或钻模等加工
assembly [əˈsembli] n. 组装件，装配；集会，集合
contour [ˈkɔntuə] n. 轮廓；电路；概要
integrate [ˈintiɡreit] v. 使一体化，集成，成为一体 adj. 整体的，完整的
metalworking [ˈmetəlˌwəːkiŋ] n. 金属加工，金属制造
retrofit [ˈretrəufit] n. 改型（装，进）；（式样）翻新
stamping [ˈstæmpiŋ] n. 冲压，冲压制品
advent [ˈædvənt] n.（事件、时期等的）出现，到来
work-holding 工件夹持
rough casting 铸造毛坯
bar stock 棒料
in-process gauging 在线检测
computer-integrated machining (CIM) 计算机集成制造
tooling industry 刀具业
prototype work 原始模型工件
mould-making 模具制作

Exercises

Ⅰ. Answer the following questions according to the passage.

1. What advantages does NC have? (State simply)
2. Is it just the part manufacturing process as we progress through the various stages of NC?
3. Where is NC widely accepted and used nowadays?

Ⅱ. Translate the following phrases into English.

1. 在线检测　2. 柔性制造　3. 第一次工业革命　4. 焊接
5. 计算机辅助制造　6. 铸造毛坯　7. 模具制作　8. 广泛接受

Unit 9 Elements of CNC Machine Tools

Text

The CNC machine tool system consists of six basic *components*. They are the part program, the machine control unit (MCU), the measuring system, the servo control system, the actual CNC machine tool (lathe, drill press, milling machine, etc.), and the adaptive control system. It is important to understand each element prior to actual programming of a numerically controlled part.

Part Programs

The part program is a detailed set of *commands* to be followed by the machine tool. Each command specifies a position in the Cartesian coordinate system (X, Y, Z) or motion (workpiece travel or cutting tool travel), machining parameters and on/off function. Part programmers should be well versed with machine tools, machining processes, effects of process variables, and limitations of CNC controls.[1] The part program is written manually or by using computer-assisted language such as APT (Automated Programming Tool).

Machine Control Units (MCU)

The machine control unit (MCU) is a microcomputer that stores the program and *executes* the commands into actions by the machine tool. The MCU consists of two main units: the data processing unit (DPU) and the control loops unit (CLU). The DPU software includes control system software, calculation *algorithms*, translation software that converts the part program into a usable format for the MCU, *interpolation* algorithm to achieve smooth motion of the cutter, editing of part program.[2] The DPU processes the data from the part program and provides it to the CLU which operates the drives attached to the machine lead-screws and receives feedback signals on the actual position and velocity of each one of the axes.[3] The CLU consists of the circuits for position and velocity control loops, deceleration and backlash take-up, function controls (such as spindle on/off).

In CNC systems, the DPU functions are always performed by the control program contained in the CNC computer. The major part of the CLU, however, is always implemented in the most sophisticated CNC systems.

Measuring Systems

The term measuring system in CNC refers to the method a machine tool uses to move a part from a reference point to a target point. A target point may be a certain location for drilling a hole, milling a slot, or other machining operation. The two measuring systems used on CNC machine are the absolute and incremental. The absolute (also called coordinate) measuring system uses a fixed reference point (origin). It is on this point that all positional information is based. In other words, all the locations to which a part will be moved must be given dimensions relating to that original fixed reference point. The incremental measuring system (also called delta) has a floating coordinating system. With the incremental system, the machine establishes a new origin or reference point each time the part is moved. Noticed that with this system, each new location bases its values in X and Y, from the preceding location. One disadvantage to this system is that any errors made will be repeated throughout the entire program, if not detected and corrected.

Servo Control Systems

CNC servomechanisms are devices used for producing accurate movement of a table or *slide* along an axis. Instead of causing motion by manually turning cranks and hand-wheels as is required on conventional machine tools, CNC machines allow motions to be actuated by *servomotors* under control of the CNC, and guided by the part program. Generally speaking, the motion type (rapid, linear, and circular), the axes to move, the amount of motion and motion rate (feed rate) are programmable with almost all CNC machine tools.

Both electric and hydraulic powers are used to achieve slide motion. There are a number of very effective, responsive, and thoroughly proved hydraulic systems currently in use. But by far the most common power source is the electric motor. Three types of motor are stepping motors, non-stepping motors and *magnetic* linear drive motors. Stepping motors are special type of motor designed so that they rotate in sequential finite steps when energized by electrical *pulses*. AC servomotors are larger than DC motors providing equivalent power, and are also more costly. However, they need less maintenance and this is a factor very much in their favor. The newest slide drive system to appear on machinery is the magnetic linear drive system. Machine movement is obtained through the activation of magnetic forces alone, using high force linear motors.

CNC machines that use an open-loop system contain no feedback signal to ensure that a machine axis has traveled the required distance. That is, if the input received is to move a particular table axis 1.000 in., the servo unit generally moves the table 1.000 in. There is no means for comparing the actual table movement with the input signal, however. The only assurance that the table has actually moved 1.000 in. is the reliability of the servo system used. Open-loop systems are, of course, less expensive than closed-loop systems.

A closed-loop system compares the actual output (the table movement of 1.000 in.) with the input signal and compensates for any errors. A feedback unit actually compares the amount the table has been moved with the input signal.[4] Some feedback units used on closed-loop systems are *transducers*, electrical or magnetic scales, and synchros. Closed-loop systems greatly increase the reliability of CNC machines.

A CNC command executed within the control (commonly through a program) tells the drive motor to rotate a precise number of times. The rotation of the drive motor in turn rotates the ball screw. And the ball screw drives the linear axis. A feedback device at the opposite end of the ball screw allows the control to confirm that the commanded number of rotations has taken place.

Actual CNC Machine Tools

The machine tool could be one of the following: lathe, milling machine, coordinate measuring machine, etc. Unlike the conventional machines, the CNC machine tools have higher strength and rigidity, better stability and vibration resistance, and lower clearances.

Adaptive Control Systems

From a machining operation, the term adaptive control denotes a control system that measures certain output process variables and uses these to control speed and/or feed. Some of the process variables that have been used in adaptive control machining system include spindle deflection or force, *torque*, cutting temperature, vibration amplitude, and horsepower. In other words, nearly all the metal-cutting variables that can be measured have been tried in experimental adaptive control systems. The motivation for developing an adaptive machining system lies in trying to operate the process more efficiently. The typical measures of performance in machining have been metal removal rate and cost per volume of metal removed.

The main benefits of adaptive control in machining are as followings: increased production rates, increased tool life, greater part protection, less operator intervention, and easier part programming.

New Words and Phrases

component [kəm'pəunənt] n. 零件；组件；部件
command [kə'mænd] n. 命令 vi. 命令，指挥 vt. 控制；远望
execute ['eksikjuːt] v. 执行（命令） vt. 处死，处决；履行
algorithm ['ælgəriðəm] n. 算法
interpolation [ɪnˌtəːpə'leɪʃn] n. 插补
slide [slaid] n. 滑块；滑板 v. 使滑动
servomotor ['səːvəʊˌməutə] n. 伺服电动机
magnetic [mæg'netik] adj. 磁铁的，磁性的；有吸引力的
pulse [pʌls] n. 脉冲 vt. 使跳动 vi. 跳动，脉跳
transducer [trænz'djusə] n. 传感器；换能器
torque [tɔːk] n. 转矩；力矩；扭矩
machine control unit (MCU) 机床控制装置
servo control system 伺服控制系统
drill press 钻床
milling machine 铣床
adaptive control system 自适应控制系统
APT (Automated Programming Tool) 自动编程工具
data processing unit (DPU) 数据处理装置
control loops unit (CLU) 回路控制装置
interpolation algorithm 插补算法
floating coordinating system 浮动坐标系统
hydraulic power 液压动力
drive motor 驱动电动机
electrical pulse 电脉冲
open-loop system 开环系统
input signal 输入信号
closed-loop system 闭环系统

Notes

1. Part programmers should be well versed with machine tools, machining processes, effects of process variables, and limitations of CNC controls.

译文：工件加工程序员应该非常熟悉机床、加工工艺、工艺变量的影响以及CNC机床控制的局限性。

说明：这是一个简单句。part programmers 是主语，谓语是 should be well versed with，其宾语是由四个并列成分，即 machine tools，…and limitations of CNC controls，来充当的。should be well versed with 意为“应该非常熟悉……”。例如：

Modern people should be well versed with computer handling. 现代人应该熟练掌握电脑操作技巧。

2. The DPU software includes control system software, calculation algorithms, translation software that converts the part program into a usable format for the MCU, interpolation algorithm to achieve smooth motion of the cutter, editing of part program.

译文：DPU 软件包含控制系统软件、计算算法、将工件加工程序转换成 MCU 可用格式的编译软件、实现刀具流畅运动的插补算法及工件加工程序编辑等。

说明：这是一个复合句。主句是主动宾结构，其宾语是由五个并列成分组成的。其中 that converts the part program into a usable format for the MCU 是定语从句，修饰 translation software；to achieve smooth motion of the cutter 是不定式短语作后置定语修饰 interpolation algorithm（插补算法）。

3. The DPU processes the data from the part program and provides it to the CLU which operates the drives attached to the machine lead-screws and receives feedback signals on the actual position and velocity of each one of the axes.

译文：DPU 处理工件程序中的数据并将结果输送到 CLU，CLU 则启动装在机床丝杠上的驱动器并接收各轴实际位置与速度的反馈信号。

说明：这是一个复合句。本句主干成分为 the DPU processes the data... and provides it to the CLU...，其中 the DPU 是主语，processes 和 provides 是并列谓语动词。宾语 the data 由后置定语 from the part program 修饰；而 which operates the drives... and receives feedback signals ... 是定语从句，修饰 the CLU；从句中谓语动词也有两个，即 operates 和 receives。attached to 意为“附属于……，系于……，安装在……”。例如：

There is a middle school attached to the institute. 这个学院附设一所中学。

4. A feedback unit actually compares the amount the table has been moved with the input signal.

译文：反馈装置将工作台实际移动量与输入信号进行对比。

说明：这是一个复合句。句子主干是 a feedback unit actually compares…with…。句中的 the table has been moved 为定语从句，修饰 the amount。amount 有两种词性，作为动词时意为“总计，合计，共计；产生……结果”；作为名词时意为“数量；总额，总数”。例如：

The bill amounts to \$102. 此账单共计 102 美元。

Exercises

Ⅰ. Translate the following phrases into Chinese or English.

1. 机床控制装置 2. 反馈信号 3. 铣床 4. adaptive control system 5. servo control system 6. open-loop system

Ⅱ. Complete the following sentences with the proper form of the words given below.

component, command, execute, slide, magnetic

1. The glasses ________ off the table onto the floor.
2. The management plan has four main ________.
3. Make sure that you ________ all movements smoothly and without jerking.
4. This dog can fetch and carry at his master's ________.
5. But her husband did not think she was ________ as she hoped.

Ⅲ. Decide whether the following statements are True (T) or False (F) according to the passage.

1. The data processing unit and the control loops unit are two parts of the machine control unit.
2. There is a disadvantage to measuring system is that any errors made will be repeated throughout the entire program, if not detected and corrected.
3. CNC machines that use a closed-loop system contain no feedback signal.
4. All of the metal-cutting variables that can be measured have been tried in experimental adaptive control systems.
5. The part programmer's task requires a much less conservative approach.

Translating Skills:

被动语态的译法

由于科技文章需要描述客观事物的性质、特征、生产工艺流程与科学实验的结果等，并且通常不需要说明从事这些活动的执行者，因而被动语态在科技英语及专业英语中被广泛使用。由于英、汉两种语言在表达上存在巨大差异，因此在将英语的被动句式翻译成汉语时应做一些改变，以适应汉语表达习惯。被动语态常见的翻译方法如下：

一、译成汉语主动句

All bodies are known to posses weight.

人们知道所有物体都有质量。(加施动者)

The lathe is being adjusted by the operator.

操作员正在调试那台车床。(将主语译成宾语)

The mechanical energy can be changed back into electrical energy by means of a generator.

利用发电机，可将机械能再次转变成电能。(无主语主动句)

In a common machining, several different cutting tools are used to produce a part.

在普通加工过程中，使用几把不同的刀具加工一个零件。(无主语主动句)

二、译成汉语被动句

为了满足句子表达的需要，用“被”“由”“受”“靠”“给”“遭”等汉语中表达被动概念的介词引导出施动者。

The saddle, which is mounted on the hardened and ground bed-ways.

滑板被安装在淬硬并经过研磨的导轨上。

The bed is usually made of high-quality cast iron.

床身通常是由优质铸铁制造成的。

The tool can be automatically called for use by the part program.

刀具由零件程序自动调用。

三、由 by 或 in 引导的状语通常可以转换为汉语句子的主语

Production costs had been reduced by the workers.

工人们已经把生产成本降下来了。

Storage capacity is generally provided in the MCU for any canned cycle.

通常，机床控制装置（MCU）提供用于固定循环的存储空间。

四、常用句型“it + 被动语态 + that...”的译法

It is said that...	据说……	It is believed that...	人们认为……
It is known that...	众所周知，……	It is expected that...	人们希望……
It is noted that...	人们注意到……	It is announced that...	有人宣称……

Reading: Safety *Precautions* of CNC Machines

In order to reduce the possibility of safety *hazard* and keep normal operation function, the following safety precautions are very important. [1]

1. For Operation

To *minimize* the risk of accidents, follow these safety precautions:

1) Wear safety shoes, safety glasses and safety hats.

2) Wear work clothes. Clothing should not be loose; in particular shirt cuffs should be buttoned or *tight* against the arm.

3) Do not operate the machine wearing gloves.

4) Do not touch the workpiece or spindle when machine is in operation.

5) *Sufficient* light should be provided for the environment of machinery, and always keep the environment clean as far as possible.

6) Do not use high-pressure air for blowing dust or cutting chips near the controller.

7) A rigid floor should be provided for the operation area.

8) The ground for the installation of machine should be rigid enough.

9) Don't operate this machine if you feel uncomfortable.

10) Don't leave the tool on the working table or board.

11) Please check the condition of machine before operation. Ensure the best *performance* of the machine.

12) Every day, after turning on the machine, the machine needs to warm-up to ensure the life of *spindle*. The minimum time required is 30 minutes with the spindle speed set at 3000 rpm.

13) Please stop all functions of the machine when installing the workpiece.

14) Don't open the *protective* door during the operation.

2. For Machine

Before operating the machine the user needs to understand this manual very well.

1) The operator or *maintenance* personnel should pay attention to all warning labels. Do not break or remove any marking label on machine.

2) Except when required for *adjustment* or repair, every door should be closed so as to keep unwanted objects away from the machine.

3) During adjustments, repairs, or maintenance, be sure to use appropriate tools.

4) Do not move or change the location of limit switch for the purpose of *modifying* the travel of machine.

5) Please stop operating immediately and press *emergency* stop button, in case of any problem of machine.

6) The following precautions should be paid attention to for daily work:

During operation, do not place any part of your body on the movable parts, such as spindle, ATC, working table, and chip conveyor.

Do not clean tool and/or cutting chip of working table with bare hand. Any cleaning work could be carried out only after the machine stops completely.

Before adjustment of cutting fluid, air and fluid nozzle, the machine should be stopped completely.

7) In normal situation, the daily work should be *terminated* as the following procedures:

Press the emergency stop button→ Turn off the power→ Clean the working table→ Spray anti-rust oil on the spindle and working table.

3. For Electricity

During *inspection* and maintenance, the following precautions should be taken:

1) Do not hit controller by force in any case.

2) Only use the cable, which is specified by the manufacturer. Only use the cable with *appropriate* length. If some of cable is laid down on the ground, it should be provided with appropriate protection.

3) Only the manufacturer and the *authorized* agent are allowed to modify the setting of parameters of controller.

4) Do not change the setting values of controller and control buttons.

5) Do not overload the socket and conductors.

6) Before inspect and maintenance electrical components, be sure to disconnect the device of controller and main power, and lock it at "OFF" position.

7) Do not use any wet tool to touch the electrical components.

8) Only use the *fuse* which approves by manufacturer; the fuse with high capacity or copper wire is *prohibited*.

9) Only trained and qualified personnel are allowed to open the door of electrical *cabinet* for maintenance, and the others are prohibited.

New Words and Phrases

precaution [prɪˈkɔːʃn] n. 预防，警惕；预防措施
hazard [ˈhæzəd] n. 危险；冒险的事 vt. 赌运气；冒……的危险，使遭受危险
minimize [ˈminimaiz] vt. 把……减至最低数量（程度）
tight [tait] adj. 牢固的；绷紧的；密集的，紧凑的 adv. 紧紧地；牢固地
sufficient [səˈfiʃnt] adj. 足够的，充足的
performance [pəˈfɒːməns] n. 表演；表现
spindle [ˈspindl] n. 纺锤，纱锭；轴
protective [prəˈtektiv] adj. 保护的，防护的
maintenance [ˈmeintənəns] n. 维持；保养；维修
adjustment [əˈdʒʌstmənt] n. 调解，调整；调节器
modifying [ˈmɒdifaiŋ] v. 修改，更改（modify 的现在分词）；改变
emergency [iˈməːdʒənsi] n. 紧急情况；突发事件 adj. 紧急的，应急的
terminated [ˈtəːmineitid] v. 使终结；解雇（terminate 的过去式）
inspection [inˈspekʃn] n. 检查，检验，视察，检阅
appropriate [əˈprəupriət] adj. 适当的；恰当的；合适的 vt. 占用，拨出
authorized [ˈɒːθəraizd] adj. 权威认可的，审定的，经授权的 v. 授权；批准；辩护
fuse [fjuːz] n. 熔丝；导火线 vi. 融合；熔化 vt. 使融合；使熔化，使熔融
prohibited [prəˈhibitid] v. 禁止，阻止 adj. 被禁止的
cabinet [ˈkæbinət] n. 内阁；柜橱

Notes

1. In order to reduce the possibility of safety hazard and keep normal operation function, the following safety precautions are very important.

译文：为减少安全隐患、保持设备的正常运行功能，落实下列安全防范措施是十分重要的。

说明：这是一个简单句。in order to 引出两个并列关系的不定式短语作目的状语。precaution 意为“预防，防备，警惕；预防措施”。例如：

We are aware of the potential problems and have taken every precaution. 我们已意识到潜在的问题，并采取了全面的预防措施。

Exercises

Translate the following sentences into English by using the words in brackets.

1. 他们团结一致以应付紧急情况。(emergency)
2. 他们有足够维持生命的食物。(sufficient)
3. 此方法减少了开发和维护成本。(maintenance)

Unit 10 CNC Programming

Text

In an NC (Numerically Controlled) machine, the tool is controlled by a code system that enables it to be operated with minimal *supervision* and with a great deal of repeatability. [1] CNC (Computerized Numerical Control) is the same type of operating system, with the exception that the machine tool is *monitored* by a computer.

The same principles used in operating a manual machine are used in programming an NC or a CNC machine. The main difference is that instead of cranking handles to position a slide to a certain point, the dimension is stored in the memory of the machine control once. [2] The control will then move the machine to these positions each time the program is run.

The operation of the VF-Series Vertical Machining Center requires that a part program be designed, written, and entered into the memory of the control. The most common way of writing part programs is off-line, that is, away from the CNC in a facility that can save the program and send it to the CNC control. The most common way of sending a part program to the CNC is via an RS-232 *interface*. The HAAS VF-Series Vertical Machining Center has an RS-232 interface that is *compatible* with most existing computers and CNCs.

In order to operate and program a CNC machine, a basic understanding of machining practices and knowledge of math are necessary. It is also important to become familiar with the control *console* and the placement of the keys, switches, *displays*, etc.

Before you can fully understand CNC, you must first understand how a manufacturing company processes a workpiece that will be produced on a CNC machine. The following items form a fairly common and logical sequence of tasks done in CNC programming. The items are only in a suggested order, offered for further evaluation.

Obtain or develop the part drawing.

Decide what machine will produce the part.

Decide on the machining sequence.

Choose the tooling required.

Do the required math calculations for the program coordinates.

Calculate the speeds and *feeds* required for the tooling and part material.

Write the NC program.

Prepare setup sheets and tool lists.

Send the program to the machine.

Verify the program.

Run the program if no changes are required.

Manual programming (without a computer) has been the most common method of preparing a part program for many years. The latest CNC controls make manual programming much easier than ever before by using fixed or repetitive machining cycles, variable type programming, *graphic* tool motion *simulation*, standard mathematical input and other timesaving features. [3] The need for improved efficiency and accuracy in CNC programming has been the major reason for development of a variety of methods that use a computer to prepare part programs. Computer assisted CNC programming has been around for many years.

New Words and Phrases

supervision [ˌsuːpəˈvɪʒn] n. 监督，管理
monitor [ˈmɒnɪtə(r)] n. 监视器 v. 监视
interface [ˈintəfeis] n. 界面；接口；接口电路
compatible [kəmˈpætəbl] adj. 兼容的
console [kənˈsəul] n. 控制台
display [diˈsplei] n. 显示；显示器 v. 显示
feed [fiːd] n. 馈送，供给；进给
verify [ˈverifai] v. 校验
graphic [ˈgræfik] adj. 图形的
simulation [ˌsimjuˈleiʃn] n. 仿真
control console 控制面板
part drawing 零件图
machining sequence 加工工序
program coordinate 编程坐标
setup sheet 设置表
tool list 刀具清单
manual programming 手动编程
fixed cycle 固定周期
variable type programming 变量编程

Notes

1. In an NC (Numerically Controlled) machine, the tool is controlled by a code system that enables it to be operated with minimal supervision and with a great deal of repeatability.

译文：在NC（数字控制）机床中，刀具由编码系统控制，从而使加工过程中人工监管达到最小化、加工操作的可重复性也很好。

说明：这是一个复合句。that enables it to be operated with minimal supervision and with a great deal of repeatability 是定语从句，修饰 a code system。operated with 意为“使用……运作、操作”。例如：

It has a metal-to-metal seat and can only be operated with an appropriate wrench. 它有一个金属件组合成的阀座，只能用合适的扳手来操作（旋动）阀门。

2. The main difference is that instead of cranking handles to position a slide to a certain point, the dimension is stored in the memory of the machine control once.

译文：主要区别在于手动机床通过转动曲柄使滑台移到指定点，而 NC 机床或 CNC 机床则是将这些尺寸信息事先储存在机床控制装置的存储器中。

说明：这是一个包含表语从句的复合句。在 that 引导的表语从句中，instead of cranking handles... certain point 可理解为方式状语，修饰 is stored。根据上下文，本句中的 the main difference 指的是 NC 机床、CNC 机床与手动机床之间的差别；once 可引申为"早已、事先"；instead of 意为"代替"，由于 of 为介词，其后如果加动词，需要转化为动名词的形式。例如：

We'd better present the facts and reason things out instead of quarreling. 我们最好摆事实、讲道理，不要争吵。

3. The latest CNC controls make manual programming much easier than ever before by using fixed or repetitive machining cycles, variable type programming, graphic tool motion simulation, standard mathematical input and other timesaving features.

译文：最新的 CNC 控制器通过使用固定或重复的加工周期、变量编程、刀具运动图形仿真、标准数学输入以及其他简捷功能，使手动编程比以往方便了很多。

说明：句中 make… much easier than ever before 的意思是"使……比以往任何时候都更容易"；根据本文内容，other timesaving features 可引申为"其他简捷功能"。the latest 意为"最后的、最新的、最先进的"。例如：

They are listening out for the latest news. 他们正期待着最新消息。

Exercises

Ⅰ. Answer the following questions according to the passage.

1. What is the difference between NC and CNC?
2. What are necessary when operating and programming a CNC machine?
3. Nowadays, what is the most common method of preparing a part program?

Ⅱ. Complete the following sentences with the proper form of words given below.

monitor, compatible, display, verify, simulation

1. To get started, we need to ________ your email address.
2. Out of habit, I slowed as I reached the window to see whether my book was on ________.
3. You need feedback to ________ progress.
4. To make the training realistic, the ________ operates in real time.
5. His plan is ________ with my intent, so we won't work together anymore.

Ⅲ. Translate the following phrases into English.

1. 控制面板 2. 程序坐标 3. 手动编程 4. 我们最好摆事实、讲道理，不要争吵。
5. 技术员开发了多种使用计算机编制工件程序的方法。

Translating Skills:

非谓语动词 V-ing 的用法

非谓语动词是动词的一种非限定形式，不可单独作谓语，其形式不受主语的人称和数目的限制。非谓语动词有三种形式：V-ing，V-ed，to-V。由于非谓语动词由动词变化而来，因此它也保留了动词的特性，可带有自己的宾语和状语。

V-ing 形式的非谓语动词通常被划分为动名词和现在分词。在使用时也可不加区分，但要准确把握其用法，弄清 V-ing 形式在句中的作用及意义。

V-ing（或 V-ing 短语）可在句子中用作主语、表语、宾语、定语和状语等。

1. 作主语

Knowing how to use computer is useful.

知道怎样使用电脑是很有用的。

Controlling a machine tool using a punched tape or stored program is known as numerical control (NC).

数字控制就是利用穿孔纸带或储存的程序来控制机床。

2. 作表语

His great fun is chatting with strangers online.

他最大的乐趣就是在网上和陌生人聊天。

3. 作宾语

They are considering changing their operation plan.

他们正在考虑改变操作计划。

The second step in minimizing costs of production is choosing the cheapest of the technical efficient alternatives.

降低生产成本的第二个措施是选用最便宜的技术上有效的替代品。

4. 作定语

单个的 V-ing 作定语时，常放在被修饰名词的前面（也可放在其后）；但 V-ing 短语作定语时一般放在被修饰名词的后面。

All moving bodies have energy.

所有运动的物体都具有能量。

A direct current is a current flowing always in the same direction.

直流电流是指电流总向同一个方向流动的电流。

5. 作状语

V-ing 短语作状语时，常表示主句谓语动词的动作发生的时间、原因、条件、结果、方式、让步、伴随情况等，可置于句首、句中或句尾。

Having never handled a machine, he met with a lot of difficulties at first.

由于他从来没操作过机器，一开始碰到很多困难。(原因状语)

When measuring current, the circuit must be opened and the meter inserted in series with the circuit or component to be measured.

当测量电流时，必须断开电路，以将万用表与待测电路或元器件相串联。(时间状语)

6. 作主语或宾语的补足语

We put a hand above an electric fire and feel the hot air rising.

我们把手放在电炉的上方，感觉到热气上升。

He found the machine stopping working.

他发现机床停止了转动。

7. V-ing 的独立主格结构

多数情况下 V-ing 形式的逻辑主语和主句的主语相一致，但有时不一致，这时 V-ing 形式的逻辑主语和 V-ing 构成其独立结构。

With Mr. Green taking the lead, they decided to set up a trading company.

以格林先生为首，他们决定成立一个贸易公司。

Reading: An Example of Program Showing

In this *particular* example, we are milling around the outside of a workpiece *contour*. Notice that we are using a one inch *diameter* endmill for machining the contour and we are programming the very center of the end mill. Later, during key concept number four, we will discuss a way to actually program the workpiece contour (not the cutter centerline path). While you may not understand all commands given in this program, concentrate on understanding what is happening in the motion commands (G00, G01, and G02/G03). With study, you should be able to see what is happening. Messages in *parentheses* are provided to document what is happening in each command.

Program

00002 (Program number)

N005 G54 G90 S350 M03 (Select coordinate system, absolute mode, and start spindle CW at 350 RPM)

N010 G00 X -.625 Y -.25 (Rapid to point 1)

N015 G43 H01Z -.25 (*Instate* tool length *compensation*, rapid tool down to work surface)

N020 G01 X5.25 F3.5 (Machine in straight motion to point 2)

N025 G03 X6.25 Y.75 R1.0 (CCW circular motion to point 3)

N030 G01 Y3.25 (Machine in straight motion to point 4)

N035 G03 X5.25 Y4.25 R1.0 (CCW circular motion to point 5)

N040 G0 1 X.75 (Machine in straight motion to point 6)

N045 G03 X -.25 Y3.25 R1.0 (CCW circular motion to point 7)

N050 G01 Y.75 (Machine in straight motion to point 8)

N055 G03 X.75 Y -.25 R1.0 (CCW circular motion to point 9)

N060 G00 Z.1 (Rapid away from workpiece in Z)

N065 G91 G28 Z0 (Go to the machine's reference point in Z)

N070 M30 (End of program)

Keep in mind that CNC controls do vary with regard to limitations with motion types. For example, some controls have strict rules governing how much of a full circle you are allowed to make within one circular command. Some require directional *vectors* for circular motion commands instead of allowing the R word. Some even incorporate automatic corner rounding and *chamfering*, *minimizing* the number of motion commands that must be given. Though you must be prepared for variations, and you must reference your control manufacturer's programming manual to find out more about your machine's motion commands, at least this *presentation* has shown you the basics of motion commands and you should be able to adapt to your particular machine and control with relative.[1]

New Words and Phrases

particular [pə'tikjulə] adj. 特殊的，特别的
contour ['kɔntuə] n. 轮廓；等高线；周线；概要 vt. 画轮廓；画等高线
diameter [dai'æmitə] n. 直径
parentheses [pə'renθisi:z] n. 圆括号（parenthesis 的复数形式）
instate [in'steit] vt. 任命
compensation [ˌkɔmpen'seiʃən] n. 补偿，赔偿金；报酬
vector ['vektə] n. 向量
chamfering ['tʃæmfəriŋ] n. 倒角，切角
minimizing ['minimaiziŋ] n. 极小化；求最小参数值
presentation [ˌprezən'teiʃən] n. 介绍，陈述
CCW (counter clock wise) 逆时针方向
circular motion 圆周运动
keep in mind 记住，谨记
adapt to 适合

Notes

1. Though you must be prepared for variations, and you must reference your control manufacturer's programming manual to find out more about your machine's motion commands, at least this presentation has shown you the basics of motion commands and you should be able to adapt to your particular machine and control with relative.

译文：尽管你必须为各种变化做好准备，也必须参考数控系统制造商提供的编程手册中关于机床动作编程命令的更多含义，但至少此文稿已经展示了运动命令的基础知识，并且你应该能够使这些指令适应你的特定机器，且控制机器时要与这些命令的要求相对应。

说明：这是一个复合句。though 在句首引出让步状语从句，意为“尽管……但是……”，需注意英文转折句后半句不出现 but。adapt to 意为“使适应于……”例如：

I want flexible people who can adapt to new systems and processes. 我需要可塑性强的人，能够适应新的制度、新的方法。

Exercises

Translate the following phrases or sentences into English.

1. 集中 2. 记住 3. 适合 4. 严格的规定 5. 圆周运动
6. 我没有任何特别的理由怀疑他们。(particular)
7. 集中精神用腹肌保持你自己挺直并屏住呼吸。(concentrate on)
8. 世界会变得不同，我们必须做好准备以适应其变化。(adapt to)

Unit 11 CAD

Text

The *advent* of the computer proved to be a *boon* to the design engineer in that it *simplified* the long, *tedious calculations* which were often involved in designing a part. [1] In 1963, the Massachusetts Institute of Technology (MIT) *demonstrated* a computer system called Sketchpad that created and displayed graphic information on a cathode-ray tube (CRT) screen. This system soon became known as CAD, and it allows the designer or engineer to produce finished engineering drawings from simple pencil sketches or from models and modify these drawings on the screen if they do not seem functional. [2] From three-drawing *orthographic* views, the designer can *transform* drawings into a three-dimensional view and, with the proper computer software, show how the part would function in use. This enables a designer to redesign a part on the screen, project how the part will operate in use, and make *successive* design changes in a matter of minutes.

CAD Components

CAD is a televisionlike system that produces a picture on the CRT screen from electronic signals received from a computer. [3] Most CAD systems consist of a desk-top computer which is connected to the main or host computer. The addition of a keyboard, light pen, or an electronic tablet and plotter enables the operator to produce any drawing or view required.

The operator generally starts with a pencil sketch and, with the use of the light pen or an electronic tablet, can produce a properly *scaled* drawing of the part on the CRT screen and also record it in the computer memory. If design changes are necessary, the designer is able to create and change parts and lines on the CRT screen with a light pen, an *electromechanical cursor*, or an electronic tablet.

After the design is finished, the engineer or designer can test the *anticipated* performance of the part. Should any design changes be necessary, the engineer or designer can make changes quickly and easily to any part of the drawing or design without having to redraw the original. [4] Once the design is considered correct, the plotter can be directed to produce a finished drawing of the part.

Advantages of CAD

CAD offers industry many advantages which result in more accurate work and greater *productivity*. [5] The following is a list of some of the more common CAD advantages: greater productivity of drafting personnel, less drawing production time, better drawing revision procedures, greater drawing and design accuracy, greater detail in *layouts*, better drawing appearance, greater parts *standardization*, better factory assembly procedures and less *scrap* produced.

New Words and Phrases

advent [ˈædvənt] n. 出现，到来

boon [buːn] n. 实惠
simplified [ˈsimplifaid] v. 使简单（简明）
tedious [ˈtiːdiəs] adj. 单调沉闷的；冗长乏味的；令人生厌的
calculation [ˌkælkjuˈleiʃn] n. 计算；估计
demonstrate [ˈdemənstreit] vt. 证明；显示，展示 vi. 示威
orthographic [ˌɔːθəˈgræfik] adj. 正字法的；正交的；投影的
transform [trænsˈfɔːm] vt. 改变，使变形；转换 vi. 变换，改变；转化
successive [səkˈsesiv] adj. 连续的，相继的
scaled [skeild] adj. 按比例缩放的
electromechanical [iˌlektrəumiˈkænikəl] adj. 电动机械的，机电的
cursor [ˈkəːsə] n. 指针，光标
anticipated [ænˈtisipeitid] adj. 预先的；预期的
productivity [ˌprɒdʌkˈtivəti] n. 生产率，生产力
layout [ˈleiˌaut] n. 布局，设计，规划
standardization [ˌstændədaiˈzeiʃn] n. 标准化，标定，规范化
scrap [skræp] n. 废料
Massachusetts Institute of Technology (MIT) 麻省理工学院 (MIT)
a matter of 大约；大概
consist of 由……组成，由……构成

Notes

1. The advent of the computer proved to be a boon to the design engineer in that it simplified the long, tedious calculations which were often involved in designing a part.

译文：计算机对设计师而言非常实用，因为它简化了零件设计中经常包含的冗长且乏味的计算过程。

说明：这是一个复合句。which 引导的定语从句修饰 calculations；in that 意为“因为”。例如：

However, these differ from refactoring in that they add new functionality. 不过，这些特性与重构不同，因为它们添加了新功能。

2. This system soon became known as CAD, and it allows the designer or engineer to produce finished engineering drawings from simple pencil sketches or from models and modify these drawings on the screen if they do not seem functional.

译文：该系统很快成为人们所熟知的 CAD，它使设计师或工程师根据铅笔草图或者模型便可绘制最终的工程图样，如果达不到预期功能要求，还可在显示屏上即时修改。

说明：本句是由 and 连接的并列复合句。第一个分句是简单句；第二个分句是复合句，其主语 it 指 CAD。动词不定式 to produce... 和 (to) modify... 作宾补。sketches 意为“草图，示意图”，是 sketch 的复数形式。例如：

The sketch was rather amazing.　那素描真是太神奇了。

3. CAD is a televisionlike system that produces a picture on the CRT screen from electronic signals received from a computer.

译文：CAD 类似于电视系统，它接收计算机发来的电子信号，并在 CRT 显示屏上生成图像。

说明：这是一个复合句。that 引导的定语从句修饰 a televisionlike system，televisionlike 是由 television 和 like 组成的合成词，意为“像电视一样的，类似于电视”，合成构词法的例子很多，又如 monkeylike 意为“像猴子一样”；sunflower 意为“太阳花，向日葵”；blackboard 意为“黑板”等。过去分词短语 received from a computer 作定语修饰 electronic signals。received from 意为“收到”。例如：

The client will print out the data received from the server. 客户机将把从服务器接收到的数据打印出来。

4. Should any design changes be necessary, the engineer or designer can make changes quickly and easily to any part of the drawing or design without having to redraw the original.

译文：若需要改进，工程师或设计师可迅速、方便地对图样或设计的任意部分进行修改，而无需重绘原图。

说明：should any design changes be necessary 是省略了 if 的虚拟预期的倒装句，正常的语序为 if any design changes should be necessary。在倒装句中，颠倒了位置的成分可以恢复原位而其句意基本不变。redraw 意为“重绘”，该词的词根为 draw，意为“绘画”，前缀 re- 意为“重复、再一、又一”。类似的词还有 rethink，review 和 rebuild 等。original 意为“原文；原型；原件”。例如：

After having talked about some household affairs for a long while, we reverted to the original topic. 谈了半天家庭琐事以后，我们又回到原来的话题上。

5. CAD offers industry many advantages which result in more accurate work and greater productivity.

译文：CAD 给工业生产带来很多优势，可使工作更加高效而精确。

说明：这是一个复合句，主句在前。which 引导的定语从句修饰 advantages。accurate 意为“精确的”。例如：

When experiments are to be made, one cannot rely too much upon the human senses to make accurate observation. 在做实验时，不能过分依靠人类的感觉而不去做精确的观察。

Exercises

Ⅰ. Decide whether the following statements are True (T) or False (F) according to the passage.

1. The CAD can help the designer and engineer to modify drawings on the screen if they dissatisfied with them.

2. With the proper computer software, the designer can transform drawings into a 3D view.

3. All of the CAD systems consist of a desk-top computer.

4. The engineer and designer can test the anticipated performance of the part during programming.

5. Redraw the original is the only way for the engineer or designer to change designs.

Ⅱ. Complete the following sentences with the proper form of the words given below.

simplified, tedious, transform, successive, standardization

1. They rebuffed him, and after three ________ rejections he got the message and gave up.
2. Life is far more important than art, but without it life is so ________.
3. My clothes can ________ into lots of styles.
4. The costs are very low if only basic ________ rules need to be applied.
5. Last night, the president presented a shorter, ________ version of his speech.

Ⅲ. Translate the following sentences into English by using the words in brackets.

1. 那么，一场真正的面试包括什么呢？(consist of)
2. 外交部部长已重新制定了一套外交政策。(redesign)
3. 一般来说，就是不要把你所有的鸡蛋放在一个篮子里。(generally)
4. 这个方法的优势在于它的简单性。(simplicity)

Translating Skills:

非谓语动词 V-ed 和 to V 的用法

Ⅰ. 动词不定式（to V）的句法功能：具有名词、形容词和副词的特征，可充当主语、宾语、宾语补足语、表语、定语、状语。例如：

It is easier **to perform quality assurance by a spot-check instead of checking all parts.**（作主语）

抽检代替全检，质量更容易得到保证。

To describe the motion one must introduce the concept of time.（作状语）

必须引入时间的概念来描述运动。

Ⅱ. 过去分词（V-ed）具有形容词和副词的特征，可充当表语、定语、宾语补足语、状语。例如：

1. 过去分词作表语，相当于一个形容词，说明主语的状态。例如：

The engine lathe is **broken.**

这个普通车床坏了。

2. 过去分词作定语，它的逻辑主语就是它所修饰的词。

Heat is the energy **produced by the movement of molecules.**

热是分子运动所产生的能量。

3. 过去分词作宾语补足语时，过去分词和宾语的关系是被动关系，说明宾语的状态，表示动作的完成。

When I entered the workshop, I saw many parts **machined by Master Li.**

走进车间时，我看见李师傅加工了许多零件。

4. 过去分词在句中可作时间、条件、原因、让步、方式等状语。

Observed from the spaceship, our earth looks like a blue ball.

从宇宙飞船上看，我们的地球像一个蓝色的球。

Ⅲ. 在翻译中要特别注意 to V 与 V-ing 的区别，否则易造成曲解。例如：

Stop **to assemble the CNC machine tool.**

停下来安装 CNC 机床。(不定式表示一次具体动作)

Stop **working to rest.**

停工休息。(现在分词表示经常性的动作)

I forgot **to oil the machine.**

我忘记要给机器加油。(不定式表示事情没有做)

I forgot **oiling it.**

我忘记给机器加过油了。(现在分词表示事情已做了)

IV. 注意 V-ing 与 V-ed 的区别

V-ing 形式通常含有主动、进行的意思；V-ed 通常有被动、完成之意，在阅读翻译时要注意区分。例如：

I heard him **cutting the workpiece.**

我听到他在切削工件。(现在分词表示动作正在进行)

I saw these parts **cut by him.**

我看到他把这些零件切削过了。(过去分词表示动作已经完成)

The most popular technology for **realizing microcircuits** makes use of MOS transistors.

用来实现微电路的最通用的技术是利用 MOS 晶体管。(现在分词表示概念性、抽象性动作)

The symbols **used for enhancement MOS transistors** are shown in figure.

被用来表示增强型 MOS 晶体管的符号如图所示。(过去分词表示被动)

Reading: Computer-Aided Manufacturing

Computer-Aided Manufacturing (CAM) can be *defined* as the use of computer systems to plan, manage, and control the operations of a manufacturing plant through either direct or indirect computer *interface* with the plant's production resources. [1] CAM was first used in 1971 for car body design and tooling.

CAM functions center around four main areas: numerical control, process planning, robotics, and factory management.

CAM makes it possible to manufacture *physical* models using CAD programs. CAM creates real life *version* of components design within a software package. CAM allows data from CAD software to be *converted* directly into a set of manufacturing instructions.

Workers will become tired and make mistakes if they work long. The machines to perform manufacturing functions do not get tired. Because of this, the *likelihood* of mistakes being made is greatly decreased.

In *conjunction* with CAD, CAM enables manufacturers to reduce the costs of producing goods by minimizing the *involvement* of human operators. [2]

All of these developments result in lower operational costs, lower end product prices and increase *profits* for manufacturers.

New Words and Phrases

define [di'fain] vt. 精确地解释；界定

interface [ˈintəfeis] n. 接口 v. 接合，连接 vi. 相互作用（或影响）；交流，交谈
physical [ˈfizikl] adj. 身体的；物质的，物理学的 n. 体格检查
version [ˈvɜːʃn] n. 版本；形式
convert [kənˈvɜːt] vt. 使转变，转换；使改变信仰 vi. 转变；皈依
likelihood [ˈlaiklihuːd] n. 可能，可能性
conjunction [kənˈdʒʌŋkʃn] n. 结合；联合
involvement [inˈvɒlvmənt] n. 参与；卷入
profit [ˈprɒfit] n. 利润，收益，盈利
allow ... to do ... 允许……做……
result in 导致，结果是
in conjunction with 连同；结合；与……协力

Notes

1. Computer-Aided Manufacturing (CAM) can be defined as the use of computer systems to plan, manage, and control the operations of a manufacturing plant through either direct or indirect computer interface with the plant's production resources.

译文：计算机辅助制造（CAM）可定义为使用计算机系统来规划、管理和控制制造工厂的运作，这个过程中与工厂生产资源的对接可通过直接或间接的计算机接口来实现。

说明：Computer-Aided Manufacturing 为主语；can be defined as 是含有情态动词的被动语态作谓语；the use of computer systems 是宾语；to plan, manage, and control the operations of a manufacturing plant 为状语。interface 意为“界面；接口”。例如：

Listening is our interface with the thoughts of others. 倾听是自我思想和他人思想交流的接口。

2. In conjunction with CAD, CAM enables manufacturers to reduce the costs of producing goods by minimizing the involvement of human operators.

译文：结合 CAD，CAM 使制造商得以通过人工操作的最小化来降低商品生产的成本。

说明：to reduce ... 为不定式短语作宾语补足语（或称复合宾语）；minimizing 为现在分词，作介词宾语。in conjunction with 意为“连同，结合，与……协力”。例如：

This method can be used in conjunction with other methods. 这种方法可以和其他方法结合使用。

Exercises

Translate the following sentences in English by using the words in brackets.

1. 这件事难下定义。(define)
2. 我要化悲痛为力量。(convert... into...)
3. 这种方法可以和其他方法结合使用。(in conjunction with)
4. 当你充满信心地去做某事，结果就会成功。(result in)

Unit 12 Flexible Machining Systems

Text

A *flexible* machining system (FMS) is a system of CNC machines, robots, and part *transfer vehicles* that can take a part from raw stock or casting and perform all necessary machining, part handling, and inspection operations to make a finished part or assembly. It is an entire unmanned, software-based, manufacturing/assembly line. An FMS consists of four major components, the CNC machines, coordinate measuring machines, part handling and assembly robots, and part/tool transfer vehicles (Fig. 12-1).

The main element in an FMS is the CNC machining or turning *center*. The automatic tool changing *capability* of these machines allows them to run unattended, given the proper support system. Tool monitoring systems built into CNC machine are used to *detect* and replace worn tools. The major *obstacles* in an FMS are not the machining centers but the support systems for the machines, such as part load/unload and part transfer.

Fig. 12-1 A Small Flexible Machining System

Inspection in an FMS is accomplished through the use of *coordinate* measuring machines. These operate much like CNC machinery in that they are programmed to move to different positions on a workpiece. [1] Instead of using a rotating spindle and a cutting tool, a coordinate measuring machine is equipped with electronic gaging probes which measure features on a workpiece. The results of the gaging are compared to *acceptable* limits programmed into the machine.

Robots are frequently used in an FMS to load and unload parts from the machines. Since robots are programmed pieces of equipment that lack the ability to make judgments, special workholding fixtures are employed on the transfer vehicles to orient the workpiece so that the robot can handle it correctly. [2] Specially designed machine fixtures and clamping mechanisms are employed to ensure correct placement and clamping of the part on the machine. All part handling must be accomplished in a specific orderly fashion, with coordination of the part transfer vehicle, the robot, and the CNC machines. Future robots will probably employ some type of artificial intelligence which will enable them to make limited judgments as to workpiece *orientation* and take the necessary corrective actions. [3]

The fourth *critical* component of an FMS is the part/tool transfer vehicles. These vehicles shuttle workpieces from machine to machine. They also shuttle tool magazines to and from the machinery to maintain an adequate supply of sharp cutting tools at each CNC machine. Transfer vehicles employed in current flexible manufacturing systems are of four major types: automatic guided vehicles (AGV), wire guided vehicles, air cushion vehicles, and hardware guided vehicles.

Automatic guided vehicles rely on onboard *sensors* and/or a program to determine the path they take. There is no hardware connecting them to the system. An advantage of AGVs is that they can be reprogrammed to take different routes, *eliminating* the need to run tracks or wires for each route change. The corresponding disadvantage of AGVs is that they are the most difficult of the part *delivery* vehicles to make function, because of the lack of hardware connection.

A wire guided vehicle uses a wire buried in the floor to define its path. A sensor on the vehicle detects the location of the wire. A major advantage of wire guided vehicles is the ability to use the wire as opposed to an AGV without the need to have a hardware system such as an overhead wire or track on the floor. [4] The disadvantage of wire guided vehicles is the necessity of installing new wire in the floor if a route change is required.

An air cushion vehicle is guided by some *external* hardware device, such as an overhead wire, but glides on a cushion of air rather than a track system. When using air cushioned vehicles, particular attention to chip removal and control must be built into the FMS. Chips in the path of an air cushion vehicle will stop its progress. These vehicles are generally used for straight paths.

Hardware guided vehicles are the most reliable but least flexible of the transfer vehicles. A track on the floor or an overhead guide rail controls the vehicle path. The advantages of these vehicles are their reliability and the ease of coordinating them with the rest of the system. The major disadvantage is, of course, the need to run new rail or track whenever a vehicle route change or new route is *deemed essay*. A large FMS may employ several different types of vehicles, depending on the requirements of different parts of the manufacturing line.

New Words and Phrases

flexible [ˈfleksəbl] adj. 灵活的；易弯曲的；柔韧的
transfer [trænsˈfəː(r)] vt. 使转移；使调动 vi. 转让；转学
vehicle [ˈviːəkl] n. 搬运装置，传送装置；车辆
capability [ˌkeipəˈbiləti] n. 才能，能力；容量；性能
detect [diˈtekt] vt. 查明，发现；洞察；侦察，侦查
obstacle [ˈɒbstəkl] n. 障碍
coordinate [kəuˈɔːdineit] vt. 使协调，使调和；整合 vi. 协调；协同
acceptable [əkˈseptəbl] adj. 可接受的；合意的
orientation [ˌɔːriənˈteiʃn] n. 方向，定位，取向

critical [ˈkritikl] adj. 批评的，爱挑剔的；关键的；严重的
sensor [ˈsensə] n. 传感器，敏感元件
eliminate [iˈlimineit] v. 排除，消除；除掉
delivery [diˈlivəri] n. 传输，传送
external [ikˈstəːnl] adj. 外面的，外部的；表面上的；外用的
deem [diːm] v. 认为，相信
essay [ˈeseɪ] n. 散文；试图；试验 vt. 尝试；对……做试验
flexible machining system 柔性制造系统
coordinate measuring machine 坐标测量机
part/tool transfer vehicles 工件/刀具传输装置
turning center 车削中心
tool monitoring system 刀具监测系统
rotating spindle 旋转主轴
clamping mechanism 夹紧装置
artificial intelligence 人工智能
tool magazine 刀库
automatic guided vehicle 自动导向车
wire guided vehicle 导线导向车
air cushion vehicle 气垫车
hardware guided vehicles 硬件导向车
overhead guide rail 悬空导轨
manufacturing line 生产线

Notes

1. These operate much like CNC machinery in that they are programmed to move to different positions on a workpiece.

译文：坐标测量机在程序控制下向工件各处移动，这一点与 CNC 机床十分相似。

说明：本句开头的 these 指文中上句提到的 coordinate measuring machines（坐标测量机）。in that 意为“在……方面”。例如：

And I hate disappointing anybody in that respect. 我不想在那方面让大家失望。

2. Since robots are programmed pieces of equipment that lack the ability to make judgments, special workholding fixtures are employed on the transfer vehicles to orient the workpiece so that the robot can handle it correctly.

译文：由于机器人是程序控制装置，缺乏判断能力，所以传输装置需使用专用夹具定位工件，以便机器人能正确处理工件。

说明：这是一个复合句。since 引导原因状语从句，that 引导的定语从句修饰 equipment。orient 意为“ 使适应、确定方向”。例如：

Walking tours are a great way to orient yourself, as well as meet other adventure-seekers discovering the same place as yourself. 徒步旅行是进行自我定位的一个好方法，而且可以在旅途

中遇到和你一样的探险者。

3. Future robots will probably employ some type of artificial intelligence which will enable them to make limited judgments as to workpiece orientation and take the necessary corrective actions.

译文：未来的机器人可能具有某种人工智能，这使它们对工件定位有一定的判断能力，且能采取必要的纠错动作。

说明：这是一个复合句。which 引导的定语从句修饰 some type of artificial intelligence；该从句中 to make... 和（to）take... 并列作宾补，由 and 连接。enable ... to do 意为“使……能够做某事”．例如：

The skills that you have enable you to do the job. 你所拥有的技能，让你能够做好你的工作。

4. A major advantage of wire guided vehicles is the ability to use the wire as opposed to an AGV without the need to have a hardware system such as an overhead wire or track on the floor.

译文：导线导向车的主要优点是，不同于自动导向车，导线导向车具有使用导线的能力，但不需要如悬空导轨或地面导轨这样的硬件系统。

说明：这是一个简单句。to use the wire 为后置定语，修饰 the ability。as opposed to an AGV…是状语，as opposed to 意为“与……相反/ 不同”。例如：

How does a poem change when you read it out loud as opposed to it being on the page? 将写在纸上的诗大声读出来会有怎样的不同?

Exercises

Ⅰ. Decide whether the following statements are True（T）or False（F）according to the passage.

1. There are four components that make up of a flexible machining system.

2. The major obstacles in an FMS are the support systems for the machines.

3. Robots are programmed pieces of equipment, so that the robot can't orient the workpiece.

4. If automatic guided vehicles reprogrammed to take different routes, they need to run tracks or wires for each route change.

5. Hardware guided vehicles are the most reliable of the transfer vehicles.

Ⅱ. Translate the following phrases into Chinese or English.

1. 柔性制造系统 2. 自动导向车 3. 生产线 4. coordinate measuring machine

5. overhead guide rail 6. hardware guided vehicles

Ⅲ. Complete the following sentences with the proper form of the words given below.

capability, obstacle, acceptable, adequate, delivery

1. His wages are ________ to support three people.

2. A lack of qualifications can be a major ________ to finding a job.

3. Air pollution in the city had reached four times the ________ levels.

4. The strike caused a great delay in the ________ of the mail.

5. As a scientist, he has the ________ of doing important research.

Translating Skills:

句子成分的转换

英、汉两种语言，由于表达方式不同，翻译时有时需要改变原文的语法结构，以达到译文逻辑正确、通顺流畅、重点突出等目的。所采用的主要方法除了词类转换之外，还有句子成分的转换。

一、将介词宾语译成主语

将英语翻译成汉语时，为了符合汉语的表达规范，有时需要将原文中的介词宾语翻译成主语，使译文重点突出，行文流畅。例如：

Our refrigerator is light in **weight**, portable in **size** and low in **power consumption**.

我们生产的冰箱质量轻、体积小、耗电低。

A motor is similar to a generator in **construction**.

电动机的结构与发电机类似。

二、将动词 have 的宾语译成主语

Levers have little **friction** to overcome.

杠杆要克服的摩擦力很小。

Evidently semiconductors have a lesser **conducting capacity** than metals.

半导体的导电能力显然比金属差。

三、将其他动词的宾语译成主语

除动词的宾语以外，其他一些及物动词的宾语有时也可译成主语，不过动词（省略不译的除外）有时需要与主语一起译成定语。例如：

Light beams can carry more **information** than radio signals.

光束运载的信息量比无线电信号运载的信息量大。

Hot-set systems produce higher **strengths** and age better than cold-set systems.

热固系统比冷固系统的强度高，而且老化情况也比较好。

四、将主语译成定语

英译汉时，由于句意表达的需要往往会更换主语，而将原文的主语译成定语。

1. 当形容词译成名词，并译作主语时，原来的主语通常译成定语。

The oxygen atom is nearly 16 times as **heavy** as the hydrogen.

氧原子的重量几乎是氢的 16 倍。

Medium carbon steel is much **stronger** than low carbon steel.

中碳钢的强度比低碳钢大得多。

2. 当动词译成名词，并译作主语时，原来的主语一般需要译成定语。

The earth acts like a big magnet.

地球的作用像一块大磁铁。

A cathode ray tube is **shaped** like a large bell.

阴极射线管的形状像个大铃铛。

五、将定语译成谓语

1. 动词宾语的定语译成谓语

Water has a greater heat capacity than sand.

水的热容比沙大。

Copper and tin have a **low** ability of combining with oxygen.

铜和锡的氧化能力低。

2. 介词宾语的定语译成谓语

在介词宾语译作主语的同时，有时还需要将该宾语的定语译成谓语，原来的谓语和主语都译成定语。例如：

Gear pumps operate on the **very simple** principle.

齿轮泵的工作原理很简单（在该句中，介词宾语 principle 译作主语，其定语 very simple 译成谓语，谓语 operate 译成定语）。

Nylon is produced by **much the same** process as rayon.

尼龙的生产过程与人造丝大体相同。

六、将定语译成汉语主谓结构中的谓语

有时出于修辞的目的，将某一名词前的形容词，即名词的定语与该名词颠倒翻译，一起译成汉语的主谓结构，在句中充当一个成分。原来作定语的形容词成为主谓结构中名词的谓语。例如：

Among the advantages of numerical control are **more flexibility, higher accuracy, quicker changes, and fewer machines down time.**

数控的优点包括适应性强、精度高、换刀快及非加工时间短。

These pumps are featured by their **simple operation, easy maintenance, low oil consumption and durable service.**

这些水泵的特点是操作简单、维修容易、耗油量少、经久耐用。

Reading: Group Technology

Group technology is a concept at the very heart of CIM[1]. CIM is supposed to give manufacturers the *flexibility* to produce *customized* products without *sacrificing* productivity. Group technology is a key *ingredient* in the larger formula of CIM that makes this possible. It amounts to making *batch* manufacturing economical.

Any new components design needs its own unique *identification* number, usually drawing number, and traditionally these were just *allocated* in sequence. With development of computer techniques, the drawings can be stored in a CAD system and stored in a database.

Group technology provides components with *meaningful* drawing number. Each drawing number is allocated by a coding system and the digit has meaning. If the code number is known many of the component features can be deduced without reference to the drawing.

Group technology involves grouping parts according to design characteristics, the processes used to produce them, or a combination of these. In Fig. 12-2, the similar parts grouped together on the basis of two design characteristics (shape and material). These similar parts fall into a family of parts. Families of parts can be produced using similar processes usually within a single flexible manufacturing cell (FMC).

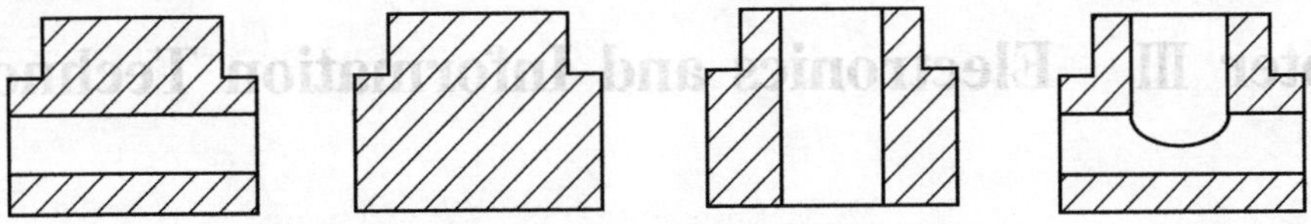

Fig. 12-2 Parts with similar design characteristics

New Words and Phrases

flexibility [fleksiˈbiliti] n. 弹性，适应性，机动性，挠性
customize [ˈkʌstəmaiz] v. 定制，用户化，按客户具体要求制造
sacrifice [ˈsækrifais] n. 牺牲，牺牲品，损失 v. 牺牲，献出
ingredient [inˈgriːdiənt] n. 成分，因素，组成部分
batch [bætʃ] n. 一炉，一批，一次所制之量
identification [aiˌdentifiˈkeiʃn] n. 识别，鉴别，辨别，鉴定，核对
allocate [ˈæləukeit] vt. 分派，分配，指定
meaningful [ˈmiːniŋful] adj. 有意义的，意味深长的
amount to 意味着；共计；发展成；折合
group technology 成组技术
CIM (Computer Integrated Manufacturing) 计算机集成制造
flexible manufacturing cell (FMC) 柔性制造单元，柔性生产单元

Notes

1. Group technology is a concept at the very heart of CIM.

译文：成组技术是计算机集成制造最为核心的一个概念。

说明：这是一个简单句。very 是一个常用副词，修饰形容词，但在本句中为形容词，意为恰好是、正是，所以 the very heart 译为“最为核心”。例如：

He knows our very thoughts. 他了解我们的真实想法。

Exercises

Ⅰ. Answer the following questions according to the passage.

1. What does the “group technology” mean to CIM?
2. What makes the larger formula of CIM possible?
3. What does group technology provide for components?
4. In what situation we can still deduce the component features without reference to the drawing?
5. How do we usually produce the family of parts?

Ⅱ. Translate the following sentences into English by using the words in brackets.

1. 这些例证是这部词典独有的。(unique)
2. 这本书按顺序阅读各章会更好。(in sequence)
3. 这个方法是经过实验推论出来的。(deduced)

Chapter Ⅲ Electronics and Information Technology

Unit 13 Alternating Current

Text

An alternating current (AC) is an electrical current whose magnitude and direction vary cyclically, as opposed to direct current, whose direction remains *constant*. [1]

1. Alternating current

Alternating currents are accompanied (or caused) by alternating *voltages*. In English the initialism AC is commonly and somewhat confusingly used for both.

Step and impulse functions are useful in determining the responses of *circuits* when they are first turned on or when sudden or irregular changes occur in the input; this is called transient analysis. However, to see how a circuit responds to a regular or repetitive input—the steady-state analysis—the function that is by far the most useful is the sinusoid.

The sine wave is the most common wave in AC and sometimes we refer to sine AC as AC in short. [2] An AC voltage v can be described mathematically as a function of time by the following *equation*

$$v(t) = V_{peak}\sin(\omega t)$$

A sine wave, over one cycle (360°) is shown in Fig. 13-1. The dashed line represents the root mean square (RMS) value at about 0.707 V_{peak}. Where V_{peak} is the peak voltage (unit: volt);

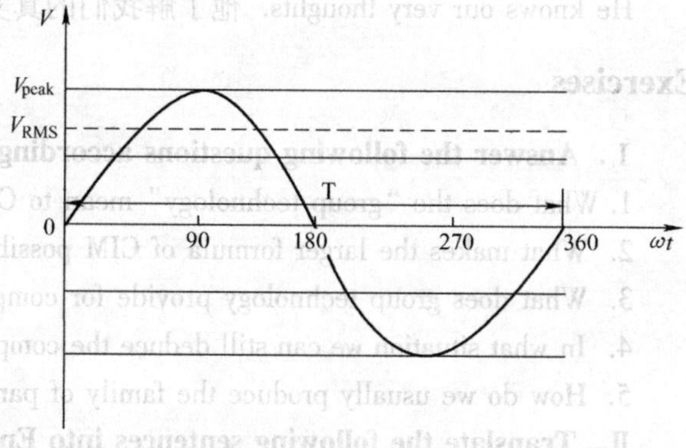

Fig. 13-1 Sine wave

ω is the angular frequency (unit: *radians* per second);

t is the time to complete one cycle, and is called period (unit: second).

The angular frequency ω is related to the physical frequency f which represents the number of oscillations per second (unit: hertz), by the equation: $\omega = 2\pi f$.

We have following conclusions about the sinusoid:

1) If the input of a linear, time-invariant circuit is a sinusoid, then the response is sinusoid of the same frequency.

2) Finding the magnitude and phase angle of a sinusoidal steady-state response can be *accomplished* with either real or complex sinusoids.

3) If the output of a sinusoidal circuit reaches its peak before the input, the circuit is a lead *network*. Conversely, it is a lag network.

4) Using the concepts of phasor and *impedance*, sinusoidal circuits can be analyzed in the frequency *domain* in a manner *analogous* to *resistive* circuits by using the phasor *versions* of KCL, KVL, *nodal* analysis, *mesh* analysis and *loop* analysis.

Though electromechanical generators and many other physical phenomena naturally produce sine waves, this is not the only kind of alternating wave in existence. Other "waveforms" of AC are commonly produced within electronic circuitry. Here are but a few sample waveforms and their common *designations* in Fig. 13-2.

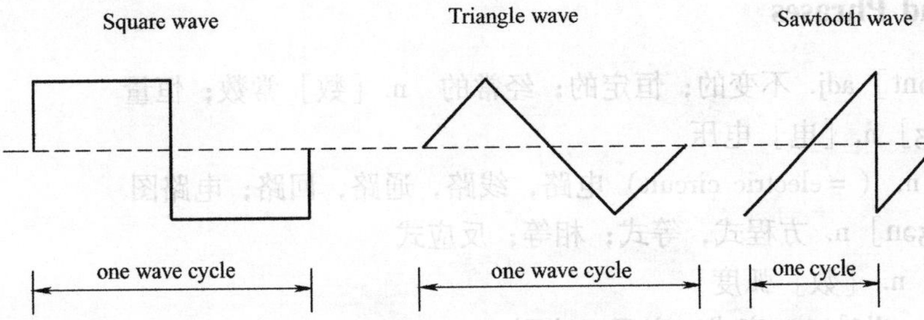

Fig. 13-2 Some common waveshapes (waveforms)

They're simply a few that are common enough to have been given *distinct* names. Even in circuits that are supposed to *manifest* "pure" sine, square, triangle, or sawtooth voltage/current waveforms, the real-life result is often a *distorted* version of the intended waveshape. Generally speaking, any waveshape bearing close *resemblance* to a perfect sine wave is termed sinusoidal, anything different being labeled as non-sinusoidal.

2. AC Electric *power*

We have previously defined power to be the *product* of voltage and current in the DC circuits. For the case that voltage and current are constants, the *instantaneous* power is equal to the average value of the power. The voltage and current are both sinusoids in AC circuits, however, the instantaneous power, which is still the product of voltage and current, changes with time and is not equal to the average power.

In alternating current circuits, energy storage elements such as *inductance* and *capacitance* may result in periodic *reversals* of the direction of energy flow.[3] The portion of power flow that, averaged over a complete cycle of the AC waveform, results in net transfer of energy in one direction is known as real power (also referred to as active power). That portion of power flow due to stored energy that returns to the source in each cycle is known as reactive power.

The relationship between real power, reactive power and apparent power can be expressed by representing the quantities as *vectors*. Real power is represented as a horizontal vector and reactive power is represented as a vertical vector (Fig. 13-3). The apparent power vector is the hypotenuse of a right triangle formed by connecting the real and reactive power vectors.

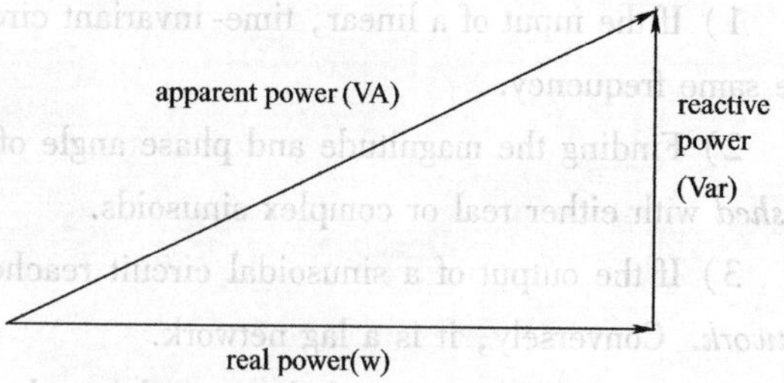

Fig. 13-3　Power triangle

This representation is often called the power triangle. Using the Pythagorean Theorem, the relationship among real, reactive and apparent power is:

$$(\text{apparent power})^2 = (\text{real power})^2 + (\text{reactive power})^2$$

The *ratio* of real power to apparent power is called power factor and is a number always between 0 and 1.

New Words and Phrases

constant [ˈkɔnstənt] adj. 不变的；恒定的；经常的　n. [数] 常数；恒量
voltage [ˈvəultidʒ] n. [电] 电压
circuit [ˈsəːkit] n. (= electric circuit) 电路，线路，通路，回路；电路图
equation [iˈkweiʒən] n. 方程式，等式；相等；反应式
radian [ˈreidiən] n. [数] 弧度
accomplish [əˈkʌmpliʃ] vt. 完成；实现；达到
network [ˈnetwəːk] n. 网络；广播网；网状物
impedance [imˈpiːdəns] n. [电] 阻抗
domain [dəuˈmein] n. 领域；域名；产业；地产
analogous [əˈnæləgəs] adj. 类似的，相似的；可比拟的 (analogous to/with)；模拟的
resistive [riˈzistiv] adj. 有抵抗力的
version [ˈvəːʃən] n. 版本；译文；倒转术
nodal [ˈnəudəl] adj. 节的；结的；节似的
mesh [meʃ] n. 网眼；网丝；圈套　vi. 相啮合　vt. [机] 啮合；以网捕捉
loop [luːp] n. 环，圈，弯曲部分　vi. 打环；翻筋斗　vt. 以环连接
designation [ˌdezigˈneiʃən] n. 指定；名称；指示；选派
distinct [disˈtiŋkt] adj. 明显的；独特的；清楚的；有区别的
manifest [ˈmænifest] vt. 证明，表明；显示，出现　n. 货单　adj. 显然的
distort [disˈtɔːt] vt. 扭曲；使失真；曲解　vi. 扭曲；变形
resemblance [riˈzembləns] n. 相似；相似之处；相似物；肖像
power [pauə] n. 功率；力量；能力；政权　vt. 激励　vi. 快速前进
product [ˈprɒdʌkt] n. 产品；结果；乘积；作品
instantaneous [ˌinstənˈteinjəs] adj. 瞬间的；即时的；猝发的

inductance [in'dʌktəns] n. 电感；感应系数；自感应
capacitance [kə'pæsitəns] n. [电] 电容；电流容量
reversal [ri'vəːsəl] n. 逆转；[摄] 反转；[法] 撤销
vector ['vektə] n. 矢量；带菌者；航线 vt. 用无线电导航
ratio ['reiʃiəu] n. 比率，比例
KCL (Kirchhoff current laws) 基尔霍夫电流定律
KVL (Kirchhoff voltage laws) 基尔霍夫电压定律

Notes

1. An alternating current (AC) is an electrical current whose magnitude and direction vary cyclically, as opposed to direct current, whose direction remains constant.

译文：不同于方向保持不变的直流电，交流电是指大小和方向都会做周期性变化的电流。

说明：as opposed to 意为“与……对照；与……对比”。例如：

As opposed to gas-liquid systems, there is an interchange of gas between bubble and continuous phase in fluidized beds. 与气-液系统不同，在流化床中气泡相和连续相之间有气体交换。

2. The sine wave is the most common wave in AC and sometimes we refer to sine AC as AC in short.

译文：正弦波是交流电最常见的波形，有时我们把正弦交流电简称为交流电。

说明：in short 意为“总之；简言之”。例如：

This is, in short, a new kind of cyber-attack. 简言之，这是一种新型的网络攻击。

3. In alternating current circuits, energy storage elements such as inductance and capacitance may result in periodic reversals of the direction of energy flow.

译文：在交流电路中，储存能量的元件，如电感和电容（只能储存或释放能量），可能引起能量流方向的周期性反转。

说明：result in 意为“产生某种作用或结果”。例如：

Traditionally, selection controls do not directly result in actions—they require an imperative control to activate. 以前，选择控件不直接引发操作，操作通常需要命令控件来触发。

Exercises

Ⅰ. Translate the following phrases into English.

1. 峰值电压 2. 线性时不变电路 3. 角频率 4. 有效值 5. 无功功率 6. 视在功率 7. 暂态分析

Ⅱ. Decide whether the following statements are True (T) or False (F) according to the passage.

1. To see how a circuit responds to a regular or repetitive input—the steady-state analysis—the function that is by far the most useful is the sinusoid.

2. If the output of a sinusoidal circuit reaches its peak before the input, the circuit is a lag net-

work. Conversely, it is a lead network.

3. As electromechanical generators and many other physical phenomena naturally produce sine waves, this is the only kind of alternating wave in existence.

4. The portion of power flow that, averaged over a complete cycle of the AC waveform, results in net transfer of energy in one direction is known as reactive power.

Translating Skills:

定语从句的译法

定语从句分为限制性定语从句和非限制性定语从句，其译法一般有三种。

一、前置译法

将较短的定语从句译成汉语的定语词组或带“的”字的结构，放在所修饰的中心词之前，从而使从句和主句合成一个句子，这就是前置译法。前置译法主要用于一些结构简单，但对先行词限制较强的定语从句，既适用于翻译限制性定语从句，也适用于翻译非限制性定语从句。例如：

The resistance of any conductor depends not alone on the material of which it is made.

任何导体的电阻不仅取决于制造该导体的材料。

The electromotive force that causes the current is measured in volts named after Volta, the Italian scientist.

产生电流的电动势是以伏特为单位来测量的，该单位是以意大利科学家 Volta 来命名的。

二、分译法

分译法指将定语从句与先行词分开翻译，把定语从句译成和主句并列的一个分句。当定语从句结构复杂，或其只对先行词进行描写、叙述或解释，而不加限制时，如果将其译成汉语的前置定语，则不符合汉语的表达习惯，此时，可采用分译法。分译法既适用于翻译限制性定语从句，也适用于翻译非限制性定语从句，尤其是后者。

Transformers designed for audio frequency use must be capable of transmitting current over a wide frequency range, in contrast to a power transformer which transmits current over a narrow frequency band.

用于音频的变压器必须能够在一个宽频带内传送电流，这点与电力变压器正好相反，后者在一个狭窄的频带内传输电流。

The pioneering work done by TRW was noticed by many computer manufacturers, who saw a large potential market for their products.

由 TRW 所做的开拓性的工作引起了许多计算机制造商的关注，他们看到了他们的产品具有很大的潜在市场。

That accords with Lenz's law, which states that an induced current always has a direction such as to oppose the cause which produces it.

这符合楞次定律，即感应电流的磁场总是阻碍产生电流的磁通量的变化。

The AF signal is amplified and then converted into sound power by the loudspeaker, which sends out sound waves.

音频信号经放大后，由扬声器转换成声音功率，并发出声波。

三、翻译成状语从句

英语中，有些定语从句兼有状语从句的职能，在逻辑上（即意义上）与主句有状语关系，如说明原因、结果、让步、假设等关系，翻译时应善于从英语原文的字里行间发现这些逻辑上的关系，将其翻译成状语从句。例如：

Liquids, which contains no free electrons, are poor conductors of heat.

液体，由于不含有自由电子，是热的不良导体。

Electronic computers, which have many advantages, cannot do creative work and replace human.

尽管电脑有很多优势，但还是不能做创造性的工作，更不能代替人。

For any substance whose formula is known, a mass corresponding to the formula can be computed.

不管什么物质，只要知道其分子式，就能求出与分子式相应的质量。

Reading: Three-Phase Circuits

A circuit that contains a source that produces (sinusoidal) voltages with different phases is called a *polyphase* system. The importance of this concept lies in the fact that most of the generation and distribution of electric power is accomplished with polyphase systems. The most common polyphase system is the balanced three phase system, which has the property that it supplies constant instantaneous power, which results in less vibration of the rotating machinery used to generate electric power.

A Y (wye)-connected three-phase source in the frequency domain is shown in Fig. 13-4. Terminals a, b and c are called the line terminals and n is called the *neutral* terminal. The source is said to be balanced if the voltages $v_{an}(t)$, $v_{bn}(t)$, $v_{cn}(t)$ (or their phasor $\dot{v}_{an}$, $\dot{v}_{bn}$, $\dot{v}_{cn}$), called phase voltages, have the same *amplitude* and sum to zero, that is, if $v_{an} = v_{bn} = v_{cn}$, and $v_{an}(t) + v_{bn}(t) + v_{cn}(t) = 0$ (or their phasor form $\dot{v}_{an} + \dot{v}_{bn} + \dot{v}_{cn} = 0$).

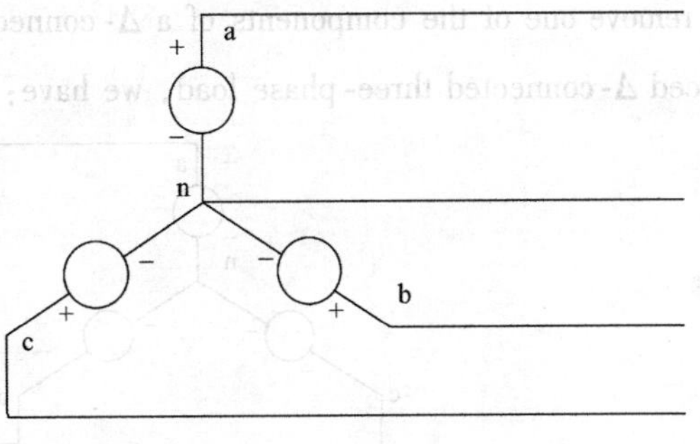

Fig. 13-4 Balanced Y-connected three-phase source

Suppose that the amplitude of the sinusoids is V, if we arbitrarily select the angle of $\dot{v}_{an}$ to be zero, that is, if $v_{an}(t) = V\sin(\omega t + 0°)$ (or $\dot{v}_{an} = V\angle 0°$), then the two situations that result in a balanced source are as follows:

Case Ⅰ

$v_{an}(t) = V\sin(\omega t + 0°)$

$v_{bn}(t) = V\sin(\omega t - 120°)$

$v_{cn}(t) = V\sin(\omega t - 240°)$

Case Ⅱ

$v_{an}(t) = V\sin(\omega t + 0°)$

$v_{cn}(t) = V\sin(\omega t - 120°)$

$v_{bn}(t) = V\sin(\omega t - 240°)$

For case Ⅰ, $v_{an}(t)$ leads $v_{bn}(t)$ by 120°, and $v_{bn}(t)$ leads $v_{cn}(t)$ by 120°. It is, therefore, called a positive or abc phase *sequence*. Similarly, Case Ⅱ is called a negative or acb phase sequence. Clearly, a negative phase sequence can be converted to a positive phase sequence simply by relabelling the terminals. Thus, we need only consider positive phase sequences.

Let us now connect our balanced source to a balanced Y-connected three-phase load as shown in Fig. 13-5, the voltages between the line terminals are called line voltages, the currents between the a and A, b and B, c and C are called line currents. In a balanced Y-connected three-phase load, we have: $V_{line} = \sqrt{3}V_p$ and $I_{line} = I_p$.

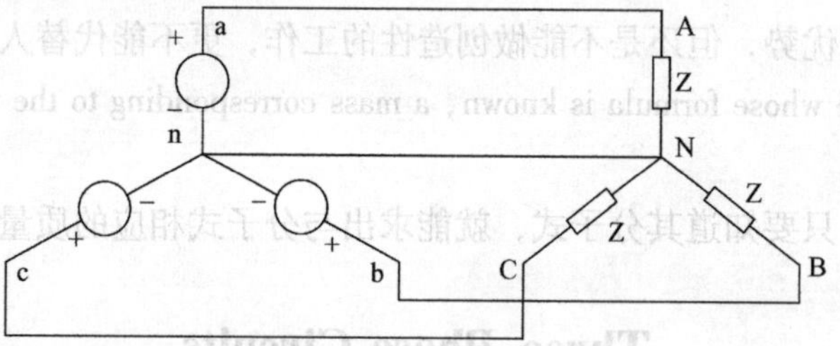

Fig. 13-5 Balanced Y-Y connected three phase circuit

If the lines all have the same impedance, the effective load is still balanced and so the neutral current is zero, and the neutral wire can be removed.

More common than a balanced Y-connected three-phase load is a Δ (delta)-connected load. A Y-connected source with a balanced Δ-connected load is shown in Fig. 13-6, we see that the individual loads are connected directly across the lines, and consequently, it is relatively easier to add or remove one of the components of a Δ-connected load than with a Y-connected load. In a balanced Δ-connected three-phase load, we have: $V_{line} = V_p$ and $I_{line} = \sqrt{3}I_p$

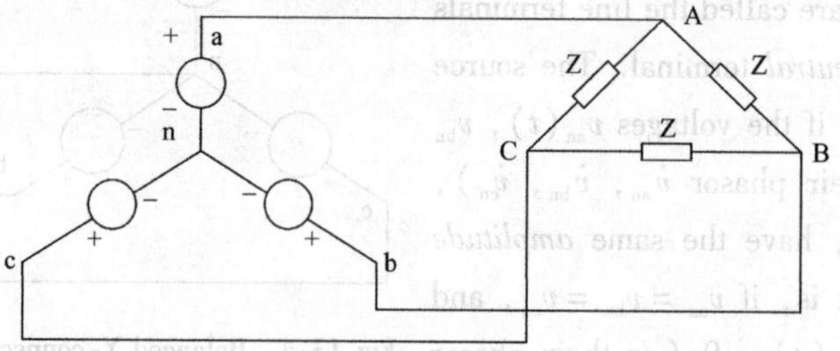

Fig. 13-6 Balanced Y-Δ connected three phase circuit

Regardless of whether a balanced load is Y-connected or Δ-connected, in terms of the line voltage, line current, and load impedance phase angle, we can use the same formula for the total power absorbed by the load

$$P=\sqrt{3}V_{\text{line}}I_{\text{line}}\cos\varphi=3V_{\text{p}}I_{\text{p}}\cos\varphi$$

New Words and Phrases

polyphase [ˈpɔlifeiz] adj. [电] 多相的
sequence [ˈsiːkwəns] n. 次序，顺序，序列
neutral [ˈnjuːtrəl] n. 中立者 adj. 中性的
amplitude [ˈæmplitjuːd] n. 广阔，充足；[物] 振幅
balanced three phase system 三相平衡系统
positive phase sequence 正相序
Δ (delta) -connected load 三角形连接的负载

Exercises

Translate the following phrases into Chinese.

1. balanced three phase system
2. positive phase sequence
3. negative phase sequence
4. line voltage
5. line current

Unit 14 Electronic Components

Text

There are a large number of symbols which *represent* an equally large range of electronic components. It is important that you can recognize the more common components and understand what they actually do[1]. A number of these components are drawn below and it is interesting to note that often there is more than one symbol representing the same type of component.

Resistors

Resistors restrict the flow of electric *current*, for example, a resistor is placed in series with a light-emitting *diode* (LED) to limit the current passing through the LED. Fig. 14-1 shows resistor example and circuit symbol.

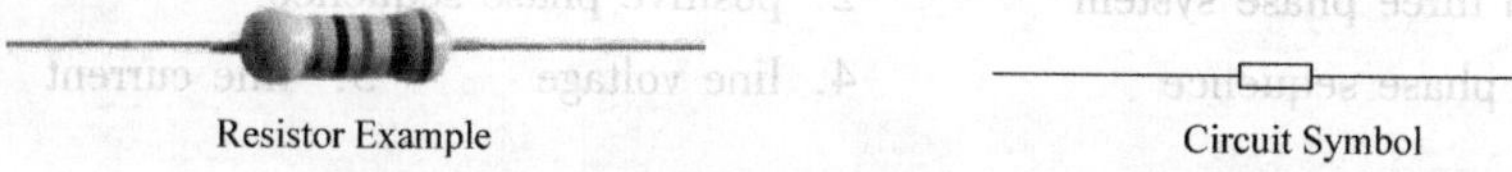

Fig. 14-1 Resistor Example and Circuit Symbol

Resistors may be connected either way round. They are not damaged by heat when soldering[2].

Capacitors

Capacitors store electric charge. They are often used in *filter* circuits because capacitors easily pass AC (changing) signals but they block DC (*constant*) signals[3]. Fig. 14-2 shows capacitor examples and circuit symbol.

Fig. 14-2 Capacitor Examples and Circuit Symbol

Inductors

An inductor is a passive electronic component that stores energy in the form of a magnetic field. An inductor is a coil of wire with many *windings*, often wound around a core made of a magnetic material, like iron. Fig. 14-3 shows inductor examples and circuit symbol.

Fig. 14-3 Inductor Examples and Circuit Symbol

Diodes

Diodes allow electricity to flow in only one direction. The arrow of the circuit symbol shows the direction in which the current can flow. Diodes are the electrical version of a *valve* and early diodes were actually called valves. Fig. 14-4 shows diode example and circuit symbol.

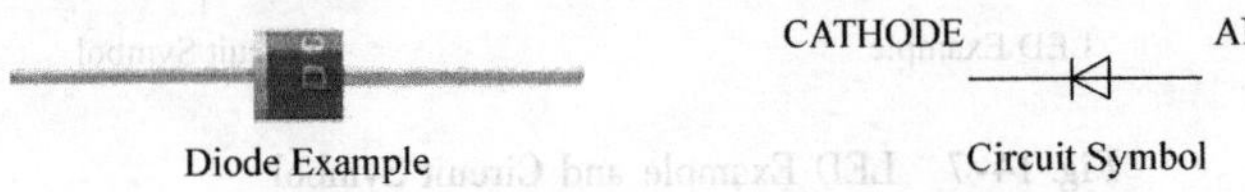

Fig. 14-4 Diode Example and Circuit Symbol

Transistors

There are two types of standard transistors, NPN and PNP, with different circuit symbols. The letters refer to the layers of *semiconductor* material used to make the transistor. Fig. 14-5 shows transistor examples and circuit symbols.

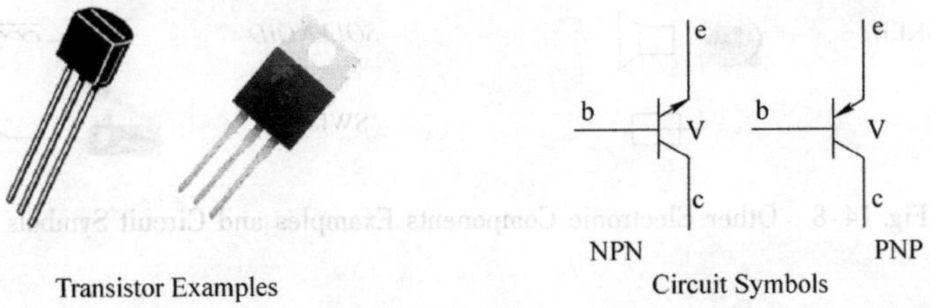

Fig. 14-5 Transistor Examples and Circuit Symbols

Integrated Circuits (*Chips*)

Integrated Circuits are usually called ICs or chips. They are complex circuits which have been *etched* onto tiny chips of semiconductor (*silicon*). The chip is packaged in a plastic holder with pins spaced on a 0.1" (2.54mm) grid which will fit the holes on *strip* board and *breadboards* [4]. Very fine wires inside the package link the chip to the pins. Fig. 14-6 shows integrated circuits example and circuit symbol.

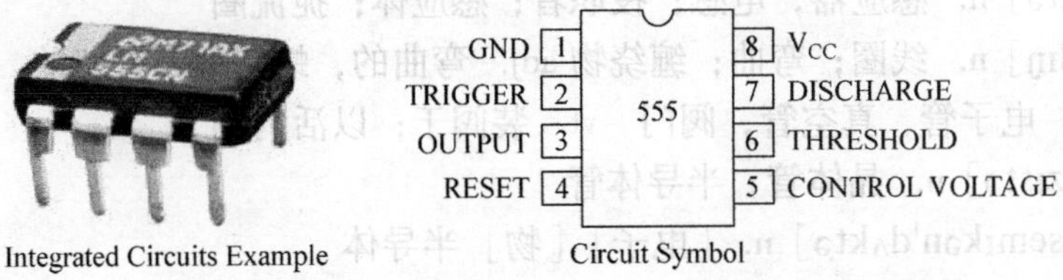

Fig. 14-6 Integrated Circuits Example and Circuit Symbol

Light Emitting Diodes (LEDs)

LEDs emit light when an electric current passes through them.

LEDs must be connected the correct way round, the *diagram* may be labelled "a" or "+" for *anode* and "k" or "−" for *cathode* (yes, it really is "k", not "c", for cathode!). The cathode is the short lead and there may be a slight flat on the body of round LEDs[5]. Fig. 14-7 shows LED example and circuit symbol.

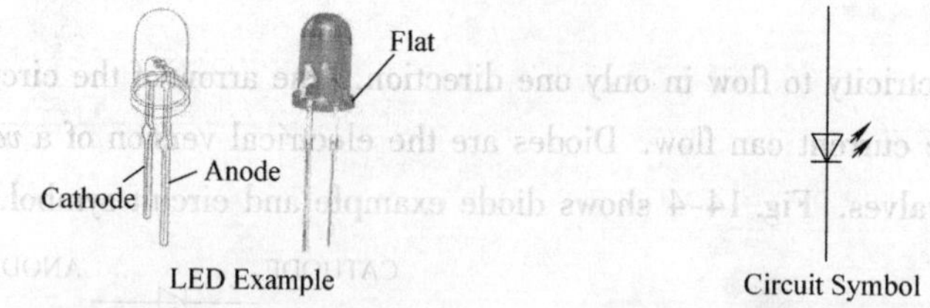

Fig. 14-7 LED Example and Circuit Symbol

Other Electronic Components (Fig. 14-8)

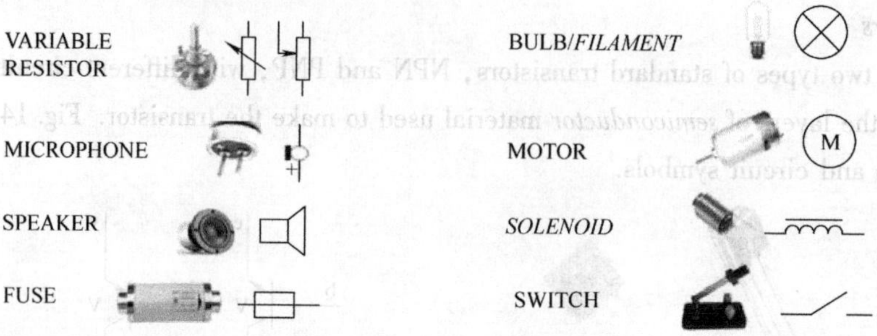

Fig. 14-8 Other Electronic Components Examples and Circuit Symbols

New Words and Phrases

represent [reprɪˈzent] v. 表现，体现，作为……的代表；描绘；回忆
current [ˈkʌr(ə)nt] n. 电流；趋势；涌流 adj. 现在的；通用的；最近的
diode [ˈdaɪəʊd] n. [电子] 二极管
capacitor [kəˈpæsɪtə] n. 电容，电容器
filter [ˈfɪltə] n. 滤波器，过滤器；筛选 vt. 过滤，渗透 vi. 滤过，渗入
constant [ˈkɒnst(ə)nt] adj. 不变的，恒定的，经常的 n. [数] 常数，恒量
inductor [ɪnˈdʌktə] n. 感应器，电感；授职者；感应体；扼流圈
winding [ˈwaɪndɪŋ] n. 线圈；弯曲；缠绕物 adj. 弯曲的，蜿蜒的；卷绕的
valve [vælv] n. 电子管，真空管，阀门 vt. 装阀于；以活门调节
transistor [trænˈzɪstə] n. 晶体管，半导体管
semiconductor [ˌsemɪkənˈdʌktə] n. [电子] [物] 半导体
chip [tʃɪp] n. 芯片；筹码；碎片，薄片 vt. 削，凿 vi. 剥落，碎裂
etched [ˈetʃid] adj. 被侵蚀的，风化的 v. 蚀刻（etch 的过去分词）
silicon [ˈsɪlɪk(ə)n] n. 硅，硅元素
strip [strɪp] n. 带，条状 vt. 剥夺，剥去
breadboard [ˈbredbɔːd] n. 擀面板；案板；电路试验板
diagram [ˈdaɪəgræm] n. 图表，图解，示意图，[数] 线图 vt. 用图表示
anode [ˈænəʊd] n. 阳极（电解）；正极（原电池）
cathode [ˈkæθəʊd] n. 阴极（电解）；负极（原电池）
filament [ˈfɪləm(ə)nt] n. 灯丝；细线；单纤维

solenoid [ˈsəʊlənɒɪd] n. 螺线管，带磁芯的电感器
light-emitting diode（LED）发光二极管
AC（Alternating Current）交流电
DC（Direct Current）直流电

Notes

1. It is important that you can recognize the more common components and understand what they actually do.

译文：能够识别更为常见的元件并掌握它们的实际用途是很重要的。

说明：这是一个复合句。it 是形式主语，that 引导的主语从句作真正的主语，is important 是系表结构作主句的谓语。

2. Resistors may be connected either way round. They are not damaged by heat when soldering.

译文：电阻可以连接在任一回路中。它们不会因焊接高温而损坏。

说明：这是两个意义关联的简单句。they 指代 resistors，第二句中 when soldering 相当于一个省略了 they are 的时间状语从句，当从句主语与主句主语相同，而且助动词是 be 动词时，往往省略。

3. They are often used in filter circuits because capacitors easily pass AC（changing）signals but they block DC（constant）signals.

译文：由于电容可使交流信号轻易通过，而阻止直流信号，因此它们经常应用于滤波电路中。

说明：这是一个复合句。because 引导原因状语从句，但从句本身包含了具有并列关系和转折含义的两个分句。

4. The chip is packaged in a plastic holder with pins spaced on a 0.1"（2.54mm）grid which will fit the holes on strip board and breadboards.

译文：该芯片被封装在一个塑料固定物上，其引脚间隔距离为 0.1 英寸（2.54mm），这样的栅格将适合带形板和电路试验板的孔距。

说明：这是一个复合句。which 引导的定语从句修饰先行词 grid，with pins spaced on a 0.1"（2.54mm）grid 是介词短语作后置定语修饰 holder。

5. The cathode is the short lead and there may be a slight flat on the body of round LEDs.

译文：负极是较短的引脚，并且在 LEDs 的圆形体上可能有窄小的平面端面。

说明：这是一个并列句，两个分句都是简单句，在第二个分句中，on the body of round LEDs 是介词短语作状语。

Exercises

Ⅰ. Decide whether the following statements are True（T）of False（F）according to the passage.

1. A resistor is placed in series with a light-emitting diode（LED）to restrict the current passing through the LED.

2. Capacitors easily pass DC signals but they block AC signals.

3. Capacitors are often used in filter circuits.

4. Integrated Circuits are usually referred to as ICs or chips.

5. There are two kinds of standard transistors, NPN and PNP.

Ⅱ. Match the following words to appropriate definition or expression.

1. resistor	a. a device for controlling the flow of a liquid which letting it move in one direction only
2. inductor	b. a piece of wire, wound into circles, which acts as a magnet when carrying an electric current
3. valve	c. a small wire or device inside a piece of electrical equipment that breaks and stops the current if the flow of electricity is too strong
4. fuse	d. a device that has resistance to an electric current in a circuit
5. solenoid	e. a component in an electric or electronic circuit which possesses inductance

Ⅲ. Complete the following sentences according to the passage.

1. ________ restrict the flow of electric current.

2. Capacitors are often used in ________ circuits because they easily pass AC signals but they block DC signals.

3. Diodes are the electrical ________ of a valve and early diodes were actually called valves.

4. The chip is packaged in a ________ holder with pins spaced on a 0. 1"(2. 54mm) grid which will fit the holes.

5. Diodes allow electricity to flow in ________ direction.

Ⅳ. Translate the following sentences into English by using the words in brackets.

1. 能够识别更为常见的电子元件并掌握它们的实际用途是非常必要的。(components)

2. 电阻器不会因焊接高温而损坏。(damaged, soldering)

3. 金属要加热到一定的温度才熔化。(melt, definite)

4. 电子元器件包括电子管、晶体管、集成电路等。(integrated circuit)

Translating Skills:

And 引导的句型的译法

And 作为连词，用来连接词、短语和句子，其基本意义相当于汉语的“和”“与”“并且”。但在实际翻译过程中，特别是在连接两个句子时，and 的译法很多，表达的意义可能相差甚远。如果不考虑 and 前后成分之间的逻辑关系，只生硬套用基本译法，难免造成理解上的错误，甚至把整个句子的意思搞错。

1. And 表示原因。

Laser is widely used for developing many new kinds of weapons, *and* it penetrates almost everything.

激光广泛用于制造各种新式武器，*因为*它的穿透力很强。

2. And 表示因果。

1) But since a digital signal is made up of a string of simple pulses, noise stands out *and* easily removed.

但由于数字信号由一组简单脉冲组成，噪声易被发现，*因而*容易排除。

2) In 1945 a new type of aeroplane engine was invented, it was much lighter and powerful than earlier engines, *and* enabled war planes to fly faster and higher than ever.

1945年发明了一种新型的飞机发动机，它比早期的发动机要轻得多，功率也要大得多，*因此*采用这种发动机的军用飞机比以往任何时候都飞得更快、更高。

3. And 表示目的。

It was later shown that the results of this work were by no means the ultimate, and further work has been put in hand *and* to provide closer control and more consistent operation in this area.

后来发现，这项研究工作的结果绝非已做定论，而进一步的研究工作已开始，*以便*在这方面提供更严密的控制和持续性更强的操作。

4. And 表示承接。

In many ways, computer is more superior than human brain, *and* human can rule it.

计算机在许多方面超过人脑，*而*人却可以控制它。

5. And 表示对照。

Motion is absolute, *and* stagnation is relative.

运动是绝对的，*而*静止是相对的。

6. And 表示递进。

The electronic brain calculates a thousand times quicker, *and* more accurately than is possible for the human being.

电脑的运算速度比人所能达到的速度要快1000倍，*甚至*更加准确。

7. And 表示转折。

There will always be some things that are wrong, *and* that is nothing to be afraid of.

错误在所难免，*但*并不可怕。

8. And 表示条件。

Even if a programmer had endless patience, knowledge and foresight, storing every relevant detail in a computer, the machine's information would be useless, *and* the programmer knew little how to instruct it in what human beings refer to as commonsense reasoning.

即使一个程序员有无限的耐心、知识和远见，把每一个相关细节都存入计算机，*如果*他不懂得按人类常识推理规律去对计算机下达指令，机器里的信息也还是没有用途的。

9. And 表示结果。

Operators found that the water level was too low so they turned on two additional main coolant pumps, *and* too much cold water flowing into the system caused the steam to condense, further destabilizing the reactor.

操作人员发现冷却水的水位过低，就起动了另外两台主冷却泵，*结果*过量的冷却水进入系统，使蒸汽冷凝，反应堆因而更不稳定。

Reading: Testing and Measuring Instruments

Once a particular component is suspected of being faulty, individual tests must be then performed. And the testing and measuring instruments often to be used are as follows:

Multimeter (Fig. 14-9)

A multimeter is sometimes called a volt-ohm *milliammeter*. Multimeters having only volt and ohm ranges are also called volt-ohmmeters. There are two main kinds of multimeters, they are mechanical handy multimeters and digital ones. It's better for learners to use handy types, because it's helpful for them to be familiar with some electronic principles.

Using a multimeter, you should first place it flat on a table and note if the meter pointer indicates exactly 0 at the extreme left end of the black *scale*. If it doesn't read 0, turn the *screw* on the meter movement slowly until the proper 0 reading is realized. [1] And then connect test-*lead* to the other. Set the RANGE switch to the proper item and range. *Insert* the test-lead *plugs* into the correct pair of *jacks*, *clip* or hold the test probes on the terminals of the part being checked, and read the meter.

With a multimeter, we can measure current, voltage, resistance, etc. And we can also *troubleshoot appliances*, test fuses, measure voltage at a wall outlet, check the wires from your roof antenna, find out whether a wall switch is broken, and perform dozens of other household jobs. In troubleshooting, voltage can be measured while the set is on, and resistance can be measured while the set is off. You get clues to the location of the trouble by comparing your multimeter readings with the normal values given in service manuals.

Oscilloscope (Fig. 14-9)

Oscilloscope is a display instrument. With it you can study the waveform of an electric signal, such as the waveform of alternating current and voltage. You can measure voltage, current, power, and frequency as well; in fact, almost any quantity that involves amplitude and waveform.

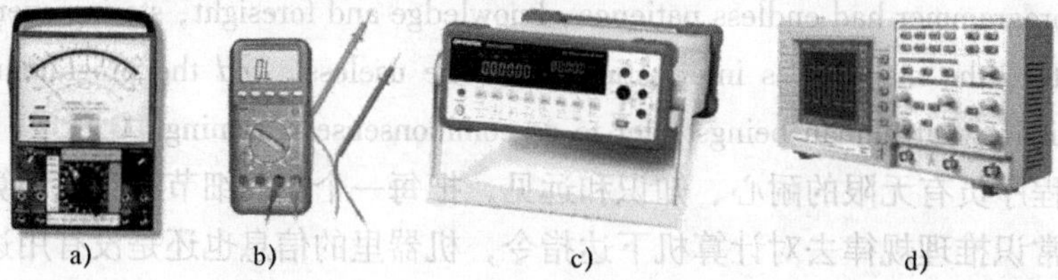

a) b) c) d)

Fig. 14-9 Testing and Measuring Instrument Examples

a) Analog Multimeter b) Digital Multimeter c) Bench-type Digital Multimeter d) Digital Storage Oscilloscope

The heart of the oscilloscope is the cathode-ray tube (CRT). This consists of the base, neck (an electron gun is included), bulb and the face-plate (screen). The electron gun consists of a cathode, a control *grid*, an anode and two sets of *deflection* plates. When the cathode is heated, it emits electrons, which form electron beam. The electron beam sweeps rapidly from left to right across the screen of a cathode-ray tube. However, recently most of the cathode-ray tubes used in oscilloscopes have been replaced by LCDs (Liquid *Crystal* Displays), which greatly decrease the weight and size of the instrument.

New Words and Phrases

multimeter [ˌmʌltiˈmitə] n. [电] 万用表
milliammeter [ˌmiliˈæmitə] n. [电] 毫安计
scale [skeil] n. 刻度；比例；天平 vi. 攀登；衡量 vt. 测量；依比例改变
screw [skruː] vt. & vi. 转动；旋，拧；压榨，强迫 n. 螺钉；吝啬鬼
insert [inˈsəːt] vt. 插入，嵌入 n. 插入物
lead [liːd] n. 铅；导线；引线 vt. & vi. 导致；领导 adj. 带头的；最重要的
plug [plʌg] n. 栓；插头；塞子 vi. 塞住；用插头接电源
jack [dʒæk] n. 千斤顶；插座；插孔 vt. 提醒；用千斤顶顶起
clip [klip] vt. 修剪；夹牢；痛打 vi. 修剪 n. 夹子；修剪；回形针
troubleshoot [ˈtrʌblʃuːt] vt. 检修，排除故障 vi. 充当故障检修员
appliance [əˈplaiəns] n. 器具，器械；装置设备；应用
oscilloscope [əˈsiləskəup] n. [电子] 示波器；示波镜
grid [grid] n. [计] 网格；格子，栅格；输电网
deflection [diˈflekʃən] n. 偏向，偏差，挠曲
crystal [ˈkristəl] n. 水晶；晶体 adj. 水晶的；透明的，清澈的
handy multimeter 指针式万用表
cathode-ray tube (CRT) 阴极射线管
control grid 控制栅（极）
electron beam 电子束
Liquid Crystal Display (LCD) 液晶显示器

Notes

1. If it doesn't read 0, turn the screw on the meter movement slowly until the proper 0 reading is realized.

译文：如果指针没指到零位，可慢慢转动万用表上的螺钉（或调整旋钮）直到（指针）归零。

说明：if 引导条件状语从句；主句省略了主语和情态动词 you can；on the meter movement 为介词短语作定语，修饰 the screw。

Exercises

Choose the best answer(s) to complete the following statements.

1. According to the text, ________ is the most common tool in repairing work.

A. oscilloscope B. multimeter C. screw D. electron gun

2. The multimeter must be put horizontally and the pointer must point to ________.

A. the left B. the end C. zero D. range

3. Different ________ must be set for different testing and measuring.

A. screw B. range switch C. probe D. plug

4. When testing or measuring, you should hold the test probes on the terminals of the ________.

A. part being checked B. meter C. jacks D. plugs

5. Oscilloscope is an instrument which can ________ the waveform of an electric signal.

A. study B. draw C. display D. generate

Unit 15 Transistor Voltage Amplifiers

Text

Amplifiers are necessary in many types of electronic equipments such as radios, oscilloscopes and record players. Often it is a small *alternating* voltage that has to be amplified. [1] A junction transistor in the common-emitter mode can act as a voltage amplifier if a suitable resistor, called the load, is connected in the collector circuit.

The small alternating voltage, the input U_i is applied to the *base-emitter* circuit and causes small changes of base current which produce large changes in the collector current flowing through the load. The load converts these current changes into voltage changes which form the alternating output voltage U_o (U_o being much greater than U_i). [2]

A transistor voltage amplifier circuit is shown in Fig. 15-1. To see just the voltage amplification occurs, consider first the situation when there is no input, i. e. $U_i = 0$, called the *quiescent* (quiet) state.

For transistor action to take place the base-emitter junction must always be forward *biased*. A simple way of ensuring this is to connect a *resistor* R_B, called the base-bias resistor. A steady (D. C.) base current I_B flows from battery " + ", through R_B into the base and back to battery " - ", via the emitter. The value of R_B, can be calculated once the value of I_B for the best amplifier performance has been decided.

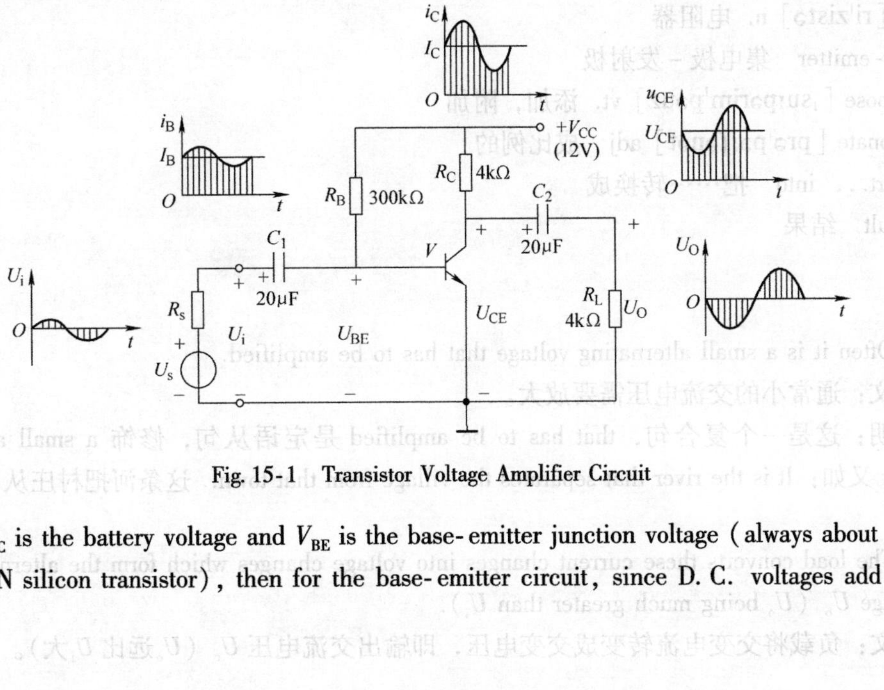

Fig. 15-1 Transistor Voltage Amplifier Circuit

If V_{CC} is the battery voltage and V_{BE} is the base-emitter junction voltage (always about +0.6V for an NPN silicon transistor), then for the base-emitter circuit, since D. C. voltages add up, we can write

$$V_{CC} = I_B R_B + V_{BE} \tag{15-1}$$

I_B causes a much larger collector current I_C which produces a voltage drop $I_C R_L$ across the load R_L. If V_{CE} is the *collector-emitter* voltage, then for the collector-emitter circuit

$$V_{CC} = I_C R_L + V_{CE} \tag{15-2}$$

When U_i is applied and goes positive, it increases the base-emitter voltage slightly (e. g. from +0.60 V to +0.61 V). When U_i swings negative, the base-emitter voltage decreases slightly (e. g. from +0.60V to +0.59V). As a result a small alternating current is *superimposed* on the quiescent base current I_B which in effect becomes a varying D. C. [3]

When the base current increases, large *proportionate* increases occur in the collector current. From equation (15-2) it follows that there is a corresponding large decrease in the collector-emitter voltage (since V_{CC} is fixed). A decrease of base current causes a large increase of collector-emitter voltage. In practice positive and negative swings of a few milli-volts in U_i can result in a fall or rise of several volts in the voltage across R_L and so also in the collector-emitter voltage. [4]

The collector-emitter voltage may be regarded as an alternating voltage superimposed on a steady direct voltage, i. e. on the quiescent value of V_{CE}.[5] Only the alternating part is wanted and capacitor C blocks the direct part but allows the alternating part, i. e. the output U_o to pass.

New Words and Phrases

amplifier [ˈæmplɪfaɪə(r)] n. 放大器；扩音器
alternating [ˈɔːltəneitiŋ] adj. 交互的；交替的 v. （使）交替
base-emitter 基极－发射极
quiescent [kwiˈesnt] adj. 不活动的，静态的，休眠的
biased [ˈbaiəst] adj. 结果偏倚的，有偏的，有偏见的
resistor [riˈzistə] n. 电阻器
collector-emitter 集电极－发射极
superimpose [ˌsuːpərimˈpəuz] vt. 添加，附加
proportionate [prəˈpɔːʃənət] adj 成比例的
to convert... into 把……转换成
as a result 结果

Notes

1. Often it is a small alternating voltage that has to be amplified.

译文：通常小的交流电压需要放大。

说明：这是一个复合句，that has to be amplified 是定语从句，修饰 a small alternating voltage。又如：It is the river that separates the village from that town. 这条河把村庄从小镇分离开来。

2. The load converts these current changes into voltage changes which form the alternating output voltage U_o (U_o being much greater than U_i).

译文：负载将交变电流转变成交变电压，即输出交流电压 U_o（U_o远比 U_i大）。

说明：这是一个复合句，主句在前。which form the alternating output voltage U_o是定语从句修饰 voltage changes。convert... into ... 意为“把……变成……”，相当于 change... into ...。例如：

At what rate does the dollar convert into pounds? 美元以什么汇率兑换成英镑？

其次，alternating 是现在分词作定语，意为“交替的、交互的”，修饰 output voltage。例如：

This is an alternating current generator. 这是一个交流发电机。

3. As a result a small alternating current is superimposed on the quiescent base current I_B which in effect becomes a varying D. C..

译文：这使得一个小的交流电流叠加在静态基极电流 I_B上，其结果是产生一个幅值变化的直流电。

说明：这是一个复合句。as a result 意为“结果”；which in effect becomes a varying D. C. 是定语从句，修饰 the quiescent base current I_B. 例如：

There was a substantial growth of industry and foreign trade increased as a result. 由于工业的大发展，对外贸易也增长了。

4. In practice positive and negative swings of a few milli-volts in $U_i$can result in a fall or rise of several volts in the voltage across R_L and so also in the collector-emitter voltage.

译文：实际上，U_i毫伏数量级的上下波动，就会导致负载 R_L 上的电压上升或下降几个伏特，集电极－发射极电压也做相应幅度的变化。

说明：这是一个复合句。in practice 意为“实际上”；result in 意为“导致……结果”；so 这里指 can also result in a fall or rise of several volts. 例如：

Tom went out to play football and so Joseph also. 汤姆出去踢足球了，约瑟也一样。

5. The collector-emitter voltage may be regarded as an alternating voltage superimposed on a steady direct voltage, i. e. on the quiescent value of V_{CE}.

译文：集电极－发射极电压可以认为是一个叠加在稳定直流电压之上的交流电压，比如叠加在静态电压 V_{CE}上。

说明：过去分词短语 superimposed on a steady direct voltage 充当定语，修饰 an alternating voltage。be regarded as 意为“被看作，被视为”。例如：

The dead student is now being regarded as a martyr. 这名死去的学生现在被视为烈士。

Exercises

Ⅰ. Decide whether the following statements are True (T) or False (F) according to the passage.

1. Amplifiers are necessary in many types of electronic equipments such as radios, oscilloscopes and record players. Often it is a large alternating current that has to be amplified.

2. A junction transistor in the common-emitter mode can act as a resistance amplifier.

3. The small alternating voltage is applied to the base-emitter circuit and causes large changes of base current which produce small changes in the collector current flowing through the load.

4. The load converts current changes into voltage changes, which form the alternating output

voltage.

5. A decrease of base current causes a large increase of collector-emitter voltage.

Ⅱ. **Match the following terms to appropriate definition or expression.**

1. load	a. a suitable resistor
2. V_{CC}	b. the alternating output voltage
3. V_{BE}	c. base-emitter junction voltage
4. U_o	d. the base-bias resistor
5. U_i	e. battery voltage
6. R_B	f. the small alternating voltage, the input

Ⅲ. **Translate the following sentences into English by using the words in brackets.**

1. 经过训练的狗可以充当盲人的向导。(act as)
2. 电容易转变成其他能量吗？(convert into)
3. 这样他们就能更好地把理论运用到实践中去。(apply to)
4. 人们把风能视为清洁能源。(regard as)

Translating Skills:

科技术语的翻译

科技术语是准确地描述科技和社科领域相关概念的词语，用来记录和表述各种现象、过程、特性、关系、状态等，是科技文献的灵魂和精华。在进行翻译时，科技术语的翻译往往是难点，也是亮点。如何用规范汉语表达术语内涵且又不失科技文体的准确性与严谨性呢？下面介绍一些常用的翻译方法：

一、意译

按照原词所表达的具体事物或概念译出其科学含意，这种译法的使用很普遍，例如：holography 全息摄影、guided missile 导弹、aircraft carrier 航空母舰、videophone 可视电话、E-mail电子邮件等。多适用于以下三类术语词汇：

1）合成词。由两个或两个以上单词构成的复合词。例如：skylab 太空实验室、moonwalk 月球漫步、friction factor 摩擦因数。

2）多义词。旧词转义，即通过赋予旧词以新义而构成术语。例如：drone（雄蜂）无人驾驶飞机、bug（臭虫）窃听器、computer 计算者→计算机。

3）派生词。在原有的词根基础上加前缀或后缀构成术语。例如：thermocouple 热电偶；voltmeter 电压表。

二、音译

根据英语单词的发音译成读音与原词大致相同的汉语。主要适用于以下两类术语词汇：

1. 计量单位词。例如：hertz 赫兹（频率单位）、bit 比特（二进位制信息单位）、lux 勒克司（照明单位）、joule 焦耳（功或能的单位）。

2. 某些材料或产品名称。例如：nylon 尼龙（聚酰胺纤维）、sonar 声呐（声波导航和测距设备）、vaseline 凡士林（石油冻）、morphine 吗啡。一般来说，意译能够明确地表达术语含义。因此，有些音译词常被意译词所取代，或二者同时使用。例如：combine（康拜因）联合收割机、laser（莱塞）激光、vitamin（维他命）维生素、penicillin（盘尼西林）青霉素。

三、形译

英语常用字母来为形状相似的物体命名。形译法多适用于以下三种情况：

1. 选用与原字母形似的汉语。例如：T-square 丁字尺、I-column 工字柱、U-bend 马蹄弯头、V-slot 三角形槽。

2. 保留原字母不译，在该字母后加“形”字。例如：A-bedplate A 形底座、D-valve D 形阀、C-network C 形网络、M-wing M 形机翼。

3. 保留原字母不译，直接代入。例如：X ray X 射线、L-electron L层电子（原子核外第二层电子）、N-region N 区（电子剩余区，即电子导电区）。

四、意、音结合译

根据具体情况意译与音译可混用。例如：logiccircuit 逻辑电路、covar 科伐合金（铁镍钴合金）、microampere 微安（培）、kilowatt 千瓦、radar-man 雷达手、Morse code 莫尔斯电码、Babbit metal 巴氏合金。

五、直译

在科技文献中，商标、牌号、型号和表示特定意义的字母均可不译，直接使用原文，只译普通名词。例如：B-52 E bomber B-52 E 轰炸机、Kubota Mobile Crane Model KM-150 库宝塔 KM-150 型流动式起重机。

六、其他注意事项

在翻译科技术语时，应避免译名不统一，不规范，以免给读者带来麻烦。首先要尽量采用已公布的统一译名，或者已为大家所公认的译名。例如 lathe 一词应译为车床，而不应译为旋床；gram 的译名有公分、克兰姆、克等，通常应译为克。对于那些已经有了习惯译法的科技术语，则不适合标新立异。例如 Nylon，现在习惯译为尼龙，如果随便译成“乃隆”，就可能被当成另一种材料。对于那些有几种习惯译法，几个译名都通用的术语，应注意保持前后译名一致。

科技术语翻译的难点主要集中于对专业词语的认知理解上，因此，在日常学习与工作中，应注意搜集相关专业知识的表达方法，包括专业术语、专业词汇以及语言特色等。并且，要对搜集到的资料及时进行归纳、整理和总结。

Reading: Integrated Circuits

An integrated circuit (IC) is a *combination* of a few *interconnected* circuit elements such as transistors, diodes, capacitors and resistors. It is a small electronic device made out of a semiconductor material. The first integrated circuit was developed in the 1950s by Jack Kilby of Texas Instruments and Robert Noyce of Fairchild Semiconductor.

The electrically interconnected components that make up an IC are called integrated elements.[1] If an integrated circuit includes only one type of components, it is said to be an *assembly* or set of components.

Integrated circuits are used for a variety of devices, including *microprocessors*, audio and *video equipment*, and *automobiles*. Integrated circuits are often *classified* by the number of transistors and other electronic components, they *contain*:

– SSI (small-scale integration): Up to 100 electronic components per chip.

– MSI (medium-scale integration): From 100 to 3, 000 electronic components per chip.

– LSI (large-scale integration): From 3, 000 to 100, 000 electronic components per chip.

– VLSI (very large-scale integration): From 100, 000 to 1, 000, 000 electronic components per chip.

– ULSI (*ultra* large-scale integration): More than 1 million electronic components per chip.

As the capability to integrate a greater number of transistors in a single integrated circuit (IC) grows, it is becoming more common that an application-specific IC (ASIC) is required, at least for high volume applications.[2] Advances in silicon technology have allowed IC designers to integrate more than a few million transistors on a chip; even a whole system of *moderate complexity* can now be *implemented* on a single chip.

The invention of IC is a great *revolution* in the electronic *industry*. Sharp size, weight reductions are possible with these techniques, and more importantly, high *reliability*, excellent *functional* performance, low cost and low power *dissipation* can be achieved. ICs are widely used in the electronic industry.

New Words and Phrases

assembly [ə'semblɪ] n. 装配；集会，集合

combination [ˌkɒmbɪ'neɪʃn] n. 结合，联合，合并

interconnect [ˌɪntəkə'nekt] v. 使互相连接，互相联系

microprocessor [ˌmaɪkrəʊ'prəʊsesə(r)] n. [计] 微处理器

equipment [ɪ'kwɪpmənt] n. 设备，装备；器材，配件

automobile ['ɔːtəməbɪːl] n. <美>汽车；驾驶汽车

contain [kən'teɪn] vt. 包含，容纳；克制；牵制

classify ['klæsɪfaɪ] vt. 分类，归类；把……列为密件

video ['vɪdɪəʊ] n. [电子] 视频　adj. 录像的，视频的

implement ['ɪmplɪmənt] vt. 实施，执行　n. 工具

ultra ['ʌltrə] adj. 极端的，过分的

moderate ['mɒdərət] adj. 有节制的，适度的，中等的

complexity [kəm'pleksətɪ] n. 复杂性，复杂的事物；复合物

revolution [ˌrevə'luːʃn] n. 革命；旋转；运行

industry ['ɪndəstrɪ] n. 工业；产业；工业界

reliability [rɪˌlaɪə'bɪlətɪ] n. 可信度；可靠性，可靠度

dissipation [ˌdɪsɪ'peɪʃn] n. （物质、精力逐渐地）消散，分散

functional [ˈfʌŋkʃənl] adj. 功能的
the circuit element 电路元件
the video equipment 视频设备
electronic component 电子元件
ultra large-scale integration (ULSI) 超大规模集成电路
the electronic industry 电子工业

Notes

1. The electrically interconnected components that make up an IC are called integrated elements.

译文：这些构成集成电路的电气连接元件称为集成元件。

说明：that 引导定语从句，修饰先行词 interconnected components。

2. As the capability to integrate a greater number of transistors in a single integrated circuit (IC) grows, it is becoming more common that an application-specific IC (ASIC) is required, at least for high volume applications.

译文：随着一个集成电路集成更多数量晶体管的能力的提高，对专用集成电路的需求已更加普遍，至少对大批量的应用来说如此。

说明：这是一个复合句。as 引导时间状语从句；主句本身也是一个复合句，it 是形式主语，真正的主语是由 that 引导的主语从句；at least 引导让步状语。

Exercises

Ⅰ. Answer the following questions according to the passage.

1. What is an integrated circuit? When was the first integrated circuit developed?
2. What are called integrated elements? What is named an assembly or set of components?
3. How many types are integrated circuits often classified into? What are they?
4. What are possibly achieved with the invention of IC?

Ⅱ. Translate the following sentences into English by using the words in brackets.

1. 他们的成功是各种环境因素偶然联系在一起的幸运产物。(combination of)
2. 他们将不得不加班弥补罢工耽误的时间。(make up)
3. 他开始细心地为他的检验结果分类。(classify)

Unit 16 Introduction to Internet

Text

The Internet is a collection of individual networks, connected by intermediate networking devices, that functions as a single large network, it is made up of millions of computers linked together around the world in such a way that information can be sent from any computer to any other 24 hours a day. [1] These computers can be in homes, schools, universities, government departments, or businesses small and large. They can be any type of computer and be single personal computers or workstations on a school or a company network. The Internet is often described as "a network of networks" because all the smaller networks of organizations are linked together into the one *giant* network called the Internet. All computers are pretty much equal once connected to the Internet, the only difference will be the speed of the connection which is dependent on your Internet Service Provider and your own modem.

History of network

The first networks were time-sharing networks that used mainframes and attached terminals. Such environments were implemented by both IBM's Systems Network *Architecture* (SNA) and Digital's network architecture[2].

Local-Area Networks (LANs) evolved around the PC revolution. LANs enabled multiple users in a *relatively* small geographical area to exchange files and messages, as well as access shared resources such as file servers and printers.

Wide-Area Networks (WANs) interconnect LANs with geographically *dispersed* users to create *connectivity*. Some of the technologies used for connecting LANs include Tl, T3, ATM, ISDN, ADSL, Frame Relay, radio links, and others. New methods of connecting dispersed LANs are appearing everyday.

Today, high-speed LANs and Internet are becoming widely used, largely because they operate at very high speeds and support such high-bandwidth applications as multimedia and videoconferencing. [3]

TCP/IP

All computers on the Internet communicate with one another using the Transmission Control Protocol/Internet Protocol suite, *abbreviated* to TCP/IP. TCP/IP is a communications protocol used to transfer digital data around the Internet. TCP and IP were developed by a Department Of Defense (DOD) research project to connect different networks designed by different vendors into a network of networks (the "Internet"). Computers on the Internet use a *client/*server architecture. This means that the remote server machine provides files and services to the user's local client machine. Software can be installed on a client computer to take advantage of the latest access technology.

Internet services

An Internet user has access to a wide variety of services: electronic mail, file transfer, vast information resources, interest group membership, interactive *collaboration*, multimedia displays, real-time broadcasting, breaking news, shopping opportunities, and much more.

The Internet represents a global resource in the form of information and knowledge and provides a platform for cooperative ventures. This has resulted in the establishment of a wide variety of information service providers. Unfortunately, the Internet has also seen the establishment of a number of unwholesome service providers, such as those dealing in child *pornography* and extremist *propaganda*. [4] Perhaps we are *oversimplifying*, but there is good and bad information. Information can also be *categorized* in another way: as open or internal information, with *security* measures required to protect the latter category.

Computers on the Internet may use one or all of the following Internet services:

1) Electronic mail (E-mail). Permits you to send and receive mail.

2) *Telnet* or remote login. Permits your computer to log onto another computer and use it as if you were there.

3) FTP or File Transfer Protocol. Allows your computer to rapidly *retrieve* complex files intact from a remote computer and view or save them on your computer.

4) *Gopher*. An early, text-only method for accessing Internet documents. Gopher has been almost entirely *subsumed* in the World Wide Web, but you may still find gopher documents linked to in web pages.

The World Wide Web (WWW or the "Web"), the largest, fastest growing activity on the Internet, *incorporates* all of the Internet services above and much more. You can retrieve documents, view images, *animation*, and video, listen to sound files, speak and hear voice.

When you log onto the Internet using Netscape or Microsoft's Internet Explorer or some other *browser*, you are viewing documents on the World Wide Web. The current foundation on which the WWW functions is the programming language called HTML. *Hypertext* is the ability to have web pages containing links, which are areas in a page or buttons or graphics on which you can click your mouse button to retrieve another document into your computer. This "click ability" using Hypertext links is the feature which is unique and revolutionary about the Web.

New Words and Phrases

giant [ˈdʒaɪənt] n. 巨人 adj. 巨大的，庞大的

architecture [ˈaːkɪtektʃə] n. 建筑，建筑学；体系机构

relatively [ˈrelətɪvlɪ] adv. 相关地

disperse [diˈspɜːs] v. （使）分散，（使）散开，疏散

connectivity [kɒnekˈtɪvɪtɪ] n. 连接；连通性

abbreviate [əˈbriːvieɪt] v. 缩写，简化，简写成

client [ˈklaɪənt] n. ［计］顾客，客户，委托人

collaboration [kəˌlæbəˈreɪʃn] n. 协作；通敌；勾结

pornography [pɒː'nəgraːfɪ] n. 色情文学，色情描写；色情
propaganda [ˌprɒpə'gændə] n. 宣传，传递；宣传运动
oversimplify ['əuvə'sɪmplɪʃaɪ] v. （使）过分地单纯化
category ['kætəgərɪ] n. 种类，类别，范畴
security [sɪ'kjuərətɪ] n. 安全；保证，担保
telnet ['telnet] 远程登录
retrieve [rɪ'trɪːv] v. 重新获得，找回 n. 取回；[计] 检索
Gopher 地鼠程序（信息检索）
subsume [səb'sjuːm] v. 包容，包含；归入
incorporate [ɪn'kɔːpəreɪt] adj. 合并的，一体化的 v. 合并
animation [ˌænɪ'meɪʃn] n. 活泼，有生气的动画
browser [brauzə(r)] n. 浏览器
hypertext ['haɪpətekst] 超文本

Notes

1. The Internet is a collection of individual networks, connected by intermediate networking devices, that functions as a single large network, it is made up of millions of computers linked together around the world in such a way that information can be sent from any computer to any other 24 hours a day.

译文：因特网是各个独立的网络通过中间网络设备连接到一起形成的一个功能独立的大型网络集合。它将世界上数百万的计算机连接在一起，网络内的任何计算机能全天候地向其他计算机传送信息。

说明：这是一个并列复合句，两个分句都是复合句。第一个分句主句在前，connected by intermediate networking devices 是过去分词短语作后置定语，其前后用逗号隔开，相当于一个非限制性定语从句修饰 individual networks；that functions as a single large network 是限制性定语从句修饰 a collection。

第二个分句的结构与第一个分句相仿，linked together around the world in such a way 是过去分词短语作后置定语修饰 computers，其前后没有用逗号隔开，与被修饰词的关系比较紧密；另外，that information can be sent from any computer to any other 24 hours a day 是一个同位语从句，对 such a way 起补充说明作用。millions of 意为数百万的；such … that 意为“如此……，以至……”，一般引导结果状语从句。例如：

He could put an idea in such a way that Alan (would) believe it was his own. 他能把观点如此表达，以至于艾伦认为那是他自己的主意。

2. The first networks were time-sharing networks that used mainframes and attached terminals. Such environments were implemented by both IBM's Systems Network Architecture (SNA) and Digital's network architecture.

译文：最早的网络是使用主机和终端的分时复用网络。此类网络主要由 IBM 的系统网络体系（SNA）和数字网络体系结构实现。

说明：这是一个复合句。that 引导的定语从句修饰 time-sharing networks，time-sharing 意为“分时”，此处指中央处理系统同时为数位使用者服务的时间安排。例如：

Each panel may include menus and dialogs to run tools on the underlying Time Sharing Option (TSO). 每个面板可以包括菜单和对话框来运行底层的分时选项（TSO）工具。

3. Today, high-speed LANs and Internet are becoming widely used, largely because they operate at very high speeds and support such high-bandwidth applications as multimedia and videoconferencing.

译文：当今，高速局域网和互联网被广泛地使用，主要因为它们可提供并支持高速率的传输和高带宽的应用，如多媒体和视频会议。

说明：这是一个复合句。主句在前，largely because 引出原因状语从句。at high speeds 意为“高速”，例如：

It also must be light, since it travels at high speeds as it moves up and down inside the cylinder. 同时它（活塞）必须很轻，因为它在气缸内以很高的速度上下移动。

4. Unfortunately, the Internet has also seen the establishment of a number of unwholesome service providers, such as those dealing in child pornography and extremist propaganda.

译文：不幸的是，在互联网上也存在一些不健康的网络服务提供者，例如那些从事儿童色情交易和极端行为宣传的。

说明：这是一个简单句。such as 意为“例如”。例如：

It should highlight what you have to offer the company, such as a specific skill or experience. 应该重点突出你能为公司贡献什么，比如特殊的技能或工作经验等。

Exercises

Ⅰ. Translate the following phrases into English.

1. 局域网	2. 高带宽应用	3. 广域网
4. 电子邮件	5. 实时广播	6. 交互式协作
7. 全球资源	8. 文件传输协议	9. 传输控制协议

Ⅱ. Translate the following phrases into Chinese.

1. a wide variety of services	2. Internet Service Provider
3. Systems Network Architecture (SNA)	4. Digital's network architecture
5. exchange files and messages	6. high-speed LANs and Internet
7. a client/server architecture	8. a relatively small geographical area
9. multimedia and videoconferencing	10. method for accessing Internet documents

Ⅲ. Translate the following sentences into English by using the words in brackets.

1. 典型的掌上电脑具有移动电话、传真发送机和个人文件夹的功能。(function as)
2. 红外线广泛应用于工业和医学科学。(widely used)
3. 这些建议是谈判的现实出发点。(represent)
4. 回流后，仔细拿开带组件的基板，并使之冷却。(allow to)

Translating Skills:

反 译 法

在英译汉的过程中常遇到这样一种情况，即原词表达的并不是其字面意义，而是其字面意义的反义，或者说是对其字面意义的否定，但这种否定又往往不出现否定词。为了符合汉语表达习惯，往往必须使用与原文意义相反的字样或句式才能确切地译出原文的含义。这种翻译方法称为“反译法”。下面分五个方面举例说明：

1. 添加否定词反译

为使译文通顺，翻译时需要使用同原文意义相反的词并加以否定，即采用“否定词+反义词”的翻译方法，在反义词前面加上否定词。例如：

Mechanical seal and ball bearing may be left assembled unless it is necessary to service them.

机械密封和滚珠轴承除非需要维修，否则就不必拆卸（left assembled 意为“让它装着”，此处译为“不必拆卸”更为通顺）。

There are many other sources in store.

还有多种其他能源尚未开发。

2. 删去否定词反译

为使译文通顺，翻译时需要删去原文中的否定词，再将被否定的词反译，从而译出原文所强调的含意。例如：

Owing to rigidity of the spindle and bearings, the fluid bearings never lose their accuracy.

由于主轴和轴承刚性良好，流体轴承能够永久保持精度。

3. 双重反译

对句中两个词（否定词或肯定词）都按相反意义翻译出来，叫作双重反译，可使汉语译文清晰、确切而严密地表达原文意义。例如：

A silicon radiation pyrometer is the only available transfer pyrometer with a stability of better than ±0.1% annually.

硅辐射高温计是唯一年不稳定性不超过±0.1%的传热高温计。

There is no material but will deform more or less under the action of force.

在压力的作用下，任何材料或多或少都会变形（but 是含有否定意义的关系代词，等于 that not；把 no 和 but 都仅译成肯定，使译文比较简明）。

4. 固定结构反译

英语中有些固定结构形似否定意为肯定，或形似肯定意为否定。译成汉语时，往往以表意为主，也可算是一种反译。例如：

We cannot be too careful in doing experiments.

我们做实验时要尽可能小心（cannot too... 是用否定的形式表示肯定意思，意为“无论怎样……也不过分”，而不是“不能太……”）。

5. 句式反译

句式反译指的是否定句和肯定句两种句式的转换翻译法。有时在翻译时必须使用与英语原文相反的句式才能确切地表达原文的意思。例如：

Don't start working before having checked the instrument thoroughly.

要对仪器彻底检查后才能开始工作。

Reading: Bill Gates

William (Bill) H. Gates is chairman and chief software *architect* of Microsoft Corporation, the *worldwide* leader in software, services and Internet technologies for personal and business computing. Microsoft had an income of $93.6 billion for the financial year ending June 2015, and employs more than 118,000 people in 60 countries.

Born on October 28, 1955, Gates and his two sisters grew up in *Seattle*. Gates attended public elementary school and the private Lakeside school. There, he discovered his interest in software and began programming computers at age of 13.

In 1973, Gates entered *Harvard* University as a freshman, where he lived down the *hall* from Steve Ballmer, now Microsoft's chief *executive* officer. While at Harvard, Gates developed a version of programming language BASIC for the first *microcomputer*—the MITS Altair.

In his junior year, Gates left Harvard to devote his energies to Microsoft, a company he had began in 1975 with his childhood friend Paul Allen. [1] Guided by a belief that the computer would be a valuable tool on every office *desktop* and in every home, they began developing software for personal computers. Gates' *anticipation* and his *vision* for personal computing have been central to the success of Microsoft and the software industry. [2]

In 1999, Gates write Business at the Speed of Thought, a book that shows how computer technology can solve business problems in *fundamentally* new ways. The book was *published* in 25 languages and is available in more than 60 countries. Business at the Speed of Thought has received wide critical *applause*, and was listed on the best-seller lists of the New York Times, USA Today, the Wall Street Journal and Amazon.com. Gates' previous book, The Road Ahead, published in 1995, held the No. 1 place on the New York Times' bestseller list for seven weeks.

In addition to his love of computers and software, Gates is interested in *biotechnology*. He is an investor in a number of other biotechnology companies. Gates also founded Corbis, which is developing one of the world's largest resources of *visual* information. In addition, Gates has *invested* with *cellular* telephone pioneer Craig McCaw in Teledesic, which is working on an *ambitious* plan to employ hundreds of low-orbit satellites to provide a worldwide two-way broadband telecommunications service.

New Words and Phrases

architect [ˈɑːkɪtekt] n. 建筑师；缔造者

worldwide [ˈwɜːldwaɪd] adj. 全世界的，世界范围的　adv. 在世界各地

Seattle [siˈætl] n. 西雅图（美国一港市）

Harvard [ˈhɑːvəd] n. 哈佛大学；哈佛大学学生

hall [hɔːl] n. 过道；食堂
executive [ɪɡˈzekjətɪv] adj. 行政的；执行的 n. 执行者
microcomputer [ˈmaɪkrəʊkəmˌpjuːtə] n. 微电脑；[计]微型计算机
desktop [ˈdesktɒp] n. 桌面；台式机
anticipation [ænˌtɪsɪˈpeɪʃn] n. 希望；预感；预支
vision [ˈvɪʒn] n. 视力；眼力；想象力 vt. 想象
fundamentally [ˌfʌndəˈmentəli] adv. 根本地，从根本上；基础地
publish [ˈpʌblɪʃ] vt. 出版；发表；发行
applause [əˈplɔːz] n. 欢呼，喝彩；鼓掌欢迎
biotechnology [ˌbaɪəʊtekˈnɒlədʒi] n. [生物]生物技术；生物工艺学
visual [ˈvɪʒʊəl] adj. 视觉的，视力的；栩栩如生的
invest [ɪnˈvest] vt. 投资；授予 vi. 投资，入股
cellular [ˈseljələ] adj. 细胞的；由细胞组成的 n. 移动电话
ambitious [æmˈbɪʃəs] adj. 野心勃勃的；热望的；炫耀的
Harvard University 哈佛大学
Steve Ballmer 史蒂夫·鲍尔默
the MITS Altair MITS Altair 电脑
the best-seller lists 畅销书目录
Ihe Road Ahead 未来之路
Amazon. com 亚马逊网站
Craig McCaw 克雷格·麦考

Notes

1. In his junior year, Gates left Harvard to devote his energies to Microsoft, a company he had began in 1975 with his childhood friend Paul Allen.

译文：大学三年级时，盖茨从哈佛退学，专心于微软公司的业务，这个公司是盖茨在1975年与他的儿时好友保罗·艾伦一起创建的。

说明：这是一个复合句。a company 是 Microsoft 的同位语，其后是定语从句修饰 a company。devote…to 意为“把……专用于……；致力于……”。例如：

He seemed to have unlimited time to devote to us. 他似乎有无限时间专门陪我们。

2. Gates'anticipation and his vision for personal computing have been central to the success of Microsoft and the software industry.

译文：盖茨对个人计算机的远见卓识是微软公司和软件行业成功的关键。

说明：central to 意为“对……至关重要”；industry 意为“行业，工业”。例如：

Education is central to a country's economic development. 教育对一个国家的经济发展至关重要。

Exercises

Choose the best answer(s) to complete the following statements.

1. Gates began to have interest in software when he was in ________.

A. elementary school B. middle school C. Harvard University D. work

2. The first microcomputer is called ________.

A. Microsoft B. BASIC C. the MITS Altair D. Teledesic

3. Gates began Microsoft with ________.

A. Steve Ballmer B. Paul Allen C. Craig McCaw D. his father

4. Business at the Speed of Thought was not listed on the best-seller lists of ________.

A. the New York Times B. USA Today

C. the Wall Street Journal D. Yahoo. com

5. Which of the following fields is NOT among the investments of Gates?

A. Education B. Biotechnology

C. Visual information D. Cellular telephone

Unit 17 4G Network Technology

Text

4G is the fourth generation of wireless communications currently being developed for high speed broadband mobile capabilities. It is characterized by higher speed of data transfer and improved quality of sound. Although not yet defined by the ITU (International Telecommunications Union), the industry identifies the following as 4G technologies:

1) WiMAX (Worldwide *Interoperability* for Microwave *Access*).

2) 3GPP LTE (3rd Generation Partnership Project Long Term *Evolution*).

3) UMB (Ultra Mobile Broadband).

4) Flash-OFDM (Fast Low-*latency* Access with *Seamless Handoff* Orthogonal Frequency Division Multiplexing).

The 4G technology is being developed to meet QoS (Quality of Service) and rate requirements that involve prioritization of network *traffic* to ensure good quality of services. [1] These *mechanisms* are essential to accommodate applications that utilize large bandwidth such as the following. [2] Wireless Broadband Internet Access, MMS (Multimedia Messaging Service), Video Chat, Mobile Television, HDTV (High Definition TV), DVB (Digital Video Broadcasting), Real Time Audio, High Speed Data Transfer.

The goal set by ITU for data rates of WiMAX and LTE is to achieve 100Mbps when the user is moving with high speed relative to the base station, and 1Gbps for fixed positions. [3]

The industry moves towards expansion of the number of 4G compatible devices. It is set to find its way to tens of different mobile devices not restricted to 4G phones or laptops, such as Video Camera, Gaming Devices, Vending Machines, and Refrigerators.

The trend is to provide wireless internet access to every portable device that could supply and incorporate the 4G *embedded* modules. The 4G technology could not only provide internet broadband connectivity but also a high level of security that is beneficial to devices that incorporate financial transactions such as vending machines and billing devices. [4]

The following key features can be observed in all suggested 4G technologies:

1) MIMO: to attain ultra-high *spectral* efficiency by means of spatial processing including multi-*antenna* and multi-user MIMO.

2) Frequency-domain-equalization (SC-FDE), for example, multi-carrier *modulation* (OFDM) in the downlink or single-carrier.

Frequency-domain-equalization in the uplink: to exploit the frequency selective channel property without complex equalization.

3) Turbo principle error-correcting codes: to minimize the required SNR at the reception side.

4) Channel-dependent scheduling: to use the time-varying channel.

5) Link adaptation: adaptive modulation and error-correcting codes.

6) Mobile-IP utilized for mobility.

7) IP-based *femtocells* (home nodes connected to fixed Internet broadband infrastructure).

New Words and Phrases

interoperability [ˌɪntərˈɒpərəbɪlətɪ] n. 互通性，互操作性
access [ˈækses] vt. 使用；存取 n. 使用权
evolution [ˌiːvəˈluːʃn] n. 演变；进化论；进展
latency [ˈleɪtənsɪ] n. 潜伏时间；延迟时间；潜在因素
seamless [ˈsiːmləs] adj. 无缝的；不停顿的；无漏洞的
handoff [ˈhændɒf] n. 切换；传送；手递手传球（美国橄榄球）
traffic [ˈtræfɪk] n. 交通 v. 交易，用……做交换
mechanism [ˈmekənɪzəm] n. 机制，原理；机械装置
embed [ɪmˈbed] vt. 使嵌入，使插入；使深深留在脑中
spectral [ˈspektrəl] adj. [光] 光谱的
antenna [ænˈtenə] n. [电信] 天线；[动] 触角；[昆] 触须
modulation [ˌmɒdjʊˈleɪʃn] n. [电子] 调制；调整
femtocell [ˈfemtəʊsel] n. 家庭基站

Technical Terms

4G 第四代移动电话通信标准
ITU 国际电信联盟
WiMAX 全球微波互联接入
3GPP 第三代合作伙伴项目
LTE 长期演进技术
QoS 服务质量技术
UMB 超移动宽带
OFDM 正交频分复用
MMS 多媒体信息服务
HDTV 高清电视
DVB 数字视频广播
MIMO (Multi-input Multi-output) 多输入多输出技术
SC-FDE 单载波频域均衡
SNR (signal-to-noise ratio) 信噪比
embedded module 嵌入模块
spectral efficiency 频谱效率
adaptive modulation 自适应调制

Notes

1. The 4G technology is being developed to meet QoS (Quality of Service) and rate requirements that involve prioritization of network traffic to ensure good quality of services.

译文：4G 技术正在被开发以满足 QoS（服务质量）和网络流量优先化的速率要求，从而保证良好的服务质量。

说明：这是一个复合句。that involve... services 是限制性定语从句修饰 requirements；meet requirements of... 意为满足……的需求。例如：

Ensure delivery date, can solve urgent problems to meet requirements of production. 保证及时供货，并能解决紧急问题，满足生产的需要。

2. These mechanisms are essential to accommodate applications that utilize large bandwidth such as the following.

译文：这些机制对于使用大宽带的应用调节是必需的。

说明：这是一个复合句。to accommodate applications 是不定式短语作状语修饰形容词 essential。that 引导定语从句修饰 applications。accommodate 意为“调节，使适应；使符合一致；调和；通融，给（某人）提供方便”。例如：

She tried to accommodate her way of life to his. 她试图使自己的生活方式与他的生活方式相适应。

3. The goal set by ITU for data rates of WiMAX and LTE is to achieve 100Mbps when the user is moving with high speed relative to the base station, and 1Gbps for fixed positions.

译文：ITU 为 WiMAX 和 LTE 设定的目标是：当用户相对于基站高速移动时，数据传输速率达 100Mb/s；位置固定时，速率达到 1Gb/s。

说明：这是一个复合句。set by... 是过去分词短语作后置定语修饰 goal；to achieve 100Mbps 为不定式短语作表语；when 引导时间状语从句修饰动词 achieve。

set by 在句中意为“设定”，此外还有“搁在一旁；抛开，撇开；留出”等意义。例如：

It's time to set our differences by and work together for a common purpose. 该是我们抛开分歧，为一个共同的目标而同心协力的时候了。

4. The 4G technology could not only provide internet broadband connectivity but also a high level of security that is beneficial to devices that incorporate financial transactions such as vending machines and billing devices.

译文：4G 技术不仅可以提供互联网宽带连接，而且具有高级别的安全性，有利于像自动售货机和计费装置等包含金融交易的设备安全运行。

说明：这是一个复合句。that is beneficial... 为定语从句，修饰 security；定语从句中 that incorporate... 为嵌套的定语从句，修饰句子中第一个 devices。例如：

It can be beneficial to share your feelings with someone you trust. 向自己信任的人倾诉感情是很有益处的。

Exercises

Ⅰ. Answer the following questions according to the passage.

1. What are the two characters of 4G technology according to the text?
2. What is the goal set by ITU for data rates of WiMAX and LTE?
3. What kind of 4G devices that incorporate financial transactions are mentioned in the text?

Ⅱ. Decide whether the following Statements are True (T) or False (F) according to the passage.

1. 4G is characterized by higher speed of data transfer and improved quality of sound.
2. The goal set by ITU for data rates of WiMAX and LTE is to achieve 10Mbps when the user is moving with high speed relative to the base station, and 100Mbps for fixed positions.
3. 4G is set to find its way to mobile devices only restricted to 4G phones or laptops.
4. 4G technology will provide the same level of security as 3G.

Ⅲ. Translate the following sentences into English by using the words in brackets.

1. 目前，欧美及日本动画片已经发展到一个相对成熟的阶段。(currently, develop)
2. 工人们要确保模具涂层质量及准时交货。(ensure, good quality)
3. 私人飞行员如何进入航线呢？(access to)
4. 几个不同的名称用来表示电磁波谱的不同部分。(characterize)

Useful Information:

零件数据库网站介绍与英文网站注册

现代的机械、电子和自动化设计早已经脱离了使用图板、圆规、直尺来绘图的手动模式，3D设计软件已经普及。很多设计工作是以先设计3D模型，然后再转化成2D工程图的形式来完成的。而现代工业的专业化分工已经很细了，从螺钉和螺母等紧固件、小的电容和电阻器、气缸电磁阀等气动部件到PLC乃至机器人等很多种类的工业产品都由很多知名的专业生产厂商来生产，这些厂商的3D和2D数字模型在这些公司的网站上基本都可以下载。但还有一个更方便的途径来获取这些数据，包括2D和3D图样、产品规格说明书等，这就是零件库（part library）。这些丰富的数据使我们无需再参照纸版说明书中的数据来重新设计已经标准化的零件，甚至机器的3D图形也可插入到我们的装配图中，从零件库网站下载的3D数字模型大大提升了我们的工作效率和质量。

1. 常用零件库网站

零件库网站多数由西方发达国家的专业公司来运营，网站的默认界面语言是英文，并可以切换到几十种语言，以方便不同国家和地区的用户使用。下面介绍几个主要的零件库网站：

（1）CADENAS零件库　德国CADENAS公司推出的零件库分为两个产品序列，即在线版和离线版，在线版的中文主站已于2010年上线，名为Linkable PARTcommunity，是CADENAS与国内的翎瑞鸿翔科技有限公司共同开发的，这个网站除了符合ISO、EN、DIN、

GB、JIS 等主要标准的标准件外，还包括数百家国内外厂商的产品模型，几乎涵盖了整个机械和自动化领域，如国内用户所熟知的工业电气领域的 ABB、魏德米勒、施耐德等，轴承行业的 FAG、NSK、WD、NTN 等，FA 自动化的 MISUMI，气动的 SMC、FESTO、AIRTAC 等，三维模型达数百万个规格。用户注册后，模型数据全部免费使用，其提供的众多主流 CAD 原始数据接口包括 Pro/E（Creo）、SolidWorks、UGS NX、Solid Edge、Inventor、CATIA、AutoCAD 等，此外提供的 STEP、STL、IGES、SAT 等中间格式接口可满足用户的不同需求。除 2D 和 3D 模型外，一些产品的规格说明书也可以提供。

而其离线版名为 PARTsolutions，除了为 CAD 系统提供上述零部件数据资源外，还提供广泛的 ERP/PDM 接口，实现与主流 PLM（Product Life Management）系统的紧密集成，如 SAP、Intralink、Windchill、Vault、Teamcenter 和 Smarteam 等。PARTsolutions 可将其上游 ERP 系统中的物料信息与模型信息实现对接，并经 CAD 系统导入 PDM（Product Data Management），从而打通企业的整个 PLM 流程。

为适应移动通信技术的发展，微信版的 PARTcommunity 已经上线。

搜索产品时，可以在 CAD model selection 页面上单击各个厂家的 Logo（商标符号）进入其产品数据库进行查找、选型和下载。如果知道产品的具体型号，也可以在 CAD model selection 页面的搜索栏直接搜索，待搜索成功后生成 3D 或 2D 数据，页面提供预览功能，以方便决定是继续完善模型还是直接下载使用。

网站也提供产品的分类索引和搜索功能，产品按照工业门类进行归档。

需要说明的是，由于国内的基础工业起步较晚，大多数厂家被归在 Chinese manufactures 这个名目下。随着国内工业的不断进步，也有厂家被逐步单独列出。

（2）TraceParts 零件库　TraceParts 拥有全球最大且使用最广泛的 3D 数字零件库，包含 1 亿多个 3D 数字零件模型，支持 CATIA、SolidWorks、AutoCAD、Inventor、Mechanical Desktop、Pro/E（Creo）、Unigraphics 和 Solid Edge 等主流 3D 建模软件格式。所有的零件库模型开放免费下载。除了 3D 模型外，还提供很多电气或者气动符号等供设计选用。

其标准零件库涉及多个主流的工业标准，包括 GB（中国国家标准）、DIN（德国工业标准）、ISO（国际标准化组织标准）、ANSI（美国国家标准）、UNI（意大利国家标准）和 JIS（日本工业标准）等。

其制造商零件库包括各个门类的众多知名制造商的产品数据，如西门子、施耐德、FAG、NSK、空中客车、米其林、ABB、阿尔斯通和 SKF 等，数量达到数百家。

除提供按厂家英文名称首字母索引的功能外，TraceParts 还提供产品的分类索引和搜索功能。包含以下分类：机械系统和通用部件、液压系统和通用部件、制造工程、能量和热传导工程、电气工程、电子元件、信息技术、办公设备、新能源、图像处理、道路和车辆、飞机和航天器、原料处理设备、货物包装和调运、农业、食品技术、采矿、石油及相关设备、冶金、橡胶和塑料、建筑和建筑物、土木工程、计量和测量、文娱和体育等。

（3）其他零件库　其他类似的零件库还有 3Dpartlib 和国内的制造云等。3Dpartlib 的内容与上述两家零件库网站类似，是 CADENAS 的合作伙伴，而制造云网站由国内公司运营，网站上有一些课程和论坛等内容，内容在某种程度上丰富一些，但标准化的 3D 资源比 TraceParts 和 PARTcommunity 要少。

需要说明的是，即使是中文界面，在 TraceParts 和 PARTcommunity 上进行选型时，有的产品参数只有英文说明，比如气缸的直径为 Bore、行程为 Stroke 等，这对使用者的英文水平提出了一定的要求，当然也是学习技术英语的好机会，科技英语水平会在工作中不知不觉地提高。

2. 零件库英文网站的注册

下面以 TraceParts 为例介绍一下零件库英文网站注册申请表（图 17-1）的填写。注册为网站的会员时会要求填写一些必要的个人信息，如姓名、电子信箱、电话、通讯地址、职业、工作单位名称和喜好等。除了购物网站等涉及金钱交易等内容的网站外，其他网站内容服务商一般情况下并不关注用户电子信箱以外的信息，所以电子信箱地址必须是真实的，否则忘记密码时无法重新激活账户。

可以参照以下内容了解零件库英文网站注册时经常需要填写的内容。下面介绍几条图 17-1 中没有包含的信息，这些信息有时也被要求填写：

Age：年龄：
Sex（Gender）：性别：
Marital Status：婚姻状况：
Religious：宗教：
Language：语言：
Preference：喜好：
Title：称呼，职务：

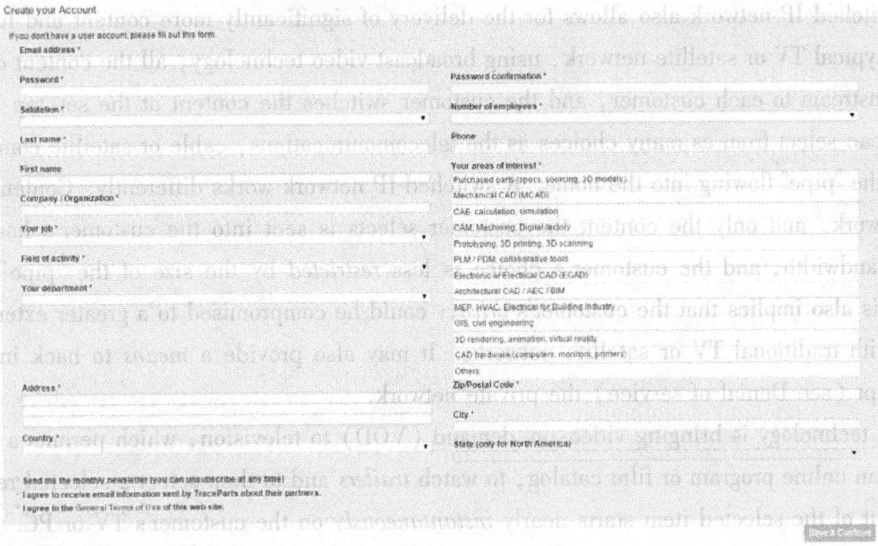

图 17-1　注册申请表

Reading:　Internet Protocol Television (IPTV)

Internet *protocol* television (IPTV) is a system through which television services are delivered using the Internet protocol suite over a packet-switched network such as the Internet, instead of being delivered through traditional terrestrial, satellite signal, and cable television formats.

IPTV services may be classified into three main groups:

1) live television, with or without interactivity related to the current TV show;

2) time-shifted television: catch-up TV (replays a TV show that was broadcast hours or days ago), start-over TV (replays the current TV show from its beginning);

3) video on demand (VOD): browse a catalog of videos, not related to TV programming.

IPTV is *distinguished* from Internet television by its on-going standardization process (e. g.: European Telecommunications Standards Institute) and *preferential* deployment *scenarios* in subscriber-based telecommunications networks with high-speed access channels into end-user premises via set-top boxes or other customer-premises equipment. [1]

IPTV is defined as the secure and reliable delivery to subscribers of entertainment video and related services. These services may include, for example, Live TV, Video On Demand (VOD) and Interactive TV (ITV). These services are delivered across an access *agnostic*, packet-switched network that employs the IP protocol to transport the audio, video and control signals. In contrast to video over the public Internet, with IPTV deployments, network security and performance are tightly managed to ensure a superior entertainment experience, resulting in a compelling business environment for content providers, advertisers and customers alike. [2]

The Internet protocol-based platform offers significant advantages, including the ability to integrate television with other IP-based services like high speed Internet access and VoIP.

A switched IP network also allows for the delivery of significantly more content and functionality. In a typical TV or satellite network, using broadcast video technology, all the content constantly flows downstream to each customer, and the customer switches the content at the set-top box. The customer can select from as many choices as the telecommunications, cable or satellite company can stuff into the "pipe" flowing into the home. A switched IP network works differently. Content remains in the network, and only the content the customer selects is sent into the customer's home. That frees up bandwidth, and the customer's choice is less *restricted* by the size of the "pipe" into the home. This also implies that the customer's *privacy* could be compromised to a greater extent than is possible with traditional TV or satellite networks. It may also provide a *means* to hack into, or at least disrupt (see Denial of service) the private network.

IPTV technology is bringing video-on-demand (VOD) to television, which permits a customer to browse an online program or film catalog, to watch *trailers* and to then select a selected recording. The playout of the selected item starts nearly *instantaneously* on the customer's TV or PC.

Technically, when the customer selects the movie, a point-to-point *unicast* connection is set up between the customer's decoder (set-top box or PC) and the delivering streaming server. The signalling for the trick play functionality (pause, slow-motion, wind/rewind etc.) is *assured* by RTSP (Real Time Streaming Protocol).

New Words and Phrases

Protocol [ˈprəʊtəkɒl] n. 礼仪；（数据传递的）协议 vt. 把……写入议定书

distinguish [dɪˈstɪŋgwɪʃ] v. 辨别，分清；辨别是非

preferential [ˌprefəˈrenʃl] adj. 优先的；特惠的
scenario [səˈnɑːrɪəʊ] n. （行动）方案；剧情概要；分镜头剧本
agnostic [æɡˈnɒstɪk] n. 不可知论者 adj. 不可知论（者）的
restrict [rɪˈstrɪkt] vt. 限制，限定，约束
privacy [ˈprɪvəsi] n. 隐私，秘密，私事
means [miːnz] n. 方法，手段；收入
trailer [ˈtreɪlə（r）] n. 拖车；追踪者；（电影或电视的）预告片
instantaneously [ˌɪnstənˈteɪnɪəslɪ] adv. 即刻，突如其来地
unicast [ˈjuːnɪkɑːst] n. 单播
assure [əˈʃʊə] vt. 向……保证；使……确信；<英>给……保险
time-shifted 时移
video on demand（VOD）视频点播（VOD）
subscriber-based 基于用户
high-speed access 高速接入
end-user premises 终端用户处所
set-top boxes 机顶盒（set-top box 的名词复数）
defined as 定义为
stuff into 把……塞入
point-to-point 点对点

Notes

1. IPTV is distinguished from Internet television by its on-going standardization process（e. g.：European Telecommunications Standards Institute）and preferential deployment scenarios in subscriber-based telecommunications networks with high-speed access channels into end-user premises via set-top boxes or other customer-premises equipment.

译文：IPTV 与互联网电视的不同主要体现在它持续的标准化制定进程（比如欧洲电信标准化组织）和基于用户的通信网络优先部署方案，这些方案中的频道能够通过机顶盒或其他客户端设备高速连接至终端用户处所。

说明：这是一个较长的复合句。be distinguished from 意为“显示出特性，将……与……区别开”；high-speed access 意为“高速接入”。例如：

Their uniforms distinguish soldiers，sailors and marines from each other. 从制服可区分陆军士兵、水手和海军陆战队战士。

2. In contrast to video over the public Internet，with IPTV deployments，network security and performance are tightly managed to ensure a superior entertainment experience，resulting in a compelling business environment for content providers，advertisers and customers alike.

译文：与公共互联网上的视频传播相比，IPTV 的部署使网络安全和性能都受到严格管理，从而保证了卓越的娱乐体验，也因此为内容提供商、广告商和客户提供了非常有吸引力的商业环境。

说明：这是一个复合句。resulting in 引导一个伴随状语从句，意为“结果是……”。in

contrast to 意为“相比之下”; superior 意为“（在质量等方面）较好的”; alike 意为“同样地；类似于”。例如：

In contrast to similar services in France and Germany, Intercity rolling stock is very rarely idle. 与法国和德国的类似铁路服务系统不同，“城际列车”很少闲置。

Exercises

Ⅰ. Decide whether the following statements are True（T）or False（F）according to the passage.

1. Internet Protocol television（IPTV）is a system through which television services are delivered through traditional terrestrial, satellite signal, and cable television formats.

2. IPTV services may be classified into live television, without interactivity related to the current TV show.

3. IPTV is regarded as the safe and trustworthy delivery to subscribers of entertainment video and related services.

4. The Internet protocol-based platform can't provide any benefits.

Ⅱ. Translate the following sentences into English by using the words in brackets.

1. 这种交互方式非常简单，用户可以定义所收到的邮件为垃圾邮件。(define as)

2. 我奉命把它送给詹姆斯先生本人。(deliver)

3. 作为篮球运动员，他的身高给他带来很大优势。(significant advantage)

Unit 18 Internet of Things

Text

Internet of Things (IOT), also known as the sensor network, refers to a variety of information sensing devices and the Internet combine to form a huge network, will enable all of the items and network connections to facilitate the *identification* and management.[1] Because of its *comprehensive* sense, reliable delivery, intelligent processing features, it is considered as another wave of the Information industry after the computer, the Internet and mobile communication network.

Touch of a button on the computer or cell phone, even thousands of miles away, you can learn the status of an item, a person's activities. Send a text message, you can turn on the fan; if an *illegal* invasion of your home takes place, you will receive automatic telephone alarm. They are not just the scenes in Hollywood sci-fi blockbusters. They are gradually approaching in our lives.

It can be achieved due to the "things" in which there is a key technology for information storage object called radio frequency identification (RFID).[2] An RFID system consists of two components (as shown in Fig. 18-1): an antenna transceiver (often combined into a reader) and a *transponder* (the *tag*). The antenna emits radio signals to activate the tag and to read and write data to it. When activated, the tag transmits data back to the antenna. The data transmitted by the tag may provide identification or location information, or specifics about the product tagged, such as price, color, date of purchase, etc. Low-frequency RFID systems (30kHz to 500kHz) have short transmission ranges (generally less than six feet). High-frequency RFID systems (850 MHz to 950 MHz and 2.4 GHz to 2.5GHz) offer longer transmission ranges (more than 90 feet). In general, the higher the frequency, the more expensive the system.

Fig. 18-1 A Basic RFID System

For example, in mobile phones, embedded RFID-SIM card, your phone "information sensing device" can be connected with the mobile network. This phone can not only confirm the user's identity but also to pay the bills for water, gas and electricity, lottery, airline tickets and other payment services.

As long as an object embedded in a specific radio frequency tags, sensors and other devices connected to the Internet will be able to form a large network systems.[3] On this line, even thousands of miles away, people can easily learn and control the information of the object (as shown in Fig. 18-2).

To speak more concretely, let's imagine a world in which a large number of things that surround us are "*autonomous*", because they have:

a name: a tag with a unique code;

a memory: to store everything that they cannot obtain immediately from the net;

a means of communication: mobile and energy-efficient, if possible;

sensors: in order to *interact* with their environment;

acquired or *innate* behaviors: to act according to logic, an objective given by its owner.

And, of course, like everything else on Earth, these things must have an electronic existence on the network.

Some experts predict that, in 10 years, "things" may become very popular, and develop into a trillion-scale high-tech market. Then, in almost all areas, such as the personal health, traffic control, environmental protection, public safety, industrial monitoring, elderly care, "things" will play a role. Some experts said that in only three to five years' time, it will change people's way of life.

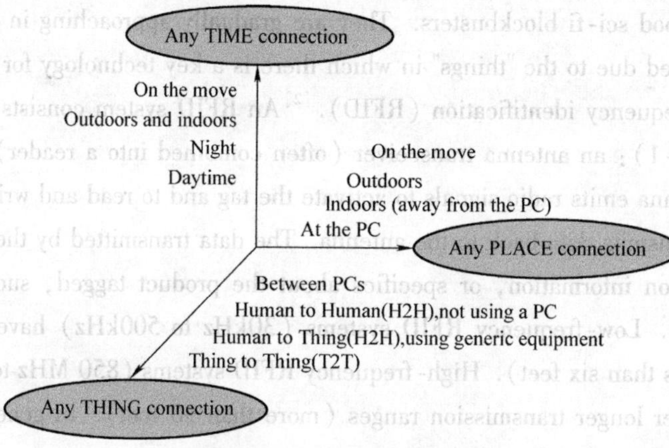

Fig. 18-2 How will Internet of Things affect our lives

The Internet of Things has great promise, yet business, policy, and technical challenges must be tackled before these systems are widely *embraced*. Early adopters will need to prove that the new sensor driven business models create superior value. Industry groups and government regulators should study rules on data privacy and data security, particularly for uses that touch on sensitive consumer information. On the technology side, the cost of sensors and *actuators* must fall to levels that will *spark* widespread use. Networking technologies and the standards that support them must evolve to the point where data can flow freely among sensors, computers, and actuators. Software to *aggregate* and analyze data, as well as graphic display techniques, must improve to the point where huge volumes of data can be absorbed by human.[4]

New Words and Phrases

identification [aɪˌdentɪfɪˈkeɪʃn] n. 鉴定，识别；身份证明

comprehensive [ˌkɒmprɪˈhensɪv] adj. 综合的；广泛的；有理解力的
illegal [ɪˈliːgl] adj. 非法的；违法的；违反规则的
transponder [trænsˈpɒndə] n. 应答器；转调器，变换器
tag [tæg] n. 标签 v. 尾随；起浑名
autonomous [ɔːˈtɒnəməs] adj. 自治的；自主的；自发的
interact [ˌɪntərˈækt] v. 相互作用，互相影响；互动
innate [ɪˈneɪt] adj. 先天的，固有的，与生俱来的
embrace [ɪmˈbreɪs] v. 拥抱；包括；接受
actuator [ˈæktjʊeɪtə] n. 执行机构；激励者；促动器
spark [spɑːk] n. 火花；闪光 v. 发火花
aggregate [ˈægrɪgət] vt. 使聚集，使积聚；总计达
Internet of Things 物联网
sensor network 传感网
mobile communication network 移动通信网络
product tagged 加标签的产品
radio frequency identification 射频识别
energy-efficient 高能效的
sci-fi blockbuster 科幻大片（sci-fi 是 Science-fiction 的缩写形式）

Notes

1. Internet of Things (IOT), also known as the sensor network, refers to a variety of information sensing devices and the Internet combine to form a huge network, will enable all of the items and network connections to facilitate the identification and management.

译文：物联网（IOT）也称传感网，是指将各种信息传感设备与互联网结合起来而形成的一个巨大网络，这个网络可使所有的物品与网络连接，以便于识别和管理。

说明：这是一个复合句。also known as the sensor network 是主句主语 Internet of Things 的同位语；而主句谓语是 refers to 和 will enable；a variety of …network 为宾语从句作 refers to 的宾语；to form a huge network 与 to facilitate the identification and management 是不定式短语作目的状语，to facilitate 意为“使容易”。例如：

To facilitate the measurements, Professor Gross and his team captured the photon in a special box, a resonator. 为了方便测量，格罗斯教授和他的团队通过一个特殊的盒子——谐振器来捕获光子。

2. It can be achieved due to the "things" in which there is a key technology for information storage object called radio frequency identification (RFID).

译文：这些能够得以实现是因为物联网里有一种叫作射频识别（RFID）的可存储物体信息的关键技术。

说明：RFID 是 radio frequency identification 的缩写形式，即射频识别，又称无线射频识别，是一种通信技术。可通过无线电信号识别特定目标并读写相关数据。识别工作无需人工干预，无需识别系统与特定目标之间建立机械或光学接触。最常见的应用有门禁控制和汽车

的 ETC 收费等。

3. As long as an object embedded in a specific radio frequency tags, sensors and other devices connected to the Internet will be able to form a large network systems.

译文：只要在物体中嵌入特定射频标签，传感器和其他与互联网连接的设备就能形成一个庞大的网络系统。

说明：这是一个复合句。as long as 引导条件状语从句，sensors and other devices 为句子主句的主语，connected to the Internet 为过去分词短语作定语修饰主语。

4. Software to aggregate and analyze data, as well as graphic display techniques, must improve to the point where huge volumes of data can be absorbed by human.

译文：收集和分析数据的软件，以及图形显示技术，必须提高至海量数据可以被人们接收的水平。

说明：这是一个复合句。software to aggregate and analyze data 为主语，as well as 引导并列主语 graphic display techniques；must improve to 为谓语，where 引导定语从句修饰宾语 the point。improve 意为"改善，增进"。例如：

China is in a transition from its planned economy to a socialist market economy. Many of its current systems remain to be improved. 中国正处于计划经济体制向社会主义市场经济体制转变的时期，各方面的制度还有待完善。

Exercises

Ⅰ. Answer the following questions according to the passage.

1. What is Internet of Things (IOT)?
2. Internet of Things is considered a wave of information industry; can you mention other so-called waves of information industry according to the text?
3. How many key components an RFID system consist of and what are they?
4. What is the relation between the transmission range and frequency of an RFID system?

Ⅱ. Match the following phrases in column A with column B.

Column A	Column B
1. Internet of Things	a. 传感网
2. mobile communication network	b. 执行机构
3. sensor network	c. 高能效的
4. radio frequency identification	d. 射频识别
5. energy-efficient	e. 标签
6. actuator	f. 应答器
7. transponder	g. 移动通信网络
8. tag	h. 物联网

Ⅲ. Translate the following sentences into English by using the words in brackets.

1. 健康饮食应由天然食物构成。(consist of)
2. 老师创建自己的幻灯片，然后嵌入到他的教学博客中。(embed in)
3. 这个男孩在上小学的时候就显示出成为运动员的潜力。(promise)

Translating Skills:

长难句的翻译

对于每一个英语句子的翻译，并不只是使用一种翻译方法，而是多种翻译方法的综合运用，这在英语长句的翻译中表现得尤为突出。长句的翻译，关键在于正确理解和分析。连词、冠词和介词等功能词的使用，和非谓语动词及谓语动词特殊结构形式的存在，使得英语长句的修饰成分相当复杂，可以是单词、短语，也可以是从句，而且这些修饰成分还可以嵌套使用。再加上英汉句子语序上的差异，如定语和状语修饰语的位置差异、句子的逻辑安排差异等，这也使得英语句子结构复杂，长句较多，有时一段可能只有一句话，但只要方法得当，长句的翻译并不难。

一、长难句分析的步骤

1. 结合具体语言环境了解全句大意。
2. 剖析全句，分清主要成分和次要成分，主句和从句。
3. 提取主干，确定句子的主、谓语，明确句子的基本构架及其所属句型。
4. 理清脉络，分清主次关系，注意上下层次、前后引衬及其逻辑关系。
5. 按汉语表达的规范，进行选词和语序的调整。
6. 反复研究译文，进行词句的锤炼和润饰，最终定稿。

例如：

For a family of four, for example, it is more convenient as well as cheaper to sit comfortably at home, with almost unlimited entertainment available, than to go out in search of amusement elsewhere.

分析：该句的骨干结构为 it is more…to do sth than to do sth else，是一个比较结构，而且是两个不定式之间的比较。该句共有三个谓语结构，它们之间的关系为：it is more convenient as well as cheaper to…为主体结构，但 it 是形式主语，真正的主语为第二个谓语结构 to sit comfortably at home，并与第三个谓语结构 to go out in search of amusement elsewhere 做比较。

综合上述翻译方法，这个句子可以翻译为：譬如，对于一个四口之家来说，舒舒服服地待在家里，就有几乎数不清的娱乐活动可以选择，这与外出去别的地方消遣相比既方便又经济。

二、长难句的四种主要翻译方法

1. 顺译法

顺译法就是按照原文的主语、动作、时间、地点及其他逻辑顺序将长句的内容再现，例如：

Even when we turn off the beside lamp and are fast asleep, electricity is working for us, driving our refrigerators, heating our water, or keeping our rooms air-conditioned.

分析：该句由一个主句、三个作伴随状语的现在分词以及位于句首的时间状语从句组成，共有五层意思：a. 即使在我们关掉了床头灯深深地进入梦乡时；b. 电仍在为我们工作；c. 帮我们运行电冰箱；d. 加热水；e. 或是维持室内空调机继续运转。上述五层意思的逻辑关系以及表达顺序与汉语完全一致，因此，我们可以采用顺序法，译为：

即使在我们关掉了床头灯深深地进入梦乡时，电仍在为我们工作：帮我们运行电冰箱；加热水；或是维持室内空调机继续运转。

2. 倒译法

在翻译某些长句时，要想使译文通顺、符合汉语表达规范，翻译时必须从后往前译，这就是倒译法。例如：

It therefore becomes more and more important that, if students are not to waste their opportunities, there will have to be much more detailed information about courses and more advice.

因此，如果要使学生充分利用他们（上大学）的机会，就得为他们提供大量关于课程的更为详尽的信息，以及更多的指导。这个问题显得越来越重要。

3. 分译法

分译法是将长句拆开，将其中从句或短语译成短句，分开来叙述。例如：

Television, it is often said, keeps one informed about current events, allows one to follow the latest developments in science and politics, and offers an endless series of programs which are both instructive and entertaining.

分析：在此长句中，有一个插入语 it is often said、三个并列的谓语结构、一个定语从句，三个并列的谓语结构尽管在结构上同属于一个句子，但都有独立的意义，因此在翻译时，可以采用分译法，按照汉语的习惯把整个句子分解成几个独立的分句。此句可译为：

人们常说，通过电视可以了解时事，掌握科学和政治的最新动态。从电视里还可以看到层出不穷的、既有教育意义又有娱乐性的节目。

4. 综合法

综合法要求我们把各种方法综合使用，先仔细分析，或按照时间的先后顺序，或按照逻辑顺序，顺逆结合，主次分明地对全句进行综合处理，以便把原文翻译成通顺的汉语句子。例如：

People were afraid to leave their houses, for although the police had been ordered to stand by in case of emergency, they were just as confused and helpless as anybody else.

尽管警察已接到命令，要做好准备以应付紧急情况，但人们不敢出门，因为警察也和其他人一样不知所措和无能为力。

Reading: How Connected Cars Might Actually Make Driving Better

Why Driving *Stinks*?

Maciej Kranz, VP and GM of Cisco's Connected Industries Group, laid out some of the *grim* statistics *plaguing* the world yet-to-be-connected cars:

Between 11% and 13% of *commuting* time is wasted in *urban* traffic *congestion*, for a total of 90 billion hours. (It just seems like of half of that is on the 101 Freeway between San Francisco[1] and San Jose[2].) Some 7% to 12% of urban congestion is caused by people looking for parking. (It just seems like all of that comes in San Francisco' s North Beach neighborhood.) Between 10%

and 17% of urban fuel is wasted at stoplights when there is no cross traffic. Eighty percent of accidents (6.3 million) are caused by driver *distraction*.

How Connected Cars Make Driving Better

For congestion, that means traffic management and *optimization* of road networks. For parking issues, connected cars can link apps identifying the closest, most affordable available parking spaces to the vehicle's *navigation*. And the vehicles can intelligently adjust driving speeds to *boost* fuel efficiency.

Intelligent stoplights, for example, would know if there were 10 cars waiting in one direction but only 1 in the other, and adjust light timing to keep traffic moving. Along straight routes, Kranz said, they can build "green waves" of traffic signals to keep *lanes* flowing efficiently.

There's also the idea that if one car knows what other vehicles, traffic lights and other road *infrastructure* are doing, they can all adjust more efficiently. For example, if your car knows that the car in front is about to make a turn or start braking, it can begin reacting even before it actually senses the action.

Cisco estimates this could lead to 7.5% less time wasted in traffic congestion and 4% lower costs for vehicle fuel, repairs and insurance. The benefits are particularly obvious in *fleet* settings, Kranz said. For example, a company with 10,000 delivery trucks would find it very valuable to be able to use connected technology to schedule *preventive* maintenance.

As for preventing accidents, vehicle-to-vehicle communications could enable a connected car to *alert* you if you get too close to the vehicle in front of you. If you don't respond, Kranz said, "at some point the car will make a decision to hit the brakes and avoid the accident."

The examples of the Internet of Things applications are *countless*. By 2015, six billion objects in the world will be connected to the Internet. While it may seem *tricky* to grasp as a concept, the Internet of Things is nothing simpler, and more *stunning*, than objects being connected to the Internet. At its most mind-blowing, these objects are learning and adapting to the *behaviour* of the user. The Internet of Things probably already influences your life. And if it doesn't, it soon will.

New Words and Phrases

Stink [stɪŋk] vi. 散发出恶臭；招人厌恶；糟透 n. 恶臭；难闻的气味
urban [ˈɜːbən] adj. 都市的；具有城市或城市生活特点的；市内
grim [grɪm] adj. 冷酷的，残忍的；严厉的；阴冷的；可怕的，讨厌的
commute [kəˈmjuːt] vi. 通勤；代偿 vt. 减刑；交换 n. 通勤来往（的路程）
plague [pleɪg] n. 瘟疫；灾害，折磨 vt. 使染瘟疫；使痛苦，造成麻烦
congestion [kənˈdʒestʃən] n. 拥挤，堵车；阻塞；稠密
distraction [dɪˈstrækʃn] n. 注意力分散；消遣；精神错乱
optimization [ˌɒptɪmaɪˈzeɪʃən] n. 最佳化，最优化；优选法
navigation [ˌnævɪˈgeɪʃn] n. 航行（学）；航海（术）；海上交通
boost [buːst] vt. 促进，提高 vi. 宣扬 n. 提高；吹捧
lane [leɪn] n. 小路，小巷；规定的单向行车道；车道

fleet [fliːt] n. 舰队；船队 adj. 快速的；敏捷的 vi. 疾驰；飞逝 vt. 使（时间）飞逝
infrastructure [ˈɪnfrəstrʌktʃə（r）] n. 基础设施；基础建设
preventive [prɪˈventɪv] n. 预防，防止；预防措施；预防药 adj. 预防的，防止的
preventive maintenance 预防性维修
alert [əˈlɜːt] adj. 警觉的，警惕的，注意的；思维敏捷的；活泼的 n. 警报；警戒状态
tricky [ˈtrɪki] adj. （形势、工作等）复杂的；机警的；微妙的
stunning [ˈstʌnɪŋ] adj. 令人晕倒（吃惊）的；出色的；令人震惊的
mind-blowing 令人兴奋的
behaviour [bɪˈheɪvjə(r)] n. 行为；举止；（人、动植物、化学药品等的）表现方式；态度

Notes

1. San Francisco：圣弗朗西斯科，或称三藩市、旧金山。是美国加利福尼亚州太平洋沿岸海港、工商业大城市，位于太平洋与圣弗朗西斯科湾之间的半岛北端。由西班牙人建于1776年，1821年归属墨西哥，1848年归属美国。19世纪中叶在采金热中迅速发展，华侨称之为“金山”，后为区别于澳大利亚的墨尔本，改称“旧金山”。

2. San Jose：圣何塞。是美国加利福尼亚州西部的城市，位于旧金山湾的南部。

Exercises

Ⅰ. Answer the following questions according to the passage.

1. Why does the author say driving stinks?
2. What an intelligent stoplight can do?
3. How to prevent accidents?
4. How many objects would be connected to the Internet by 2015?

Ⅱ. Decide whether the following statements are True (T) or False (F) according to the passage.

1. People waste 13% time on commuting.
2. Drivers’ distraction caused many accidents.
3. Connected cars can adjust traffic more efficiently.
4. The examples of the Internet of Things applications can be counted.

Chapter Ⅳ Application Technology

Unit 19 Programmable Logic Controllers (PLC)

Text

A programmable logic controller, PLC, or programmable controller is a small computer used for automation of real-world processes, such as control of machinery on factory assembly lines. [1] The PLC usually uses a microprocessor. The program can often control complex sequencing and is often written by engineers. The program is stored in battery-backed memory and/or EEPROMs.

PLCs have the basic structure, shown in Fig. 19-1.

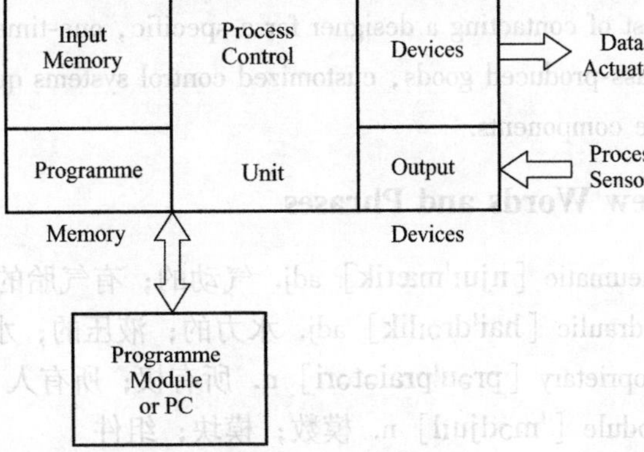

Fig. 19-1 Schematic of a PLC

From the figure, the PLC has four main units: the Programme memory, the Data memory, the Output devices and the Input devices. The Programme memory is used for storing the instructions for the logical control sequence. The status of switches, interlocks, past values of data and other working data are stored in the Data memory.

The main differences from other computers are the special input/output arrangements. These connect the PLC to sensors and actuators. PLCs read limit switches, temperature indicators and the positions of complex positioning systems. Some even use machine vision. On the actuator side, PLCs drive any kind of electronic motor, *pneumatic* or *hydraulic* cylinders or diaphragms, magnetic relays or solenoids. The input/output arrangements may be built into a simple PLC, or the PLC may have external I/O *modules* attached to a *proprietary* computer network that plugs into the PLC. [2] PLCs were invented as less expensive replacements for older automated systems that would use hundreds or thousands of relays and cam timers. Often, a single PLC can be programmed to replace thousands of relays. Programmable controllers were initially adopted by the automotive manufacturing industry, where software revision replaced the rewiring of hard-wired control panels.

The functionality of the PLC has evolved over the years to include typical relay control, sophisticated motion control, process control, Distributed Control Systems and complex networking.

The earliest PLCs expressed all decision making logic in simple ladder logic inspired from the electrical connection diagrams. The electricians were quite able to trace out circuit problems with schematic diagrams using ladder logic. This was chosen mainly to reduce the *apprehension* of the existing technicians.

Today, the PLC has been proven very reliable, but the programmable computer still has a way to go. With the IEC 61131-3 standard, it is now possible to program these devices using structured programming languages, and logic elementary operations.

A graphical programming notation called Sequential Function Charts is available on certain programmable controllers.

However, it should be noted that PLCs no longer have a very high cost (often thousands of dollars), typical of a "generic" solution. Modern PLCs with full capabilities are available for a few hundred USD. There are other ways for automating machines, such as a custom microcontroller-based design, but there are differences among both: PLCs contain everything needed to handle high power loads right out of the box, while a microcontroller would need an electronics engineer to design power supplies, power modules, etc. Also a microcontroller based design would not have the flexibility of in-field programmability of a PLC. That is why PLCs are used in production lines, for example. These typically are highly customized systems so the cost of a PLC is low compared to the cost of contacting a designer for a specific, one-time only design. On the other-hand, in the case of mass-produced goods, customized control systems quickly pay for themselves due to the lower cost of the components.

New Words and Phrases

pneumatic [njuːˈmætik] adj. 气动的；有气胎的；充气的 n. 气胎
hydraulic [haiˈdrɔːlik] adj. 水力的；液压的；水力学的
proprietary [prəuˈpraiətəri] n. 所有权；所有人 adj. 所有的；专利的
module [ˈmɔdjuːl] n. 模数；模块；组件
apprehension [ˌæpriˈhenʃən] n. 理解；逮捕；恐惧；忧惧
trace out 描绘出

Technical Terms

battery-backed memory 电池备份存储器
EEPROM 电可擦写可编程只读存储器
Distributed Control Systems 分布式控制系统
IEC 国际电工委员会
ladder logic 梯形图
Sequential Function Charts 顺序功能图
power supply 电源

Notes

1. A programmable logic controller, PLC, or programmable controller is a small computer used for automation of real-world processes, such as control of machinery on factory assembly lines.

译文：可编程序逻辑控制器，PLC，或者称可编程序控制器，是一种小型计算机，可用于实际流程的自动化生产，比如控制工厂的装配线机械等。

说明：used for automation of real-world processes 是过去分词短语作后置定语，修饰 computer。

2. The input/output arrangements may be built into a simple PLC, or the PLC may have external I/O modules attached to a proprietary computer network that plugs into the PLC.

译文：输入/输出装置可能安装在简单的 PLC 内部，或者 PLC 可通过外部 I/O 模块与接入 PLC 的专用计算机网络相连。

说明：attached to a proprietary computer network 是过去分词短语作后置定语来修饰 external I/O modules；that plugs into the PLC 是定语从句，that 在句中作主语，先行词是 computer network（专用的计算机网络）。

Exercises

Ⅰ. Answer the following questions according to the passage.

1. How many parts does the PLC consist of?
2. What functions can PLCs perform?
3. What is the programming language of PLC?

Ⅱ. Decide whether the following statements are True (T) or False (F) according to the passage.

1. The instructions for the logical control sequence are stored in the Data memory.
2. The I/O modules only can be built into PLCs.
3. Now, we can use structured programming languages, and logic elementary operations to program.
4. The simple ladder logic inspired from the electrical connection diagrams.
5. The programmability of a PLC is flexible.

Ⅲ. Translate the following sentences into English.

1. 可编程序控制器通常使用微处理器进行控制。
2. PLC 能读取限位开关、温度指示器和复杂定位系统的位置信息。
3. 最早的 PLC 采用简单的梯形图来描述其逻辑程序。
4. PLC 包含所有能直接操作大功率负载的输出功能。

Practical English:

如何阅读机电产品的英文说明书

工业生产离不开机电设备，安全高效地使用它们对用户而言相当重要，但机电产品的英文说明书与大众英语不同，很多地方不能够从基础英语的语法角度来理解，特别是一些进口机电产品的说明书，理解的难度会更大。因此，我们需要了解机电产品英文说明书的特点，常用句型，涉及内容等，从而更好地阅读和理解它们。

“产品说明书”在英语中通常有3种说法，即Instruction（使用指导）、Direction（指示，用法说明）、Description（说明书）。

一、机电产品说明书的特点

机电产品说明书是机电产品标志的重要组成部分，是在产品或包装上用于识别产品或其特征、特性的各种表达和指示的统称。产品说明书主要用文字、符号、标志、标记、数字及图案等表示，给销售者、购买者提供有关产品的信息，帮助他们了解产品的性能、质量状况，说明产品的使用、保养条件，起到引导消费的作用。许多产品说明书的内容都标在产品或产品的包装上，让人一目了然，为用户提供了方便。

二、机电产品说明书中的常用句型

句型是语言结构的要素，无论是英译汉还是汉译英都离不开对句型的分析。熟悉机电产品英文说明书中的常见句型对阅读及良好地翻译原文十分必要。

1. 情态动词/be＋形容词/过去分词＋目的状语

该句型主要用于文章的开头，说明该产品的用途。例如：

The material-piling machine is mainly used in the piling and homogenizing of material on round shape material stacking site, meeting the requirement of continuous production. 取料机主要用于圆形料场原料的均匀收取，以满足连续生产的需要。

类似常见句型还有：

be used to...

be used as...

be designed to...

be suitable to be used in...

be available for...

may be used to...

be adapted for...

be designed to be... so as to...

be capable of...

2. 情态动词/be＋介词短语

该句型用于说明物体的特征、状态和范围，以及计量单位等。例如：

The type CYJ15-18-18 oil pumping machine is of simple and compact construction.

CYJ15-18-18型抽油机的结构紧凑。

The wiring should be in good condition and core flex should not be exposed.

这些配线必须完好无损，中心导体不得裸露。

3. 现在分词+名词

该句型用于说明维修或操作程序，及有关技术要求等。例如：

Simultaneously cutting two pieces of sheet is strictly forbidden.

绝对禁止同时冲剪两块板材。

4. 名词+过去分词

Trouble: the stem sticky

故障：阀杆运动不灵活

Reasons: the stem bent; the spring broken; gland packing pressed too tightly

原因：阀杆弯曲；弹簧损坏；压盖填料压得太紧

三、机电产品说明书的组成

机电产品说明书通常由标题和正文两大部分构成，机电产品说明书主要用于帮助使用者掌握该设备的操作方法以及用途、产品规格、注意事项、维护及保养等。

1. 对用途的说明示例

These versatile wheeled tractors are reliable enough to tackle a variety of field jobs such as ploughing, harrowing, seed drilling, cultivating, harvesting, etc.

这些多用途轮式拖拉机适用于犁、耙、播种、耕作、收割等多种田间作业。

2. 对产品规格的描述示例

Processing range: 2.4 ~ 6.5mm output: 3.5 ~ 5t/shift

加工范围：2.4 ~ 6.5mm 产量：3.5 ~ 5 吨/班

3. 对设备特点的说明示例

Simple construction, easy operation and maintenance, and comparatively high productivity.

结构简单，操作容易，维修方便，生产率较高。

4. 故障排除示例

Trouble: The shaver does not work when the ON/OFF button is pressed.

问题：按下开/关按钮后剃须刀不工作。

Solution: Replace the batteries. If the shaver still does not work, see "Guarantee & Service".

解决方法：更换电池。如果剃须刀仍然不能工作，请参阅"保证及维修服务"。

5. 安全警示说明示例

Prevent the appliance and the wire from getting wet.

确保设备与电线保持干燥。

Remove the batteries from the appliance if you are not going to use it for quite some time.

长时间不使用本设备，请取出电池。

6. 操作说明示例

Switch the appliance on by pressing the switch lock and pushing the ON/OFF button upwards.

按下开关键并把开/关按钮向上推，便可起动本设备。

Reading: PLC Programming

The first PLCs were programmed with a technique that was based on relay logic wiring *schematics*. This eliminated the need to teach the electricians, technicians and engineers how to program a computer. But this method has stuck and it is the most common technique for programming PLCs today. An example of ladder logic can be seen in Fig. 19-2. To interpret this diagram we can imagine that the power is on the vertical line on the left hand side, we call this the hot rail. On the right hand side is the neutral rail. In the figure there are two rungs, and on each rung there are inputs or combinations of inputs (two vertical lines) and outputs (circles). If the inputs are opened or closed in the right combination the power can flow from the hot rail, through the inputs, to power the outputs, and finally to the neutral rail. An input can come from a sensor, or switch. An output will be some device outside the PLC that is switched on or off, such as lights or motors. In the top rung the contacts are normally open and normally closed, which means if input A is on and input B is off, then power will flow through the output and activate it. Any other combination of input values will result in the output X being off.

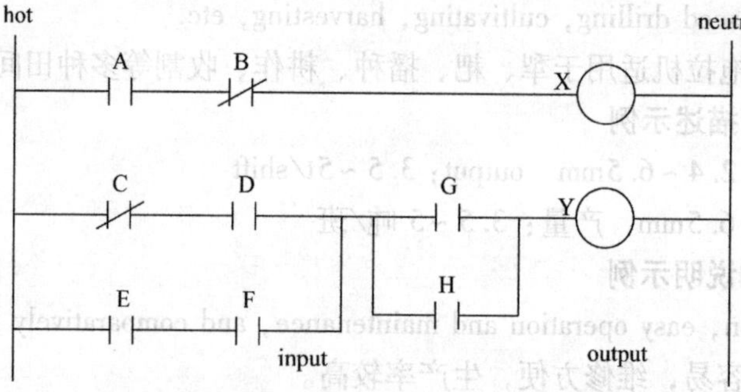

Fig. 19-2 An example of ladder logic diagram

The second rung of Fig. 19-2 is more complex, there are actually *multiple* combinations of inputs that will result in the output Y turning on. On the left most part of the rung, power could flow through the top if C is off and D is on. Power could also (and simultaneously) flow through the bottom if both E and F are true. This would get power half way across the rung, and then if G or H is true the power will be delivered to output Y.

There are other methods for programming PLCs. One of the earliest techniques involved *mnemonic* instructions. These instructions can be derived directly from the ladder logic diagrams and entered into the PLC through a simple programming terminal. An example of mnemonics is shown in Fig. 19-3. In this example the instructions are read one line at a time from top to bottom. The first line 0 has the instruction LDN (input load and not) for input 00001. This will examine the input to the PLC and if it is off it will remember a 1 (or true), if it is on it will remember a 0 (or false). The next line uses an LD (input load) statement to look at the input 00002. If the input is off it remembers a 0, if the input is on it remembers a 1. The AND statement recalls the last two numbers remembered and if they are both true the result is a 1, otherwise the result is a 0. The process is repeated for lines 00003 and 00004, the AND in line 5 combines the results from the last LD instructions. The OR instruction takes the two numbers now remaining and if either one is a 1 the result is a 1, otherwise the result is a 0. The last instruction is the ST (store output) that will look at the last value stored and if it is 1, the output will be turned on, if it is 0 the output will be turned off.

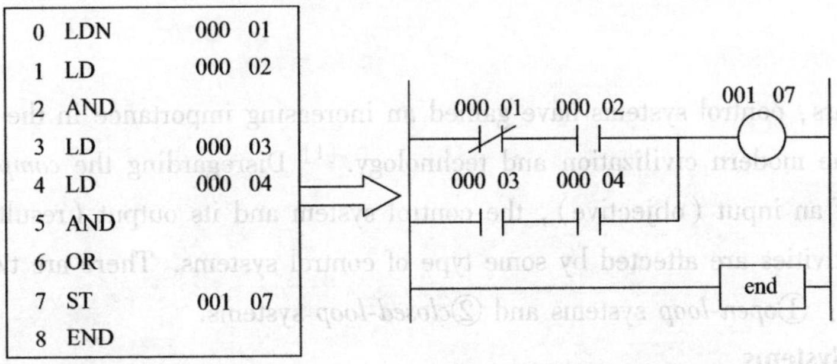

Fig. 19-3 Mnemonic and equivalent ladder logic program

The ladder logic program in Fig. 19-3 is *equivalent* to the mnemonic program. Even if you have programmed a PLC with ladder logic, it will be converted to mnemonic form before being used by the PLC.

New Words and Phrases

schematic [skiː'mætik] adj. 图解的；概要的 n. 图解视图；原理图

hot rail 火线

neutral rail 零线

multiple ['mʌltipl] adj. 多样的；许多的；多重的 n. 并联；倍数

mnemonic [niː'mɔnik] adj. 记忆的；助记的；记忆术的

equivalent [i'kwivələnt] adj. 等价的，相等的；同意义的 n. 等价物

Exercises

Ⅰ. Answer the following questions according to the passage.

1. What is the first programming language of PLC?
2. How many rails are there in the ladder logic?
3. What is the input signal of the PLCs?
4. What is the output signal of the PLCs?
5. What is the different of ladder logic and mnemonic instruction?

Ⅱ. Translate the following sentences into Chinese.

1. This eliminated the need to teach the electricians, technicians and engineers how to program a computer.
2. An output will be some device outside the PLC that is switched on or off, such as lights or motors.
3. One of the earliest techniques involved mnemonic instructions.
4. These instructions can be derived directly from the ladder logic diagrams and entered into the PLC through a simple programming terminal.

Unit 20 Automatic Control Systems

Text

In recent years, control systems have gained an increasing importance in the development and advancement of the modern civilization and technology. [1] Disregarding the *complexity* of the system, it consists of an input (objective), the control system and its output (result). [2] Practically our day-to-day activities are affected by some type of control systems. There are two main branches of control systems: ①*open-loop* systems and ②*closed-loop* systems.

Open-loop Systems

The open-loop system is also called the non-feedback system. This is the simpler of the two systems.

In this open-loop system, there is no way to ensure the actual speed is close to the desired speed automatically. [3] The actual speed might be way off the desired speed because of the wind speed and/or road conditions, such as hill or down hill, etc.

Closed-loop Systems

The closed-loop system is also called the feedback system. It has a mechanism to ensure the actual speed is close to the desired speed automatically.

Feedback Loop

A feedback loop is a common and powerful tool when designing a control system. Feedback loops take the system output into consideration, which enables the system to adjust its performance to meet a desired output response. [4]

There are two main types of feedback control systems: negative feedback and positive feedback. In a positive feedback control system the *set-point* and output values are added. In a negative feedback control system the set-point and output values are *subtracted*. As a rule negative feedback systems are more stable than positive feedback systems. Negative feedback also makes systems more *immune* to random variations in component values and inputs.

The control *function* can be defined in many ways. Recall that the system *error* is the difference between the set-point and actual output. When the system output matches the set-point the error is zero. Larger differences between the set-point and output will result in larger errors. For example if the desired *velocity* is 50 mph and the actual velocity is 60 mph, the error is −10 mph, and the car should be slowed down. The rules in the figure give a general idea of how a control function might work for a cruise control system.

When talking about control systems, it is important to keep in mind that engineers typically are given existing systems such as actuators, sensors, motors, and other devices with *set parameters*, and are asked to adjust the performance of those systems. [5] In many cases, it may not be possible to open the system (the "*plant*") and adjust it from the inside: modifications need to be made external to the system to force the system response to act as desired. This is performed by adding controllers, *compensators*, and feedback structures to the system.

New Words and Phrases

complexity [kəm'pleksəti] n. 复杂性
open-loop [ˌəupənl'uːp] adj. 开环的
closed-loop [kˌləuzdl'uːp] adj. 闭环的
set-point [ˌsetp'ɔɪnt] n. 设定值
subtract [səb'trækt] v. 减
immune [ɪ'mjuːn] adj. 免疫的；免于……的，免除的；不受影响的；无响应的
function ['fʌŋkʃn] n. 函数 v. 起作用
error ['erə(r)] n. 偏差；故障
velocity [və'lɒsəti] n. 速率
set [set] n. 设置
parameter [pə'ræmɪtə(r)] n. 参数
plant [plɑːnt] n. 被控对象
compensator ['kɒmpenseɪtə] n. 补偿器
automatic control system 自动控制系统
non-feedback system 非反馈系统
feedback system 反馈系统
output response 输出响应
positive feedback 正反馈
component value 分量值
control function 控制函数

Notes

1. In recent years, control systems have gained an increasing importance in the development and advancement of the modern civilization and technology.

译文：近些年来，控制系统在现代文明和科技的发展和进步中变得日益重要起来。

说明：这是一个简单句。句中 gained an increasing importance 等同于 became more important，原文使用"动词 + 形容词 + 名词"的结构，是为了满足科技文体客观性的需要。例如：

Ipad has gained an increasing popularity among teenagers in recent days. 近来，平板电脑日益受到年轻人的欢迎。

2. Disregarding the complexity of the system, it consists of an input (objective), the control system and its output (result).

译文：无论多复杂的系统，都是由输入（目标）、控制系统及输出（结果）组成的。

说明：这是一个简单句。现在分词 disregarding 置于句首，相当于 regardless of，那么 disregarding the complexity of the system 是非谓语动词作状语，表示一种让步关系。例如：

Taking these points into account, consider these as preferable options. 将这些因素考虑进来，再考虑将这些作为更佳选择。

3. In this open-loop system, there is no way to ensure the actual speed is close to the desired speed automatically.

译文：在开环系统中，没有办法保证实际速度自动地接近期望速度。

说明：这是一个简单句。there be 句型表示一种存在性；to ensure 修饰主语，并引导出一个宾语从句，that 被省略；desired 用来修辞名词，意为"期望的……"。例如：

Everyone has a desire to get a desired salary. 大家都渴望得到一份理想的工资。

4. Feedback loops take the system output into consideration, which enables the system to adjust its performance to meet a desired output response.

译文：反馈回路会参考系统输出，使系统能够调节其性能来实现期望的输出响应。

说明：这是一个复合句。在此句中，前一句是主句；后一句为 which 引导的非限制性定语从句，which 指代的是前面的整个句子。此外，take... into consideration 意为"将……考虑在内"，常用于正式场合。例如：

When choosing a diet, take into consideration the peculiarities of your health and environment: weight, age, profession and so on. 当你选择减肥方法的时候，要考虑你自身的健康情况和外界因素，包括体重、年龄、职业等。

5. When talking about control systems, it is important to keep in mind that engineers typically are given existing systems such as actuators, sensors, motors, and other devices with set parameters, and are asked to adjust the performance of those systems.

译文：当谈及控制系统时，需要记住的是，工程师通常已经有一些给定参数的系统，如执行机构、传感器、电动机和其他装置等，并且他们要对这些系统的性能进行调节。

说明：这是一个复合句。句中 when 为时间状语从句的省略形式。在由 that 引导的从句中，第一个 and 连接 such as 列举的最后两个事物；第二个 and 连接两个谓语 be given 和 be asked，使整个从句成为一个并列句。performance 在句中意为"性能"。例如：

How do you then profile your system and database performance? 那么您会如何配置系统和数据库性能呢？

Exercises

Ⅰ. Decide whether the following statements are True（T）or False（F）according to the passage.

1. The non-feedback system, which is also called the open-loop system, is simpler than the feedback system.

2. In an open-loop system, there is a mechanism which ensures that the actual speed is way off the desired speed.

3. Set-point and output values are added in a negative feedback control system.

4. It's possible for engineers to modify the system from the outside.

Ⅱ. Translate the following phrases into Chinese or English.

1. basic component 2. closed-loop system
3. negative feedback 4. control function
5. 正反馈 6. 分量值
7. 参数设置 8. 开环系统

Ⅲ. Translate the following sentences into English by using the words in brackets.

1. 她的双腿现在已经丧失正常功能了。(function)
2. 养在户外笼子里的鸟对空气传播的病毒有免疫力。(immune to)
3. 该空气加湿器性能优良，可以根据室内的湿度、温度变化进行自动调节。(performance, adjust)
4. F1 方程式赛车运用了高端的速度传感器和弯道反馈系统。(speed sensor, feedback system)

Practical English:

机电产品说明书范例（中英文对照）

机电产品说明书是机电产品标识的重要组成部分，良好的说明书阅读能力是我们从事相关行业的基本条件，对我们来说就是能够通过阅读机电产品的说明书掌握相应产品的特征、性能、质量状况，以及使用、保养条件等内容。下面是两例机电产品说明书的节选。

Ⅰ. Instruction Book For Refrigerator 电冰箱说明书

Features of product 产品特点

Luxury wide door series, European optimized man-machine design. No door handle, open it by pulling of the door edge.

豪华宽门系列，欧洲最优化人机工程设计。无门把手，拉动门边开门。

High energy saving. Phasing in high energy-saving compressor, three-cycling system and thick insulation get the appliance having high cooling efficiency and better energy saving.

高效节能，采用高效节能压缩机、三循环制冷系统及加厚发泡层，制冷效率更高、更节能。

Safety information 安全注意事项

Make sure to use a separate earthed socket.

务必使用单独的、带接地线的三芯电源插座。

Pull out the mains plug when you repair or clean the machine.

维修或清洁冰箱时，必须拔下电源插头。

Do not store inflammable or explosive materials in appliances to avoid explosion or fire.

禁止将易燃、易爆物品放在冰箱中，以防爆炸和火灾。

Transportation and placement 搬运和放置

Do not move the appliance by holding a door or door handle.

不可抓着门或门把手来搬运冰箱。

You should lift it from the bottom.

应从底部抬起冰箱。

You can not lie down or reverse the appliance to move.

不可将冰箱横抬或倒置。

Connecting the appliance 电器连接

The rated voltage of the appliance is 220V alternating current and the rated frequency is 50Hz. It can vary from 180V to 220V.

冰箱的额定电压为交流220V，额定频率为50Hz，电压波动范围为180 ~220V。

Ⅱ. Instruction Book For 3 Phase Low Voltage Induction Motor
低压三相异步电动机说明书

Reception check 收货检查

Immediately upon receipt check the machine for external damage and if found, inform our factory through telephone or fax number without delay (tel: 0086 -21 -6463 -1777, fax: 0086 -21 -6463 -7180).

Check all rating plate data, especially voltage and winding connection (star or delta). Turn shaft by hand to check free rotation.

收到电动机后立即检查是否有损坏，如有发现请立即拨打电话0086 -21 -6463 -1777，或发传真至0086 -21 -6463 -7180。

检查铭牌数据，特别是额定电压、接线方式（星形连接或三角形连接）是否与合同订货要求相符，用手转动电动机轴检查转动是否灵活。

Insulation resistance check 绝缘电阻检查

Measure insulation resistance before commissioning and when winding dampness is suspected. Resistance, measured at 25℃, shall exceed the reference value, i. e. where:

$$R \geqslant (20 \times U)/(1000 + 2P)$$

在调试电动机前或怀疑线圈受潮时应测量电动机的绝缘电阻，在25℃温度条件下的电阻值应超过以下表达式的值：

$$R \geqslant (20 \times U)/(1000 + 2P)$$

R: resistance, ohms (measured with 500V DC Megger);

R 表示绝缘电阻，单位为兆欧姆 MΩ（用500V直流电阻表测量）；

U: voltage, Volts;

U 表示电压；单位为伏特 V；

P: output power, kW;

P 表示输出功率；单位为千瓦 kW。

Warning 警告:

Windings should be discharged immediately after measurement to avoid risk for electric shock.

测量完绝缘电阻后应立即将线圈断电，以免线圈再次受到高电压冲击。

Insulation resistance reference value is halved for each 20℃ rise in ambient temperature.

环境温度每升高20℃，绝缘电阻就降低一半。

If the reference resistance value is not attained, the winding is too damp and must be oven dried.

如果绝缘电阻达不到参考值，表明线圈已经受潮，需进行烘干处理。

Oven temperature should be 90℃ for 12 ~ 16 hours followed by 105℃ for 6 ~ 8 hours.

先在90℃温度下烘12 ~ 16小时，再在105℃温度下烘6 ~ 8小时。

Operating conditions 运行条件

The machines are intended for use in industrial drive applications. Normal ambient temperature limits −25℃ to +40℃. Maximum altitude 1000 m above sea level.

电动机用于驱动其他机械，通常环境温度为 −25 ~ +40℃，高度为海拔1000米以下。

Storage 储存

All machine storage should be done indoors, in dry, well ventilation and dust free conditions. Unprotected machined surfaces (shaft-ends and flanges) should be given anti-corrosion coatings.

所有电动机应储存于干燥、通风以及无尘的室内。裸露的加工表面，如轴伸和法兰面，应涂防锈脂。

Installation Foundation 安装基础

The purchaser bears full responsibility for preparation of the foundation.

买方负责电动机安装的准备工作。

Metal foundations should be painted to avoid corrosion.

金属结构的基础应涂防锈漆防止被腐蚀。

Foundations shall be even, and sufficiently rigid to withstand possible short circuit forces.

电动机基础应平整，并有足够的刚度承受突然短路产生的应力。

They shall be dimensioned as to avoid the occurrence of vibration due to resonance.

应计算基础部件的尺寸以避免共振引起振动。

Alignment 校正

Correct alignment is essential to avoid bearing failures, vibrations and possible fractured shaft extensions.

校正电动机安装中心有利于避免轴承故障、振动和可能发生的轴伸断裂事故。

Connection 接线

The normal machine design is with terminal box on the right, when viewing the shaft face at the machine drive end. Availability of these solutions is described in the product catalogues.

通常出线盒在电动机的右侧（即电动机驱动端与轴面相对），还有其他可供选择位置已在产品目录中说明。

Besides the main winding and earthing terminals, the terminal box can also contain connections for standstill heating elements, bimetallic switches, or PT 100 resistance elements.

除了绕组和接地，出线盒还可以接停机加热器、双金属温度开关或PT100测温元件等。

Motor starting 电机起动

(1) Direct-on-line or star/delta starting 单速电动机起动（直接起动或Y/Δ起动）

The terminal box on standard single speed machines normally contains 6 winding terminals and at least one earth terminal. Earthing shall be carried out according to local regulations before the machine is connected to the supply voltage. The voltage and connection are stamped on the nameplate.

标准单速三相异步电动机的出线盒内通常有6个接线端和至少1个接地端，接地须根据操作规程在电动机接通电源前进行，接线方式和电压应根据铭牌指示选择。

Terminals and direction of rotation 出线与旋转方向

Direction of rotation is clockwise when viewing the shaft face at the machine drive end, when the line phase sequence L1, L2, L3 is connected to the terminals.

当相序为 L1、L2、L3 时，由电动机驱动端看向轴面，电动机的旋转方向是顺时针方向。

（2）Start-up of variable speed motor 变速电动机起动

The variable speed motor starts at low-speed status and then shifted to high-speed. For variable speed motor with many sets of windings, the speed shift shall be fulfilled when one winding has lost power.

变速电动机一般应低速起动后切换到高速，具有多套绕组的变速电动机在不同转速档切换时，应在一绕组失电后才能向另一绕组送电。

（3）Start-up of variable frequency motor 变频电动机起动

When the variable frequency motor starts, the variable frequency blower shall starts simultaneously. 变频电动机起动时，变频风机也必须同时起动。The power supply of variable frequency motor shall be of invariable frequency, and the rotating direction shall be as specified in the nameplate of blower. 变频电动机电源必须是恒频电源，电动机转向必须与风机外壳上铭牌的指示相同。

Maintenance 保养与维护

General inspection 一般检查

- Inspect the machine at regular intervals.
- 定期检查电动机。
- Keep the machine clean and ensure free ventilation air-flow.
- 保持电动机清洁，确保通风良好。
- Check the condition of shaft seals and replace if necessary.
- 检查轴封状态，如有必要即进行更换。
- Check the condition of connections and mounting and assembly bolts.
- 检查各连接、地脚及装配螺栓的状态。
- Check the bearing condition by listening for unusual noise, vibration measurement, bearing temperature, inspection of spent grease.
- 通过听异常噪声、测量振动、轴承温度以及检查润滑油是否过期来检测轴承的状态。

Lubrication 更换润滑油

（1）Machines with sealed bearings 装有密封轴承的电动机

Machines up to frame size 160 are normally fitted with sealed bearings. It needn't regrease during operation.

中心高在160mm 以下的电动机装有密封轴承，此类电动机在运行期间不必加润滑油。

（2）Machines fitted with grease nipples 带有加油口的电动机

Machines frame from 180 to 355 is normally fitted with grease nipples. Lubricate the machine while running.

中心高180~355mm的电动机通常配有加油口，可以在电动机运行时加润滑油。

Reading: Control System Components

It is well known that there are three basic *components* in a closed-loop control system. They are:

1. The Error Detector. This is a device which receives the low-power input signal and the output signal which may be of different physical natures, converts them into a common physical quantity for the purposes of *subtraction*, performs the subtraction, and gives out a low-power error signal of the correct physical nature to actuate the controller[1]. The error detector will usually contain "*transducers*"; these are devices which convert signals of one physical form to another.

2. The Controller. This is an amplifier which receives the low-power error signal, together with power from an external source. A controlled amount of power (of the correct physical nature) is then supplied to the output element.

3. The Output Element. It provides the load with power of the correct physical nature in accordance with the signal received from the controller.

Other devices such as gear-boxes and "compensating" devices are often featured in control systems, but these can usually be considered to form part of one of the other *elements*[2].

New Words and Phrases

component [kəm'pəunənt] adj. 组成的，构成的 n. 成分；元件；组件
subtraction [səb'trækʃn] n. 减少；减法，差集
transducer [trænz'djuːsə] n. 传感器；变换器；换能器
element ['elimənt] n. 元素，成分，要素；原理；自然环境

Notes

1. This is a device which receives the low-power input signal and the output signal which may be of different physical natures, converts them into a common physical quantity for the purposes of subtraction, performs the subtraction, and gives out a low-power error signal of the correct physical nature to actuate the controller.

译文：该装置接收低功率输入信号和具有不同物理性质的输出信号，并将它们转换成同一个常见物理量来求差，求差后输出一个低功率的具有正确物理性质的误差信号来驱动控制器。

说明：这是一个复合句。第一个 which 引导的定语从句修饰 device；第二个 which 引导的定语从句修饰 output signal，该从句具有三个并列谓语动词，即 may be of、converts... into 和 gives out，之后是相应的宾语。subtraction 意为"减法；减少；差集"。例如：

One of these is the subtraction of water from the ocean by means of evaporation-conversion of liquid water to water vapor. 其中一部分是通过蒸发转换的方式，即使液态水转化为水蒸气，

减少海洋中的水分。

2. Other devices such as gear-boxes and "compensating" devices are often featured in control systems, but these can usually be considered to form part of one of the other elements.

译文：齿轮箱、“补偿”设备等其他设备经常出现在控制系统中，但它们通常被认为是形成其他部件的一部分。

说明：这是一个并列句，前后是转折关系。devices 意为“设备”；be featured in 意为“出现在……”；be considered to 意为“被认为做……”。例如：

Because it can tone down the gray hair color, it may be considered to be a real hair dying product. 由于它能够染黑花白的头发，所以可能会被认为是一种真正的染发产品。

Exercises

Ⅰ. Complete the following sentences according to the passage.

1. A closed-loop control system usually contains ( ) basic components.

2. The controller is ( ) which receives the low-power error signal, together with power from an external source.

3. The error detector will usually contain "( )"; these are devices which convert signals of one physical form to another.

4. The output element provides the load with power of the correct physical nature in accordance with the signal received from the ( ).

Ⅱ. Translate the following sentences into Chinese.

1. It is well known that there are three basic components in a closed-loop control system.

2. It provides the load with power of the correct physical nature in accordance with the signal received from the controller.

3. Honesty is a vital element of her success.

Unit 21 Basic Robots

Text

General Information

The word *robot* originated from the Czech word robota, meaning work. Webster's dictionary defines robot as " an automatic device that performs functions ordinarily ascribed to human beings". A definition used by the Robot Institute of America gives a more *precise* description of industrial robots: " A robot is a reprogrammable *multi-functional manipulator* designed to move materials, parts, tools, or *special-variety* of tasks. " In short, a robot is a programmable *general-purpose* manipulator with external sensors that can perform various assembly tasks.

With this definition, a robot must possess intelligence, which is normally due to computer *algorithms* associated with its control and sensing systems. [1] Robots are the general-purpose, computer-controlled manipulator consisting of several *rigid* links connected by joints into an open *kinematic* chain. Joints are typically rotary (revolute) or linear (*prismatic*) (Fig. 21-1). A revolute joint is like a *hinge* and allows relative *rotation* between two links. A prismatic joint allows a linear relative motion between two links. The *convention* R is used to represent revolute joint and convention P is used to denote prismatic joints. The joints are shown in the figure 21-1. Each joint represents the interconnection between two links, $l_i + l_{i+1}$. Similarly, the axis of rotation of a revolute joint, or the axis along which a prismatic joint *slides* is represented by z_i if the joint is the interconnection of links l_i and l_{i+1}. [2] The joint variables, denoted by "θ_i" for a revolute and "d" for a prismatic joint, represent the relative *displacement* between *adjacent* links.

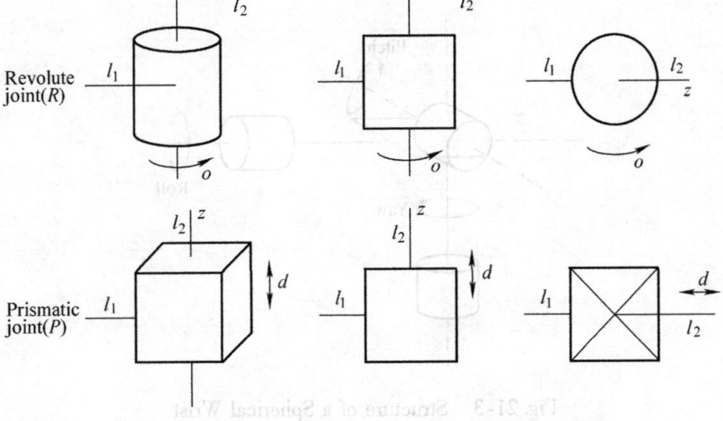

Fig. 21-1 Symbolic Representation of Robot Joints

Wrist* and *End-Effectors

Mechanically, a robot is composed of an arm and a wrist *subassembly* plus a tool. It is designed to reach a work-piece located within its work volume. The work volume is the *sphere* of influence of a robot whose arm can deliver the wrist subassembly unit to any point within the sphere. The arm subassembly generally can move with three degrees of freedom. The combination of the movements positions the wrist unit at the work-piece. The wrist subassembly unit usually consists of three rotary motions. The combination of these motions *orients* the tool according to the configuration of the object for ease in pick-up. These last three motions are often called *pitch*, yaw, and roll. Hence, for a six-jointed robot, the arm subassembly is the position mechanism, while the wrist subassembly is the *orientation* mechanism. These concepts are illustrated by the Unimation PUMA robot arm as shown in the Fig. 21-2 and Fig. 21-3.

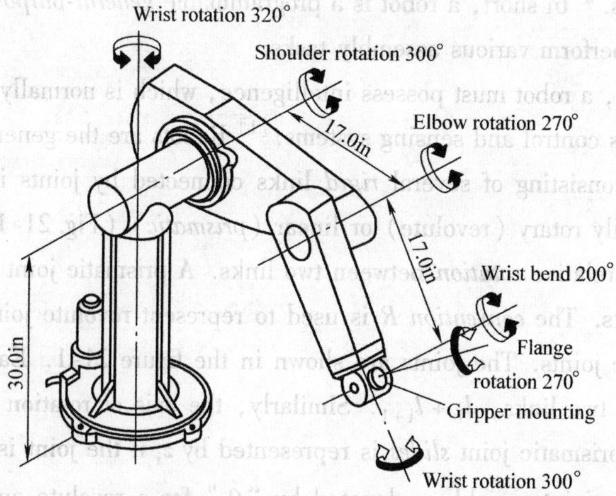

Fig. 21-2 PUMA 560 Series Robot Arm

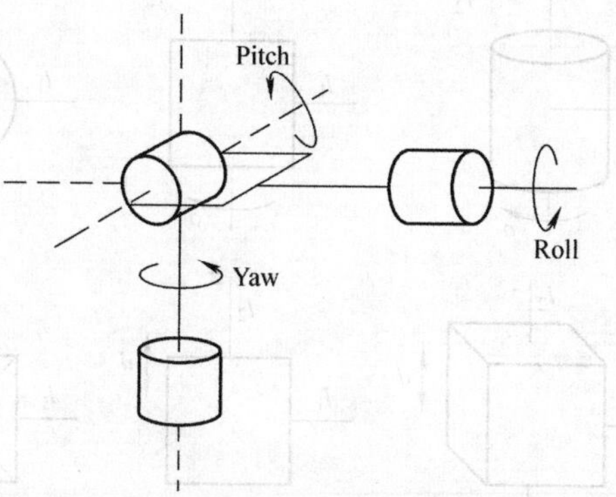

Fig. 21-3 Structure of a Spherical Wrist

Notes

1. With this definition, a robot must possess intelligence, which is normally due to computer algorithms associated with its control and sensing systems.

译文：根据这个定义，机器人必须智能化，这通常由与其控制和传感系统有关的计算机算法实现。

说明：这是一个复合句。主句主语是 a robot，谓语是 must possess，其宾语是 intelligence；句首的 with 短语是介词短语作状语，修饰 possess。例如：

With the growth of tourism, there arises a new urge to travel abroad. 随着旅游业的发展，人们产生了出国旅游的新欲望。

2. Similarly, the axis of rotation of a revolute joint, or the axis along which a prismatic joint slides is represented by z_i if the joint is the interconnection of links l_i and l_{i+1}.

译文：同样，如果关节使得 i 和 $i+1$ 杆件相互连接的话，那么转动副关节的旋转轴或移动副关节的滑动轴由 z_i 表示。

说明：这是一个包含有条件状语从句的复合句。if 引导的条件状语从句位于句末，英语中的条件状语从句可位于句首，也可以位于句末。主句主语是 the axis of rotation of a revolute joint, or the axis... joint slides，其中 along which a prismatic joint slides 是定语从句修饰 the axis。例如：

An oil seal was replaced, along with both front wheel bearings. 换了一个油封和两个前轮轴承。

New Words and Phrases

robot [ˈrəubɒt] n. 机器人；遥控装置
precise [prɪˈsaɪs] adj. 精确的；正规的；精密的
multi-functional [ˌmʌltiˈfʌŋkʃənl] adj. 多功能的
manipulator [məˈnɪpjuleɪtə] n. 操纵者；操纵器；翻钢机
special-variety [ˌspeʃlvəˈraɪəti] adj. 特殊品种
general-purpose [ˌdʒenrəlˈpɜːpəs] adj. 通用；通用的
algorithm [ˈælgərɪðəm] n. 运算法则；演算法；计算程序
rigid [ˈrɪdʒɪd] adj. 严格的；僵硬的；（规则、方法等）死板的
kinematic [ˌkɪnɪˈmætɪk] adj. 运动学的，运动学上的
prismatic [prɪzˈmætɪk] adj. 棱镜的
hinge [hɪndʒ] n. 铰链，折叶；转折点
rotation [rəuˈteiʃn] n. 旋转，转动；循环
convention [kənˈvenʃn] n. 会议；国际公约；惯例
slide [slaɪd] v. 滑落；下跌 n. 幻灯片
displacement [dɪsˈpleɪsmənt] n. 替代；停职；［化］置换
adjacent [əˈdʒeɪsnt] adj. 邻近的，毗邻的；（时间上）紧接着的
wrist [rɪst] n. 腕，手腕；腕关节

end-effector [endɪˈfektə] n. 终端执行器
subassembly [ˌsʌbəˈsemblɪ] n. 部件，组件
sphere [sfɪə] n. 球（体）；（兴趣或活动的）范围；势力范围
orient [ˈɔːrient] v. 标定方向；使……向东方；以……为参照
pitch [pɪtʃ] n. 场地；最高点 vt. 扔，投
orientation [ˌɔːriənˈteiʃn] n. 方向，定位，取向
sensing system 传感系统
open kinematic chain 开放运动链
relative motion 相对运动
axis of rotation 旋转轴
work volume 加工区域
degree of freedom 自由度
position mechanism 定位机构
orientation mechanism 定向机构

Exercises

Ⅰ. Decide whether the following statements are True (T) or False (F) according to the passage.

1. A robot is considered as "an automatic device", meaning work.
2. A rotary joint is like a ring and allows relative rotation between two joints.
3. Generally speaking, a robot consists of an arm, a wrist subassembly and a tool.
4. For a six-jointed robot, the arm subassembly is the position mechanism the same as the wrist subassembly.

Ⅱ. Translate the following phrases into Chinese or English.

1. 传感系统　2. 相对运动　3. 自由度　4. 加工区域
5. 定向机构　6. 定位机构　7. 公法线相对运动

Ⅲ. Translate the following sentences into English by using the words and phrases in brackets.

1. 我家乡的房子过去多由木料制成。(be composed of)
2. 各种煤都是由植物的朽烂形成的。(originate from)
3. 这些建议是谈判的现实出发点。(represent)
4. 母亲告诫孩子们别和坏人交往。(associate with)

Practical English:

如何阅读英文招聘广告

英文招聘广告是较常见和较实用的应用文体之一，它具有一般广告的共同特点，但又有别于产品广告、非商业广告和政治广告，它是一种特定的，以招募劳动力为目的的文体，因

此在语言和结构上也具有自己的特点。英文招聘广告的阅读属于应用性阅读，要实现高效阅读，必须掌握英文招聘广告的特点和一些实际的阅读技巧。英文招聘广告具有以下特点：

一、结构清晰，层次分明

招聘广告通常由广告标题 、公司简介 、工作职责 、资格要求 、薪金待遇 、联系方式六部分组成，虽然某些部分也可能省略，但每一部分通常都独立成段，各部分之间的界限清晰明了。

二、标题言简意赅，精确具体

英文招聘广告往往以招聘单位名称或招聘职位作标题，使求职者一目了然。例如：

Avon（Shanghai）Company Limited　上海雅芳有限公司

R&D software Engineer　研发软件工程师

Business advisor　商务顾问

三、多用省略句

为了节省广告的篇幅和广告的费用，同时使读者一目了然，英文招聘广告经常省略句子的主语、动词、定冠词、不定冠词和从句。由于有特定的上下文，读者也十分清楚，不会产生误会或歧义。例如：

Excellent written and verbal communication skills in English and Mandarin.

良好的口头和书面中英文表达能力。（原文省略了实义动词 have）

Ability to react quickly and efficiently to crisis.

能够快速有效地应对危机。（原文省略了冠词 the）

Qualifications needed：

所需资格：（原文省略了从句 which are）

四、多用短语或短句

英文招聘广告在描述工作职责和应聘资格时广泛使用名称短语、形容词短语和介词短语以缩短广告的篇幅。短语的使用既方便记忆，又能节约版面，同时突出了关键词，容易引起求职者的注意。例如：

Good analytic skills，problem solving and conflict management skills .

良好的分析能力、问题解决能力和冲突管理能力。

Familiar with general office software.

熟悉办公软件。

Under the age of 40.

40 岁以下。

With 2 ~ 3 years experience in telecom.

2 ~ 3 年电信行业工作经验。

五、多用祈使句

祈使句一般用来说明应聘方法。为表示礼貌，使读者感到亲切，都用“请”字开头。例如：

Please apply in your own handwriting with full English resume and resent photo to：

请亲自誊写详细的英文简历并附上近照寄：

Please call 0377 – 63457867 for interviewing.

请致电 0377－63457867 商定面试事宜。

六、采用缩略词

为了节省广告撰稿人的撰稿时间与读者的阅读时间，并节省费用，英文招聘广告中能进行缩略的词语都尽可能缩略。例如：

Dept：Department 部门
CV：Curriculum Vitae 个人简历
JV：joint venture 合资
G. M.：General Manager 总经理
Ad：advertisement 广告
add：address 地址
P/T：part time 兼职（的）
refs：references 推荐信
exp'd：experienced 有经验的
Wpm：words per minute 打字/每分钟
req：required 需要
CAD：computer assisted design 计算机辅助设计

因此，在阅读英文招聘广告时，首先要浏览广告标题，确定招聘单位和招聘职位。然后浏览公司简介，掌握招聘公司的一些基本情况。接下来，要细读工作职责和资格要求，确定自己是否能够胜任这份工作，注意辨别一些省略句和缩略词。最后，要留意联系方式，必要时要动手记下具体的联系方式。

七、英文招聘广告（范例）

Sample 1

Machine Operator

Responsibilities:

—Basic operation of machine, including start-up, shutdown and test to machine.

—Work according to production plan and program or process drawings.

—Replace and maintain the machine parts.

Requirements:

—College degree or above, majored in mechanical engineering or relevant.

— Excellent in reading and interpreting machining work instructions.

— Experience in an industrial production environment is a plus.

—Good sense of responsibility and working attitude.

We offer attractive salary dependent on qualification and experience. Interested parties please send full resume with expected salary, photo, contact address and tel. no. to Personnel Dept. before June 18, 2018. The address of our company: 106 Fang-yuan Road, Guangzhou. Our postcode: 510370

招聘机床操作员

工作职责：

1. 掌握机床的基本操作，包括机床的起动、关闭及调试。
2. 依据生产计划、加工程序或工艺图样进行加工。
3. 更换及维护机床零件。

职位要求：

1. 机械工程或相关专业大学及以上学历。
2. 能正确解读机床加工工作指令。
3. 有工厂实际操作经验者优先。

4. 责任心强，工作态度认真。

我们将根据应聘者的学历和经验提供富有吸引力的薪酬。有兴趣的人员请将包含期望待遇、照片、联系地址和电话的完整简历于 2018 年 6 月 18 日前寄到我公司人事部。公司地址：广州市方圆大道 106 号，邮编：510370。

Sample 2

Marketing Assistant

Responsibilities:

—Responsible for the local management of marketing and sales activities according to instructions from head office.

—Collect related information for the head office.

Requirements:

—College degree or above with good English (speaking and writing).

—Develop relationship with local media and customers.

—With basic idea of sales and marketing, related experience is preferred.

—Working experience in an international organization is a must.

—Good communication and presentation skills.

The right candidate will receive an appropriate and attractive compensation package. Interested candidates please apply in writing in both English and Mandarin as follows:

1. Qualifications, job experience with salary history and expected compensation package.

2. Recent photo.

And send them to the address below by July 31:

Room 302, International Hotel, 1S. St. Beijing 100088.

All employments will be done through FESCO.

No telephone call will be entertained.

招聘销售助理

工作职责：

1. 根据总公司的要求负责区域销售管理及销售活动。

2. 收集相关信息资料上报总公司。

职位要求：

1. 大学及以上学历，英语口语流利，书面表达能力强。

2. 开发所负责区域内的媒体与客户关系。

3. 了解市场，熟悉销售相关业务工作者优先。

4. 国际公司的工作经验是必备的条件。

5. 具有良好的沟通及表达能力。

合适的应聘者将获得合理而富有吸引力的薪酬待遇。有意者请用中英文书写以下应聘材料：

1. 包含学历、工作经验、以往工作待遇和期望薪酬待遇的简历。

2. 准备近期照片一张。

请于 7 月 31 日前，将应聘材料寄往：

北京南街 1 号国际宾馆 302 房间，邮编：100088。
所有招聘都将通过外国企业人力资源服务有限公司进行。
恕不接受电话咨询。

Reading: The Robot *Applications*

In 1995 about 700,000 robots were operating in the industrialized world. Over 500,000 were used in Japan, about 120,000 in Western Europe, and about 60, 000 in the United States. Many ro-bot applications are for tasks that are either dangerous or unpleasant for human beings.[1] In medical laboratories, robots handle potentially dangerous materials, such as blood or urine samples (尿样). In other cases, robots are used in *repetitive*, unchangeable tasks in which human performance might *degrade* over time. Robots can perform these repetitive, *high-precision* operations 24 hours a day without *fatigue*.

A major user of robots is the automobile industry. General Motors Corporation uses *approxi mately* 16, 000 robots for tasks such as spot welding (定位焊), painting, machine loading, parts transfer, and assembly. Assembly is one of the fastest growing industrial applications of robotics. It requires higher precision than welding or painting and depends on low-cost sensor systems and powerful inexpensive computers.[2] Robots are used in electronic assembly where they load *microchips* on circuit boards.

Activities in environments that cause great danger to humans, such as locating sunken ships, clean-up of nuclear waste, exploring for underwater mineral deposits, and active volcano exploration, are ideally suited to robots.[3] Similarly, robots can explore distant planets. NASA's Galileo, an unmanned space probe, travelled to Jupiter in 1996 and performed tasks such as determining the chemical content of the Jovian atmosphere.

Robots are being used to assist surgeons in installing *artificial* hips, and very high-precision robots can help surgeons with *delicate* operations on the human eye. This robot travels around hospitals to deliver medical supplies, medication, food trays, or just about anything to *nursing* stations. Once it is finished it goes back to its charging station and waits for its next task.

Head There, Inc. has introduced a *telepresence* robot that can be moved around its location by remote control using the Internet. The robot enables a user to hear, see, speak, and be seen at a far away location. In a sense, the robot acts as a stand-in for the user.

For education in schools and high schools and *mechatronics* training in companies robot kits are becoming more and more popular. Robots historically used in education include the Turtle robots (strongly associated with the Logo programming language) and the Heathkit HERO series.

New Words and Phrases

application [ˌæplɪˈkeɪʃn] n. 应用，运用；申请
repetitive [rɪˈpetətɪv] adj. 重复的，啰嗦的
degrade [dɪˈgreɪd] vt. 降低；使降级；降低……身份

high-precision [haɪprɪˈsɪʒn] adj. 高精密度；高精确度
fatigue [fəˈtiːg] adj. 疲劳 vt. 使疲劳，使疲乏
approximately [əˈprɒksɪmətlɪ] adv. 近似地，大约
microchip [ˈmaɪkrəutʃɪp] n. 晶片；微型集成电路片
artificial [ˌɑːtɪˈfɪʃl] adj. 人造的，人工的；虚假的
delicate [ˈdelɪkət] adj. 微妙的；熟练的；易损的
nursing [ˈnɜːsɪŋ] n. 护理，看护；养育
telepresence [ˈteliprezns] n. 远程监控
mechatronics [ˌmekəˈtrɔniks] n. 机电一体化

Notes

1. Many robot applications are for tasks that are either dangerous or unpleasant for human beings.

译文：许多机器人应用于危险或人类不愿意做的工作任务中。

说明：这是一个复合句。that 引导定语从句，修饰 tasks。同时，从句中的复合结构 either... or... 用于连接句子中的两个并列成分，表示选择关系，意为“要么……要么……，不是……就是……”。例如：

The alternative option is a structured bankruptcy of either or both firms. 替代性选择是结构性破产一方或双方公司.

2. It requires higher precision than welding or painting and depends on low-cost sensor systems and powerful inexpensive computers.

译文：它比焊接或涂装要求的精确度更高，并且依赖于低成本的传感器系统和强大的廉价计算机即可。

说明：这是一个简单句。requires 与 depends on 是并列谓语动词。require 意为“要求”，其后可接名词或从句；例如：

Today's complex buildings require close teamwork between the architect and the builders. 如今的建筑结构复杂，需要建筑师和施工人员密切协作。

3. Activities in environments that cause great danger to humans, such as locating sunken ships, clean-up of nuclear waste, exploring for underwater mineral deposits, and active volcano exploration, are ideally suited to robots.

译文：一些对人类来说很危险的环境中的活动，如寻找沉船、清理核废料、探索水下矿藏和勘探活火山，非常适合机器人。

说明：这是一个复合句。activities 是主句主语，in environments 是介词短语作后置定语，that 引导定语从句修饰 environments。such as 意为“例如”，引出 activities 的几个同位语。suit to 意为“使与……相适合；适合于……”。例如：

Environment of my province geography is superior, suit to foster race most. 我省地理环境优越，最适合培养赛马。

Exercises

Choose the best answer(s) to complete the following statements.

1. Which of the following is the main idea of this article?

A. Robots B. Uses of Robots C. Development of Robots D. Advantages of Robots

2. According to the article, which of the following is NOT true?

A. In 1995 about 700, 000 robots were operating in the industrialized world.

B. Why people use the robots is that they are neither dangerous nor unpleasant.

C. In medical laboratories, robots can be used to handle potentially dangerous materials, such as blood or urine samples.

D. In other cases, robots are used in repetitive, unchangeable tasks in which human performance might degrade over time.

3. In the sentence "Robots can perform these repetitive, high-precision operations 24 hours a day without fatigue. The word "fatigue" can be explained as ________.

A. rest B. paying C. powering D. being exhausted

4. The robots can do many jobs EXCEPT ________.

A. spot welding, painting B. drinking, eating and shaving

C. machine loading, parts transfer, and assembly D. exploring distant planets

5. Sometimes which kind of doctor may NOT need robots' help?

A. Physician B. Surgeon C. Brain specialist D. Dentist

Unit 22 3D Printing[1]

Text

Additive Manufacturing (AM) is an appropriate name to describe the technologies that build 3D objects by adding layer-upon-layer of material, whether the material is plastic, metal, *concrete* or one day... human *tissue*.[2]

Common to AM technologies is the use of a computer, 3D modelling software (Computer Aided Design or CAD), machine equipment and layering material. Once a CAD *sketch* is produced, the AM equipment reads in data from the CAD file and lays downs or adds *successive* layers of liquid, *powder*, sheet material or other, in a layer-upon-layer fashion to *fabricate* a 3D object.

The term AM *encompasses* many technologies including *subsets* like 3D Printing, Rapid *Prototyping* (RP), Direct Digital Manufacturing (DDM), layered manufacturing and additive fabrication.

AM application is limitless. Early use of AM in the form of Rapid Prototyping focused on pre-production *visualization* models. More recently, AM is being used to fabricate end-use products in aircraft, *dental* restorations, medical *implants*, automobiles, and even fashion products.

While the adding of layer-upon-layer approach is simple, there are many applications of AM technology with degrees of sophistication to meet diverse needs including: a visualization tool in design, a means to create highly customized products for consumers and professionals alike, as industrial tooling, to produce small lots of production parts, one day... production of human organs.

At MIT, where the technology was invented, projects *abound* supporting a range of forward-thinking applications from multi-structure concrete to machines that can build machines;[3] while work at *Contour* Crafting supports structures for people to live and work in.

Some *envision* AM as a compliment to foundational *subtractive* manufacturing (removing material like drilling out material) and to lesser degree forming (like forging). Regardless, AM may offer consumers and professionals alike, the accessibility to create, *customize* and/or repair product, and in the process, redefine current production technology.

Whether simple or sophisticated, AM is indeed amazing and best described in the adding of layer-upon-layer, whether in plastic, metal, concrete or one day... human tissue.[4]

Examples of Additive Manufacturing

SLA (Stereo lithography Appearance)

This is a very high end technology utilizing laser technology to cure layer-upon-layer of *photopolymer* resin (*polymer* that changes properties when exposed to light).

The build occurs in a pool of resin. A laser beam, directed into the pool of resin, traces the cross-section *pattern* of the model for that particular layer and cures it. During the build cycle, the platform on which the build is repositioned, lowers by a single layer thickness. The process repeats until the build or model is completed and fascinating to watch. Specialized material may be needed to add support to some model features. Models can be machined and used as patterns for injection molding, *thermoforming* or other casting processes.

FDM (Fused *Deposition* Modelling)

Process oriented involving use of *thermoplastic* (polymer that changes to a liquid upon the application of heat and solidifies to a solid when cooled) materials injected through *indexing nozzles* onto a platform.[5] The nozzles trace the cross-section pattern for each particular layer with the thermoplastic material hardening prior to the application of the next layer. The process repeats until the build or model is completed and fascinating to watch. Specialized material may be needed to add support to some model features. Similar to SLA, the models can be machined or used as patterns.

MJM (Multi-Jet Modelling)

Multi-Jet Modelling is similar to an inkjet printer in that a head, capable of *shuttling* back and forth (3 dimensions—x, y, z), *incorporates* hundreds of small jets to apply a layer of thermopolymer material, layer-by-layer.

3DP (Three-Dimensional Printing)

This involves building a model in a container filled with powder of either *starch* or *plaster* based material. An inkjet printer head shuttles and applies a small amount of *binder* to form a layer. Upon application of the binder, a new layer of powder is swept over the prior layer with the application of more binder.[6] The process repeats until the model is complete. As the model is supported by loose powder there is no need for support. Additionally, this is the only process that builds in colours.

SLS (*Selective* Laser *Sintering*)

Somewhat like SLA technology Selective Laser Sintering (SLS) utilizes a high powered laser to fuse small particles of plastic, metal, *ceramic* or glass. During the build cycle, the platform on which the build is repositioned, lowers by a single layer thickness. The process repeats until the build or model is completed. Unlike SLA technology, support material is not needed as the build is supported by unsintered material.

New Words and Phrases

additive [ˈædɪtɪv] adj. 附加的；[数] 加法的 n. 添加剂，添加物
concrete [ˈkɒŋkriːt] adj. 混凝土的；实在的，具体的 n. 混凝土
tissue [ˈtɪʃuː] n. 组织；纸巾
sketch [sketʃ] n. 素描；略图，梗概
successive [səkˈsesɪv] adj. 连续的，继承的，依次的，接替的
powder [ˈpaʊdə] n. 粉；粉末 vt. 使成粉末；撒粉 vi. 搽粉；变成粉末
fabricate [ˈfæbrɪkeɪt] vt. 制造；伪造；装配
encompass [ɪnˈkʌmpəs; en－] vt. 包含，包围，环绕；完成
subset [ˈsʌbset] n. 子集
prototype [ˈprəʊtətaɪp] n. 原型；标准，模范
visualization [ˌvɪzjʊəlaɪˈzeɪʃən] n. 形象化；清楚地呈现

dental [ˈdent(ə)l] adj. 牙科的；牙齿的，牙的
implant [ɪmˈplɑːnt] vt. 种植；嵌入 vi. 被移植 n. [医] 植入物；植入管
abound [əˈbaʊnd] vi. 富于；充满
contour [ˈkɒntʊə] n. 轮廓；等高线；周线；概要 vt. 画轮廓；画等高线
envision [ɪnˈvɪʒn] vt. 想象；预想
subtractive [səbˈtræktɪv] adj. 减去的；负的；有负号的
customize [ˈkʌstəmaɪz] vt. 定做，按客户具体要求制造
photopolymer [ˌfəʊtəʊˈpɒlɪmə] n. 光聚合物，光敏聚合物；感光性树脂
polymers [ˈpɔliməs] n. [高分子] 聚合物
pattern [ˈpæt(ə)n] n. 模式；图案；样品
thermoforming [θəməˈfɔːmɪŋ] n. 热成型；热压成型
deposition [ˌdepəˈzɪʃ(ə)n; diː-] n. 沉积物；矿床；革职
thermoplastic [θɜːməʊˈplæstɪk] adj. 热塑性的 n. [塑料] 热塑性塑料
indexing [ˈɪndeksɪŋ] n. 指数化；[机械学] 分度，转位
nozzle [ˈnɒz(ə)l] n. 喷嘴；管口；鼻
shuttle [ˈʃʌtl] n. 梭子；航天飞机 vt. & vi. 穿梭般来回移动
incorporate [ɪnˈkɔːpəreɪt] vt. 组成公司；包含 vi. 包含；吸收；合并
starch [stɑːtʃ] n. 淀粉
plaster [ˈplɑːstə(r)] n. 灰泥，涂墙泥；石膏；膏药
binder [ˈbaɪndə(r)] n. 黏合剂；包扎物，包扎工具；装订工
selective [sɪˈlektɪv] adj. 精心选择的，不普遍的；淘汰的
sintering [ˈsɪntərɪŋ] n. 烧结 v. 烧结；使熔结
ceramic [sɪˈræmɪk] adj. 陶瓷的；陶器的 n. 陶瓷；陶瓷制品

Technical Terms

Additive Manufacturing 增材制造
Rapid Prototyping (RP) 快速成型
Direct Digital Manufacturing (DDM) 直接数字制造
Selective Laser Sintering 选择性激光烧结
Fused Deposition Modelling 熔融沉积成型
Multi-Jet Modelling 多重喷射成型
Contour Crafting 轮廓工艺
Three-Dimensional Printing 三维打印
cross-section 剖面

Notes

1. 3D Printing 是 Three-Dimensional Printing（三维打印）的缩写形式，是近年来对 Additive Manufacturing（增材制造）更为流行的通俗叫法。

2. Additive Manufacturing (AM) is an appropriate name to describe the technologies that build

3D objects by adding layer-upon-layer of material, whether the material is plastic, metal, concrete or one day... human tissue.

译文：增材制造（AM）是描述3D物体生成技术的一个恰当的名称，这种技术通过一层一层地添加材料来生成物体，这些材料可以是塑料、金属、水泥，甚至有一天可能会是人体组织。

说明：这是一个复合句。that build 3D... material 是定语从句，修饰 the technologies；whether 则引导状语从句，修饰 build。

3. At MIT, where the technology was invented, projects abound supporting a range of forward-thinking applications from multi-structure concrete to machines that can build machines.

译文：在AM技术的发明地——麻省理工学院，有大量的项目来支持一系列的前瞻性应用，从多结构混凝土到可以制造机器的机器。

说明：这是一个复合句。主句主语是 projects；abound 是谓语动词，意为"富于，充满"。at MIT 为地点状语，where 引导的定语从句修饰 MIT。supporting a range of... machines 为分词短语作状语，修饰动词 abound；该分词短语中 that 引导的定语从句修饰第一个 machines。

4. Whether simple or sophisticated, AM is indeed amazing and best described in the adding of layer-upon-layer, whether in plastic, metal, concrete or one day... human tissue.

译文：无论简单还是复杂，AM确实令人惊叹，层层叠加的方式对这个工艺进行了最好呈现，无论材料是塑料、金属、混凝土或者有一天是人体组织。

说明：这是一个包含有两个让步状语的简单句。AM是主语；is 为系动词；amazing and best described 分别为形容词和动词过去分词作表语；whether（the material adopted is）in plastic, metal, concrete or one day... human tissue 为让步状语，可以看作在 whether 后边省略了 the material adopted is；而句首的 whether simple or sophisticated，可以看作在 whether 后面省略了（it is）的让步状语，用来修饰主句谓语 is indeed amazing。

5. Process oriented involving use of thermoplastic (polymer that changes to a liquid upon the application of heat and solidifies to a solid when cooled) materials injected through indexing nozzles onto a platform.

译文：这种工艺定位于使用热塑性材料（加热时变成液体并在冷却时固化成固体的聚合物），通过转位喷嘴将其注射到平台上。

说明：去掉括号内容，这是一个包含主、动、宾结构的简单句。process 是主语，oriented 是谓语动词，宾语是 involving use... materials，之后的 injected through... 是过去分词短语作定语，修饰 materials。括号中由 that 引导的定语从句修饰 polymer，而 polymer 是 thermoplastic 的同位语，是对 thermoplastic 起解释说明作用的。注意 indexing 在句中的用法，index 在机械技术英语中一般意为"转位的，分度的"。比如：indexing table，分度盘；indexable insert，可转位刀片。

6. Upon application of the binder, a new layer of powder is swept over the prior layer with the application of more binder.

译文：当一层黏合剂施加完成，马上在其上撒一层粉末，然后施加下一层黏合剂。

说明：这是一个比较复杂的简单句。upon application of the binder 在句首作时间状语，

upon 意为“当……时候，一……就……”，意思与 when 相当，但 when 是副词，而 upon 是介词，后面只能跟名词结构。

Exercises

Ⅰ. Answer the following questions according to the passage.

1. What is Additive Manufacturing?

2. What is common to AM technologies?

3. What is the focus of early use of AM technology in the form of Rapid Prototyping?

4. Where did the AM technology invented?

5. What kind of material can be used in the process of SLA?

Ⅱ. Match the following phrases in column A with column B.

Column A	Column B
1. polymer	a. 剖面
2. laser	b. 直接数字制造
3. cross-section	c. 增材制造
4. Direct Digital Manufacturing	d. 选择性激光烧结
5. Additive Manufacturing	e. 激光
6. Selective Laser Sintering	f. ［高分子］聚合物
7. Rapid Prototyping	g. 快速制造
8. Rapid Fabrication	h. 快速造型

Ⅲ. Translate the following sentences into Chinese.

1. Everybody recognizes the term 3D printing nowadays but what many people mean when talking about 3D printing is actually one of the several Additive Manufacturing (AM) processes.

2. Individual Additive Manufacturing processes differ depending on the material and machine technology used.

3. You will have to prepare a 3D digital model before it is ready to be 3D printed.

4. The term "3D printing" 's origin sense is in reference to a process that deposits a binder material onto a powder bed with inkjet printer heads layer by layer.

Practical English:

如何写英文个人简历

英文简历一般主要包括个人资料，教育背景，工作经历，兴趣爱好四个部分。

1. 个人资料（Personal Information）

（1）姓名　英文名字一般是名在前，姓在后。但是中国人的名字译成英文时，现在更为时兴的形式是保持汉语拼音拼写，只需把姓和名的首字母大写就可以了。如李阳 Li Yang，高凤林 Gao Fenglin。

（2）地址　省市名之后要写中国（China）。邮编的位置应在省市名与国名之间，即在

China之前。

（3）电话　电话号码一定加地区号，如86－10；电话号码每隔4位数字加一个“－”，如6505－2126；地区号后的括号和号码间加空格，如（86－10）6505－2126。

2. 教育背景（Education Background）

（1）时间顺序　在英文简历中，求职者受教育的情况从求职者的最高教育层次写起。至何种层次为止，则无具体规定，可根据个人实际情况安排。

（2）校名和地名　学校名称建议大写并加粗，便于招聘者迅速识别。地名右对齐，同样建议大写并加粗，地名后一定写中国（China）。

（3）社会工作　担任班干部应写明具体职务；参加有社团、协会应写明组织名称和所任职务，如果未担任任何职务，写 member of club（s）。

（4）语言和计算机水平　个人的语言水平和计算机能力应在此单列说明。

3. 工作经历（Work Experience）

（1）时间顺序　工作经历应从当前的工作岗位写起，直至求职者的第一个工作岗位为止。也可按技能类别分类写，这主要是为了强调个人的某种技能。

（2）公司名和地名　公司名称建议大写并加粗。若全称太复杂，可以用缩略形式代替，比如 International Business Machines Corporation 可简写为 IBM。

（3）公司简介　对于新公司、小公司或招聘者不甚熟悉的某些行业公司，简略介绍一下公司情况。

（4）职务与部门　建议在公司名称之后的第2行开始写，职务与部门建议加粗，每个单词的第1个字母要大写，如 Manager，Finance Department。

（5）工作内容

1）多用点句（bullet point）。避免大段文字，点句的长度以1行为宜，句数以3～5句为佳；点句以动词开始，目前的工作用一般现在时，以前的工作用过去时。

2）主要职责与主要成就。初级工作以及开创性不强的工作应把主要职责放在前面，而较高级或开创性较强的工作则应把主要成就写在前面。工作成就尽量数字化、精确化。在同一公司的业绩中，应秉持“重要优先”的原则。

3）工作培训。接受过的培训可放在公司名称的后面，因为培训是公司内部的，与公司业务有关。

4. 兴趣爱好（Hobbies and Interests）

一般写2～3项强项即可，弱项一定不要写，不具体的爱好不写。

除上述要点外，写英文简历在用词上也要多加斟酌，比如表达有闯劲，用 energetic，enterprising 或者 can-do spirit 比较通俗易懂，而 aggressive 和 ambitious 也都有“上进心强的”意思，但语气逐渐加强，对初级岗位的适用性不如前几个词好。

总之，一份好的英文简历，要目的明确、语言简练，切忌拖沓冗长，词不达意。下面给出一份英文简历的示例。

Example

Personal Information:

Name: Xu Wenbin

Date of Birth: July 12, 1998

Place of Birth: Nanyang

Sex: Male

Marital Status: Unmarried

Telephone: 139316××××××

E-mail: Wyn. xu@ 163. com

Education:

Period: August 2014 to July 2017

School: Henan Polytechnic Institute

Major: Mechanical Engineering

Achievements & Activities:

Class President and Founder of CAM Committee

Representative in the Student Association

Honors & Awards:

2015 Henan Polytechnic Institute, Third class scholarship

2016 Henan Polytechnic Institute, Outstanding class cadre

Certificates:

National College English Test 4 (CET4)

BEC Vantage level

National Computer Rank Examination 2

Professional Skills:

Have a good command of both spoken and written English.

Have a good command of CAD & CAM software—AutoCAD, Solidworks, Mastercam.

Be skillful of operation of Microsoft Office system.

Work Experience:

Period: June 2017 to date

Company: Sandvik Tools Co., Ltd

Title: CNC Line Engineer of workshop

Job Responsibilities:

To debug CNC metal cutting programming, operate milling and boring machine center.

To maintain metal cutting machine and carry out work site 5S activities.

Trouble shooting of process failures and feedback to R&D department.

Period: june 2016 to May 2017

Company: Luoyang Bearing Technology Co., Ltd

Title：Technician of metal cutting workshop

Job Responsibilities：

CNC metal cutting machine operating，including lathe and milling machines.

Bench work on metal parts.

Reference：

Xu GuoWei，Chairman of Mechanical Department of Henan Polytechnic Instiute.

Others：

Energetic，with teamwork and can-do spirit，be able to work under pressure，well prepared to face challenge work. Have very good communication skills. Strong interests in new technologies.

Reading： American Scientists Work on Printing of Living Tissue Replacements

Three-dimensional printers work very much like ordinary desktop printers. But instead of just putting down ink on paper，they *stack* up layers of living material to make 3-D shapes. The technology has been around for almost two decades，providing a shortcut for dentists，*jewelers*，*machinists* and even chocolate producers who want to make custom pieces without having to create molds. [1]

In the early 2000s，scientists and doctors saw the potential to use this technology to construct living tissue，maybe even human organs. They called it 3-D bioprinting，and it is a *red-hot* branch of the *burgeoning* field of tissue engineering.

In laboratories all over the world，experts in chemistry，biology，medicine and engineering are working on many paths toward an *audacious* goal：to print a *functioning* human *liver*，*kidney* or heart using a patient's own cells. That's right—new organs，to go. If they succeed，*donor* waiting lists could become a thing of the past.

Bioprinting technology is years and possibly decades from producing such complex organs，but scientists have already printed skin and *vertebral* disks and put them into bodies. So far，none of those bodies have been human，but a few types of printed replacement parts could be ready for human trials in two to five years.

Scientists say the biggest technical challenge is not making the organ itself，but *replicating* its *intricate* internal network of *blood vessels*，which *nourishes* it and provides it with oxygen. [2]

Many tissue engineers believe the best bet for now may be printing only an organ's largest connector vessels and giving those vessels' cells time，space and the ideal environment in which to build the rest themselves；[3] after that，the organ could be implanted.

New Words and Phrases

stack [stæk] vt. & vi. 堆成堆；堆起来或覆盖住

jeweler [ˈdʒuːələ] n. 珠宝商；宝石匠

machinist [məˈʃiːnɪst] n. 机械师

red-hot [redhɔt] adj. 赤热的；激烈的，恼怒的；近期的，新的

burgeoning [ˈbɜːdʒənɪŋ] adj. 迅速成长的 v. 迅速发展；发（芽）
audacious [ɔː ˈdeɪʃəs] adj. 大胆的；鲁莽的；大胆创新的
functioning [ˈfʌŋkʃənɪŋ] v. 起作用（function 的现在分词）；正常工作
liver [ˈlɪvə(r)] n. 肝脏
kidney [ˈkɪdni] n. 肾，肾脏
donor [ˈdəʊnə(r)] n. 捐赠者
vertebral [ˈvɜːtibrəl] adj. 椎骨的；脊椎的
replicate [ˈreplɪkeɪt] vt. 复制，复写；[生] 复制
intricate [ˈɪntrɪkət] adj. 错综复杂的；难理解的；曲折；盘错
nourish [ˈnʌrɪʃ] vt. 滋养，施肥于；抚养，教养；使健壮
blood vessel [blʌd ˈvesəl] n. 血管

Notes

1. The technology has been around for almost two decades, providing a shortcut for dentists, jewelers, machinists and even chocolate producers who want to make custom pieces without having to create moulds.

译文：这项技术已经有将近20年的历史，为牙医、珠宝商以及机械师提供了快捷的工作手段，甚至还为那些希望不用打造模具就可以为人们定制巧克力的生产商实现了梦想。

说明：这是一个复合句。主句的主语是 the technology；has been 为系动词的完成形式；around for almost two decades 是表语部分。providing a shortcut... to create moulds 是现在分词短语作伴随状语，其中 who 引导定语从句修饰 producers。

2. Scientists say the biggest technical challenge is not making the organ itself, but replicating its intricate internal network of blood vessels, which nourishes it and provides it with oxygen.

译文：科学家称最大的技术挑战并不是制造器官本身，而是复制器官内部错综复杂的血管网络，这些血管起到了滋养器官、为器官提供氧气的作用。

说明：这是一个包含有宾语从句的复合句。scientists 是主语；say 是谓语动词；the biggest technical... with oxygen 前面省略了 that，是宾语从句。在宾语从句中使用的 is not... but 意为“不是……而是……”，which 引导的非限制性定语从句修饰 blood vessels，对先行词起补充说明作用。

Exercises

Ⅰ. Decide whether the following statements are True (T) or False (F) according to the passage.

1. 3D printing is a technology which was invented five years ago.

2. Scientists and doctors saw the potential to use this technology to print human tissues 10 years ago since the technology developed not very fast.

3. There has already been a successful case to print human organs by 3D printers.

4. The biggest technical challenge to print a human organ is to replicate its intricate internal

network of blood vessels.

Ⅱ. Translate the following sentences into English.

1. 3D 打印机从 20 世纪八十年代就已经存在并被广泛应用于快速成型和研发用途。

2. 3D 打印已经被考虑作为能够生成人类活体新组织和器官的干细胞移植的方法。

3. STL 文件从 20 世纪八十年代中期起就是增材制造设备和设计程序间传递信息的行业标准。

Unit 23　Industry 4. 0 Introduction

Text

Industry 4. 0's *provenance* lies in the *powerhouse* of German manufacturing. [1] However the *conceptual* idea has since been widely adopted by other industrial nations within the European Union, and further *afield* in China, India, and other Asian countries. The name Industry 4. 0 refers to the fourth industrial revolution, with the first three coming about through mechanization, electricity, and IT.

The fourth industrial revolution, and hence the 4. 0, will come about via the Internet of Things and the Internet of services becoming integrated with the manufacturing environment. [2] However, all the benefits of previous revolutions in industry came about after the fact, whereas with the fourth revolution we have a chance to *proactively* guide the way it transforms our world.

The vision of Industry 4. 0 is that in the future, industrial businesses will build global networks to connect their machinery, factories, and warehousing facilities as *cyber*-physical systems (CPS), which will connect and control each other intelligently by sharing information that *triggers* actions. [3] These cyber-physical systems will take the shape of smart factories, smart machines, smart storage facilities, and smart supply chains. This will bring about improvements in the industrial processes within manufacturing as a whole, through engineering, material usage, supply chains, and product lifecycle management. These are what we call the horizontal value chain, and the vision is that Industry 4. 0 will deeply integrate with each stage in the horizontal value chain to provide tremendous improvements in the industrial process. [4]

At the center of this vision will be the smart factory, which will alter the way production is performed, based on smart machines but also on smart products. It will not be just cyber-physical systems such as smart machinery that will be intelligent; the products being *assembled* will also have embedded intelligence so that they can be identified and located at all times throughout the manufacturing process. The *miniaturization* of RFID tags enables products to be intelligent and to know what they are, when they were manufactured, and crucially, what their current state is and the steps required to reach their desired state. [5]

This requires that smart products know their own history and the future processes required to transform them into the complete product. This knowledge of the industrial manufacturing process is embedded within products and this will allow them to provide alternative routing in the production process. For example, the smart product will be capable of instructing the conveyor belt, which production line it should follow as it is aware of it current state, and the next production process it requires to step through to completion. Later, we will look at how that works in practice.

For now, though, we need to look at another key element in the Industry 4. 0 vision, and that is the integration of the vertical manufacturing processes in the value chain. The vision held is that the embedded horizontal systems are integrated with the vertical business processes, (sales, *logistics*, and finance, among others) and associated IT systems. They will enable smart factories to control the end-to-end management of the entire manufacturing process from supply chain through to services and lifecycle management. This merging of the Operational Technology (OT) with Information Technology (IT) is not without its problems, as we have seen earlier when discussing the Industrial Internet. However, in the Industry 4.0 system, these entities will act as one.

Smart factories do not relate just to huge companies, indeed they are ideal for small- and medium-sized enterprises because of the flexibility that they provide. For example, control over the horizontal manufacturing process and smart products enables better decision-making and *dynamic* process control, as in the capability and flexibility to *cater* to last-minute design changes or to alter production to *address* a customer's preference in the products design. [6] Furthermore, this dynamic process control enables small lot sizes, which are still profitable and *accommodate* individual custom orders. These dynamic business and engineering processes enable new ways of creating value and innovative business models.

In summary, Industry 4.0 will require the integration of CPS in manufacturing and logistics while introducing the Internet of Things and services in the manufacturing process. This will bring new ways to create value, business models, and *downstream* services for small medium enterprises.

New Words and Phrases

provenance [ˈprɒvənəns] n. 出处，起源
powerhouse [ˈpaʊəhaʊs] n. 精力旺盛的人；发电所，动力室；强国
conceptual [kənˈseptjuəl] adj. 概念上的
afield [əˈfiːld] adv. 在战场上；去野外；在远处；远离
proactively [ˌprəʊˈæktɪvlɪ] adv. 主动地
cyber [ˈsaɪbə] adj. 计算机（网络）的，信息技术的
trigger [ˈtrɪɡə(r)] vt. 引发，引起；触发 n. 扳机
assemble [əˈsemb(ə)l] vt. 集合，聚集；装配；收集 vi. 集合，聚集
miniaturization [ˌmɪnətʃəraɪˈzeɪʃn] n. 小型化，微型化
logistic [ləˈdʒɪstɪk] adj. 后勤学的
dynamic [daɪˈnæmɪk] adj. 动态的；动力学的；有活力的 n. 动态；动力
cater [ˈkeɪtə(r)] vt. 投合，迎合；满足需要；提供饮食及服务
address [əˈdres] vt. 演说；写地址；向……致辞；处理 n. 地址；致辞
accommodate [əˈkɒmədeɪt] vt. 容纳；使适应；调解 vi. 适应；调解
downstream [ˌdaʊnˈstriːm] adv. 下游地；顺流而下 adj. 下游的
come about 发生；产生；改变方向
bring about 引起；使调头
end-to-end 端对端；首尾相连
Operational Technology 经营技术；操作工艺

cyber-physical system（CPS）信息物理系统；网宇实体系统
RFID abbr. 无线射频识别（radio frequency identification devices）
lot size 批量

Notes

1. Industry 4.0's provenance lies in the powerhouse of German manufacturing.

译文：工业4.0起源于制造强国德国。

说明：这是一个简单句。powerhouse一般意为“精力旺盛的人；发电所”，这里取其引申意义“强国”。将英语翻译成中文时，措辞要尽量符合汉语习惯。

2. The fourth industrial revolution, and hence the 4.0, will come about via the Internet of Things and the Internet of services becoming integrated with the manufacturing environment.

译文：第四次工业革命，也因此称之为工业4.0，将通过物联网和服务互联网与制造环境的集成而实现。

说明：这是一个比较复杂的简单句。the fourth industrial revolution 为主语；and hence the 4.0 为插入语；will come about 为谓语动词的将来时态；via... environment 为介词短语作方式状语来修饰 come about。在介词短语结构中，the Internet of Things and the Internet of services 是 becoming integrated with the manufacturing environment 这个动名词短语的逻辑主语。即 via doing 的复合结构作状语。

3. The vision of Industry 4.0 is that in the future, industrial businesses will build global networks to connect their machinery, factories, and warehousing facilities as cyber-physical systems (CPS), which will connect and control each other intelligently by sharing information that triggers actions.

译文：工业4.0的远景是，将来的工业企业将建立全球网络来连接他们的机器、工厂和仓储设施从而形成一个信息物理系统（CPS），这个系统将通过分享引发操作的信息智能地连接各个部分并进行相互控制。

说明：这是一个复合句。the vision 为主语中心词，is 为系动词，其后的 that 引导表语从句。表语从句本身也是一个复合句，industrial businesses 为主语；will build global networks 是动宾结构作谓语、宾语；to connect... as cyber-physical systems 是不定式结构作目的状语；而 which 引导的非限制性定语从句对其先行词 cyber-physical systems 的功能和作用起补充说明作用。

4. These are what we call the horizontal value chain, and the vision is that Industry 4.0 will deeply integrate with each stage in the horizontal value chain to provide tremendous improvements in the industrial process.

译文：上述便是我们所说的横向价值链，对横向价值链的展望是，工业4.0将与横向价值链的每个阶段进行深度集成从而使工业生产过程发生巨大的变化。

说明：这是一个并列复合句。第一个分句是主系表结构的复合句，其表语是由 what 引导的表语从句；第二个分句也是主系表结构的复合句，vision 作主句主语，is 是系动词，that 引导表语从句。stage 意为阶段，指代前面提到的形成 value chain 的 engineering, material usage, supply chains, and product lifecycle management 等。to provide... 是不定式短语作目的状语。

5. The miniaturization of RFID tags enables products to be intelligent and to know what they are, when they were manufactured, and crucially, what their current state is and the steps required to reach their desired state.

译文：微型的无线射频识别装置标签使产品变得智能，以便能够知道它们是什么、什么时候被生产的，更为关键的是，能够知道它们目前的状态和达到期望状态还需要的步骤。

说明：这是一个复合句。the miniaturization of RFID tags 为主句的主语，enables 即主句谓语动词，products 在主句中作宾语，to be intelligent and to know... 是两个并列的不定式短语作宾语 products 的补语。to know 后面依次由 what、when 和 what 引导三个宾语从句。

6. For example, control over the horizontal manufacturing process and smart products enables better decision-making and dynamic process control, as in the capability and flexibility to cater to last-minute design changes or to alter production to address a customer's preference in the products design.

译文：例如，对横向制造过程和智能产品的控制使得我们更好地进行决策制定和动态过程控制，在能力和灵活性方面来适应最新的设计变更或者改变生产来满足顾客在产品设计方面的偏好。

说明：这是一个复杂的简单句。control 为名词，作主语中心词，其后的 over... products 为介词短语作后置定语修饰 control；enables 是谓语动词，后面有两个并列的宾语成分 better decision-making and dynamic process control；as in the capability and flexibility 后面的 to cater to 和 to alter 可以理解为 enable 的复合谓语动词的组成部分；to address a customer's preference in the products design 则为 production 的宾语补足语。

Exercises

Ⅰ. Answer the following questions according to the passage.

1. Where the Industry 4.0 concept originated from?
2. By which technologies the first three industry revolutions come about?
3. Via what kinds of technologies the fourth industry will come about?
4. What is the horizontal value chain of industrial process?
5. What are the vertical business processes or vertical value chain?

Ⅱ. Match the following phrases in column A with column B.

Column A	Column B
1. powerhouse	a. 标签
2. miniaturization	b. 二者选一
3. tag	c. 信息物理系统
4. alternative	d. 缩微化
5. Internet of Things	e. 首尾相连
6. end-to-end	f. 发电所；强者
7. cyber-physical system	g. 物联网

Ⅲ. Translate the following sentences into Chinese.

1. The First Industrial Revolution mobilized the mechanization of production using water and steam power.

2. The Second Industrial Revolution dates from Henry Ford's introduction of the assembly line in 1913, which resulted in a huge increase in production.

3. The Third Industrial Revolution resulted from the introduction of the computer onto the factory floor in the 1970s, giving rise to the automated assembly line.

4. The vision of Industry 4.0 is for "cyber-physical production systems" in which sensor-laden "smart products" tell machines how they should be processed.

Practical English:

如何写英文求职信

1. 求职信的内容

1）写信的目的或动机。求职信中一定要说明写信的缘由和目的。

2）个人资料。写信人应说明自己的基本情况，包括年龄（或出生年月）、教育背景，尤其是和应征职位有关的训练或教育科目、工作经验或特殊技能。如无实践经验，略述在学类似经验亦可。

3）结尾。求职信的结尾可以表达希望未来的雇主给予面谈的机会，因此，信中要说明可以面谈的时间。

2. 求职信的客观表达

写求职信要有客观的表达，下列几点可供参考。

1）陈述事实，避免表达个人意见。

2）不要批评他人。

3）不要过分渲染自我。

3. 求职信的格式

1）求职信的第一段说明写信的缘由和目的，不宜采用分词从句，下面的示例表述较为恰当；

In your advertisement for an accountant, you indicated that you require the services of a competent person, with thorough training in the field of cost accounting. Please consider me an applicant for the position. Here are my reasons for believing I am qualified for this work.

2）求职信中提到待遇时，不必过分谦虚，可参考以下示例：

I feel it is presumptuous of me to state what my salary should be. My first consideration is to satisfy you completely. However, while I am serving my apprenticeship, I should consider ¥2,600 a month satisfactory compensation.

3）求职信结尾部分应尽量避免陈腐的表述，也不要表现得太过自信。

4. 求职信的语气

求职信要发挥最大的效果，语气须肯定、自信、有创意而不过分夸张。可以参考以下示例：

I am confident that my experience and references will show you that I can fulfill the particular requirements of your bookkeeping position. 相信我的经验和推荐人可以证明，我能够满足贵公司簿记员一职的特定要求。（自信简洁）

I feel quite certain that as a result of the course in filing which I completed at the Crosby School of Business, I can install and operate efficiently a filing system for your organization. 我相信在克洛斯比商业实习班修完档案管理这门课后，我能够替贵公司高效地设置并且操作一套档案分类系统。（比较谦虚）

要写好英文求职信，除上述要点外，还要注意格式规范，语法、标点和单词拼写要正确无误。下面给出一份英文求职信的示例：

April 6, 2010
P. O. Box 3
Henan Polytechnic Institute
Nanyang 473009, China

Dear Sir/Madame:

Your advertisement for a Network Maintenance Technician in the April 10 Student Daily interested me, because the position that you described sounds exactly like the kind of job I am seeking.

According to the advertisement, your position requires a good college student, Bachelor or above in Computer Science or equivalent field and proficient in Windows NT 4. 0 and Linux System. I feel that I am competent to meet the requirements. I will be graduating from Henan Polytechnic Institute this year with my B. S. degree and BEC Vantage. My studies have included courses in computer control and management and I designed a control simulation system developed with Microsoft Visual Studio and SQL Server.

During my education, I have grasped the principles of my major subject area and gained practical skills. Not only have I passed CET-Band 4 and BEC Vantage, but more importantly I can communicate fluently in English. I am proficient in English writing and speaking.

I would welcome an opportunity to attend you for an interview.

Enclosed is my resume and if there is any additional information you require, please contact me (Tel: ×××××××××).

Yours faithfully
×××

Reading: Made in China 2025 and *Industrie* 4.0 *Cooperative* Opportunities

Following the Chinese government's *issuance* of its **Made in China 2025** strategy, which outlines plans to upgrade the mainland's industries, its 13th Five-Year Plan, adopted in March 2016, sets out to deepen the implementation of this strategy in the next five-year period (2016－2020). [1] While this has *aroused* interests as regards the development direction of Chinese industry, some industry observers have drawn parallels with Germany's **Industrie 4.0** strategy, which was designed to enhance the efficiency of German industry.

It is worth noting that some have raised concerns that the two strategies may lead to intensified competition between Chinese and German industries. [2] Nevertheless, the two countries signed a *memorandum* of understanding to step up cooperation in the development of smart manufacturing technology in July 2015. And further to an October meeting between Premier Li Keqiang and the visiting German *Chancellor* Angela Merkel in Beijing last year, both sides have also agreed to expand strategic cooperation in tapping newly-emerging opportunities in line with both **Made in China 2025 and Industrie 4.0**.

Indeed, the relative industrial development of the two countries, coupled with different strategic development priorities, reveal more opportunities for cooperation than competition, including in the area of industrial robots. Moreover, the different positions held by Chinese and German industries in the global supply chain also hint at further opportunities for relevant players *stemming* from Sino-foreign cooperation projects.

Essentially, Germany's **Industrie 4.0** *advocates* the adoption of state-of-the-art information and communication technology in production methods as a means to further enhance industrial efficiency. This strategy is developed on the basis that Germany's strong machinery and plant manufacturing industry, its IT competences and expertise in embedded systems and automation engineering make it well placed to *consolidate* its position as a global leader in the manufacturing engineering industry.

Industrie 4.0 aims for intelligent production by connecting the current embedded IT system production technologies with smart processes in order to transform and upgrade industry value chains and business models. This will require Germany to enhance its research and development efforts in areas such as further integrating manufacturing systems. New industry and technical standards will be required to enable connections between the systems of different companies and devices, while data security systems will need to be upgraded to protect information and data contained in the system against misuse and unauthorized access. All of these developments are expected to enhance the efficiency and innovative capacity of German industry, while saving resources and costs.

As regards **Made in China 2025**, the focus is on innovation and quality, as well as guiding Chinese industries to move away from low value-added activities to medium- and high-end manufacturing operations, rather than pursuing expansion of production capacity. The strategy is also aimed at eliminating inefficient and outdated production capacity, and helping enterprises to conduct more own-design and own-brand business. These objectives are to be facilitated by actions including the establishment of manufacturing innovation centers, strengthening *intellectual* property rights protection, building up new industrial standards, and facilitating the development of priority and strategic sectors.

New Words and Phrases

Industrie n. 工业（德语）

cooperative [kəʊ'ɒpərətɪv] adj. 合作的；协助的；共同的

issuance [ˈɪʃjuːəns] n. 发布，发行
arouse [əˈraʊz] vt. 引起；唤醒；鼓励 vi. 激发；醒来
memorandum [meməˈrændəm] n. 备忘录；便笺
chancellor [ˈtʃɑːnsələ(r)] n.（德、奥等国的）总理；（英）大臣；校长
stem [stem] n. 干；茎；血统 vt. 阻止 vi. 阻止；起源于某事物；逆行
advocate [ˈædvəkeɪt] vt. 提倡，主张，拥护 n. 提倡者；支持者；律师
consolidate [kənˈsɒlɪdeɪt] vt. 巩固，使固定；联合 vi. 巩固，加强
intellectual [ˌɪntəˈlektʃʊəl] adj. 智力的，聪明的，理智的 n. 知识分子
in line with 符合；与…… 一致
hint at 暗示
state-of-the-art 最先进的；已经发展的；达到最高水准的

Notes

1. Following the Chinese government's issuance of its **Made in China 2025** strategy, which outlines plans to upgrade the mainland's industries, its 13th Five-Year Plan, adopted in March 2016, sets out to deepen the implementation of this strategy in the next five-year period (2016－2020).

译文：随着中国政府对提升中国大陆工业的纲要规划《中国制造2025》战略的发布，2016年3月通过的“十三五”规划对该战略在下一个五年（2016－2020年）的实施进行了部署。

说明：这是一个复合句。主句的主语是 its 13th Five-Year Plan；sets out 是主句谓语动词；后面的不定式短语 to deepen... 作宾语；following... 是现在分词短语作时间状语，其中 which 引导非限制性定语从句，对 strategy 起补充说明作用。adopted in March 2016 是过去分词短语作 13th Five-Year Plan 的后置定语。

2. It is worth noting that some have raised concerns that the two strategies may lead to intensified competition between Chinese and German industries.

译文：值得一提的是，已经有人开始担心这两个战略将使中国和德国之间的工业竞争加剧。

说明：这是一个含有主语从句的复合句。第一个 that 引导主语从句，some have raised concerns 作后置主语，it 为形式主语。第二个 that 引导定语从句，修饰 concerns。

Exercises

I. Decide whether the following statements are True (T) or False (F) according to the passage.

1. There is no cooperation between China and German governments in updating their industry because of concerns that the two strategies may lead to intensified competition between Chinese and German industries.

2. The relative industrial development of China and Germany, coupled with different strategic development priorities, render more opportunities for cooperation than competition.

3. Germany has strong competences and expertise in embedded systems and automation engineering.

4. For better communication and connecting between different systems, some new industrial and technical standards need to be developed.

5. The priorities of Germany's Industrie 4.0 and China's **Made in China 2025** strategies are different.

Ⅱ. Translate the following sentences into English.

1. 工业4.0还没有成为现实，它还只是一个理念，但它可能会带来深远的变化和影响。

2. 预测所有的环境变化是不可能的，而这些变化又是控制系统动态响应所必需的，所以可编程逻辑将变得非常重要。

3. 无论是革命性的变化还是不断演变，工业化生产将变得更加高效。

Unit 24 National Craftsmen

Text

China's government put forward the concept of "spirit of the *craftsman*" in the year of 2015 as a measure to upgrade its industry so as to face the fierce competition from other countries. [1] The widely accepted concept of craftsman spirit is that one should be devoted to make things with *perfection*, *precision*, *concentration*, *patience* and *persistence*.

From the year of 2015 and on, to *advocate* craftsman spirit, China's government and organizations conferred honorable titles to many dedicated workers and publicized their model deeds on many sites and media.

Below is a brief introduction to some of these national craftsmen.

Gao Fenglin—Making the 'heart' for rocket

Gao Fenglin is a senior technician from China *Aerospace* Science and Technology Corporation. He has spent the past 35 years in welding rocket engine nozzles. There are several thousand void tubes on the engine nozzles of the Long March 5 Series Launch Vehicle, each of which requires a very *meticulous* and skillful *welder* to work on, the diameter of some tubes is only about 0.16 mm. As a craftsman who has high *proficiency* in special welding technology, he has once been invited by Nobel *laureate* Chao Chung Ting to be the special expert of NASA to lead a sub-project of antimatter probe, the *sophisticated* structure of which requires high welding skills. [2]

More than 130 Long March series rockets have been propelled into the space by the engines that he welded, and this number accounts for more than half of the total number of Long March rockets.

Pang Huiyong—Doctor Craftsman of Steel industry

Pang Huiyong, who is 39 years old in the year of 2017, is a senior engineer from HBIS WUSTEEL Company. As a material science doctor, he started his career as a common technician in steel shop and worked there for 3 years so as to get familiar with the *process* of steel-making. Many times, he practiced the operation in person. Because of his dedicated performance, professional ability and down to root attitude toward work, he was appointed as a project team leader to develop some special steels to *substitute* the imported ones. His team developed more than 10 kinds of steels which have filled the blanks of some steel products in China.

When the development of ultra-low temperature steel entered the key phase, he had to apologize to his newly married wife and her relatives because he couldn't go to her home town to visit her parents as spring festival was coming. [3] When the clock of the New Year clicked, the trial production of new steel succeeded! This is the first time for China to produce this kind of steel which is used for LNG ship construction.

The success of the special steel for nuclear island is a memorial event in his history. After 24 times of experiments in nearly 4 months, they succeeded the production of this special steel, the price of which is only 25 percent of that of the imported one, which is 80, 000RMB/ton! By then, all kinds of steels of the third generation nuclear power station can be produced ourselves. Before that, there is only one French supplier who can provide this kind of steel in the world.

Zhang Dongwei—Generalist of LNG ship welding

Zhang Dongwei is a welder from Hudong-Zhonghua shipbuilding Co., Ltd of CSSC. He is famous for his high skills of LNG ship body welding.

There is one kind of special steel used in LNG ship body structure——the Invar steel which requires very critical welding conditions and high welding skills of the operator. The most widely used welding technologies are carbon-dioxide arc welding and argon arc welding.

Zhang was only 34 years old in 2015 but he had worked as a welding technician for almost 15 years. Zhang overcame many obstacles and spared no effort to learn every kind of welding technologies, and finally become a master in this field. Now, Zhang Dongwei has trained and instructed over 40 workers on many kinds of ship building and Invar steel welding technologies, of which 30 workers have become skilled workers in this field and got the certificates from Classification Society. [4]

Ning Yunzhan—*Paramount locksmith* of railway industry

Ning Yunzhan is a locksmith of CSR Qingdao Sifang Co., Ltd. He is famous for *lapping* or *grinding* skills in high speed railway industry. The first time Ning showed up *prominently* was in the year of trial production of an introduced high speed train. He developed a new method to manually grind the *bogie* positioning arm in trial production. Later he repeated tests by trial and error, developed "pneumatic grinder manual grinding method" which has been used in high volume production. There is a unique pattern on the part surface produced by him, and the precision of part is also very high. He produced zero defect bogie positioning arms in the past ten years with the method he has developed and with his serious attitude toward work.

Ning is not only a locksmith but also a generalist who is good at welding, maintaining, repairing and electrical technologies. He developed many maintaining & repairing methods and tools with which big value has been created. He is also the owner of two patents.

Guan Yan'an—Deep sea locksmith master

Guan was *nicknamed* the "number one deep sea locksmith" after the completion of Hong Kong-Zhuhai-Macao Bridge which is the biggest bridge in the world.

For historical reasons, China didn't have the key *know-how* of immerse tube installation in the past before the construction of the bridge. Chinese engineering team lead by Lin Ming developed the installation method themselves since they don't like to take the super high *quotation* from foreign company just only for technical instruction. Both the immerse tube jointing and the outfitting on ship requires very careful work, strict management methodology and high skills. Outfitting means to assemble several system equipments on the ship before sinking the immerse tube into the sea; it is the most complicated work section of *tunnel* construction. Since some methods were newly developed by Chinese themselves, there is no established working instruction to follow up, Guan *rehearsed* several times of the very method before finalizing it to a standard process method, and many times, he took the life risk to perform the test under very harsh conditions. The installation of the last immerse tube is the most complicated work in traditional way invented by companies of developed countries, and the work time is normally 8 ~ 10 months. Thanks to the method invented by engineering team lead by Lin and hard work of front line installation team lead by Guan, the work time was shortened from 8 ~ 10 months to within one day![5] Guan received high appraisal for his can-do spirit, team spirit and *curiousness* to new technologies.

Gu Qiuliang—Chief bench worker of Underwater Vehicle

Gu Qiuliang is a senior locksmith from China Shipbuilding Industry Corporation. He was conferred the title of "national craftsman" for his high skills of deep sea underwater vehicle assembling and can-do spirit, he has worked in the ship building industry for almost 40 years.

As a senior locksmith, Gu contributed a lot to the successful building of many equipments of national key projects. He showed high skills in the assembling process of deep sea underwater vehicle "Jiaolong manned *submersible*" and developed many methods and tools to secure the assembly to reach the most critical requirement. He can manually grind the surface on which the window glass will be mounted with a very high precision which *coplanarity* tolerance is only 0.02 millimeters.[6] This observation window can stands up to 700 standard atmospheric pressures. To improve his filing and grinding skills, he exercised many times with file and grinders; feeling the deviation of part dimension by hands time and time again; finally he improved his skills to a very high level at the cost that he can't register himself in the attendance machine because his finger print has disappeared!

New Words and Phrases

craftsman [ˈkrɑːf(t)smən] n. 工匠；手艺人；技工
perfection [pəˈfekʃ(ə)n] n. 完善；完美
precision [prɪˈsɪʒ(ə)n] n. 精度，精密度；精确度
concentration [kɒns(ə)nˈtreɪʃ(ə)n] n. 浓度；集中；浓缩；专心；集合
patience [ˈpeɪʃ(ə)ns] n. 耐性，耐心；忍耐，容忍
persistence [pəˈsɪst(ə)ns] n. 持续；固执；存留；坚持不懈；毅力
advocate [ˈædvəkeɪt; –ət] vt. 提倡，主张，拥护 n. 提倡者；支持者
aerospace [ˈeərəspeɪs] 航空宇宙；[航] 航空航天空间
meticulous [məˈtɪkjələs] adj. 一丝不苟的；小心翼翼的
welder [ˈweldə] n. 焊接工
proficiency [prəˈfɪʃ(ə)nsɪ] n. 精通，熟练
laureate [ˈlɒrɪət; ˈlɔː–] adj. 戴桂冠的；荣誉的 n. 得奖者 vt. 使戴桂冠
sophisticated [səˈfɪstɪkeɪtɪd] adj. 复杂的；精致的；久经世故的
process [prəˈses; (for n.) ˈprəʊses] vt. 处理；加工 n. 过程，进行；方法
substitute [ˈsʌbstɪtjuːt] vt. & vi. 代替，替换，代用 n. 代替者；替补
paramount [ˈpærəmaʊnt] adj. 最重要的；至高无上的 n. 最高统治者
locksmith [ˈlɒksmɪθ] n. 钳工；修锁工，锁匠
lapping [ˈlæpiŋ] n. 研磨；抛光；搭接

grind [graɪnd] vt. 磨碎；磨快 vi. 磨碎；折磨 n. 磨；苦工作
prominently [ˈprɔminəntli] adv. 显著地
bogie [ˈbəʊgɪ] n. [铁路] 转向架；小车；妖怪；可怕的人
nickname [ˈnɪkneɪm] n. 绰号；昵称 vt. 给……取绰号
know-how [ˈnəuhau] n. 诀窍；实际知识；专门技能
quotation [kwə(ʊ)ˈteɪʃ(ə)n] n. [贸易] 报价单；引用语；引证
tunnel [ˈtʌnl] n. 隧道；坑道
rehearse [rɪˈhɜːs] vt. 排练；预演 vi. 排练；演习
curiousness [ˈkjuəriəsnis] n. 好学；好奇；不寻常
submersible [səbˈmɜːsɪb(ə)l] adj. 能潜水的；能沉入水中的
coplanarity [kəupləˈnærəti] n. 平面度
LNG abbr. 液化天然气（Liquefied Natural Gas）
antimatter probe 反物质探测器
argon arc welding 氩弧焊
carbon-dioxide arc welding 二氧化碳气体保护焊
Classification Society 船级社
Invar steel 殷瓦钢
show up 露面；露出；揭露
trial and error 反复试验
immerse tube [建筑] 沉管
outfitting 码头舾装；舾装设备
can-do spirit 敢于担当的精神
team spirit 团队精神
attendance machine 考勤机
nuclear island 核岛

Notes

1. China's government put forward the concept of " spirit of the craftsman" in the year of 2015 as a measure to upgrade its industry so as to face the fierce competition from other countries.

译文：中国政府在2015年提出了“工匠精神”的概念，作为提升工业水平的措施之一，以应对来自其他国家的激烈竞争。

说明：这是一个主干为主动宾结构的简单句。the concept是宾语中心词，of " spirit of the craftsman" 是后置定语，其后的成分分别是时间状语、方式状语和目的状语，修饰put forward。

2. As a craftsman who has high proficiency in special welding technology, he has once been invited by Nobel laureate Chao Chung Ting to be the special expert of NASA to lead a sub-project of antimatter probe, the sophisticated structure of which requires high welding skills.

译文：作为一名精通特种焊接技术的工匠，他曾受诺贝尔奖获得者丁肇中邀请，作为美国宇航局的一位特邀专家，领导反物质探测器项目的一个子项目，该探测器的结构非常复杂，对焊接技师焊接技能的要求也非常高。

说明：这是一个复合句。He has once been invited by Nobel laureate Chao Chung Ting to be... probe 是主句；who has high proficiency in special welding technology 是定语从句，修饰 craftsman；the sophisticated structure of which requires high welding skills 是定语从句，修饰 probe；as a craftsman who... 为条件状语；to be the special expert of NASA to lead a sub-project of antimatter probe 为目的状语。proficiency 意为“熟练，精通，娴熟”。例如：

Do you think the TOEFL score is a good yardstick for English proficiency? 你认为托福成绩是一种衡量英语流利程度的好标准吗?

3. When the development of ultra-low temperature steel entered the key phase, he had to apologize to his newly married wife and her relatives because he couldn't go to her home town to visit her parents as spring festival was coming.

译文：当超低温钢的开发进入关键阶段的时候，结婚后的第一个春节即将到来，他不得不向新婚妻子及其亲友道歉，因为他不能去她的家乡看望父母了。

说明：这是一个复合句。第一个 he 为主句主语，had to apologize to 是主句的复合谓语，其后是宾语。句首的 when 引导时间状语从句；后面的 because 引出原因状语从句。

4. Now, Zhang Dongwei has trained and instructed over 40 workers on many kinds of ship building and Invar steel welding technologies, of which 30 works have become skilled workers in this field and got the certificates from Classification Society.

译文：目前，在多种船体建造和殷瓦钢焊接技术方面，张东伟已经训练和指导了 40 多名工人，其中 30 多人已成为这一领域的熟手并获得船级社颁发的资格证书。

说明：这是一个复合句。主句在前，属于主动宾结构的完成式；of which 引导的非限制性定语从句是对 40 workers 的补充说明。因 40 workers 被看作一个整体，这里不用 whom 作代词引导词。

5. Thanks to the method invented by engineering team lead by Lin and hard work of front line installation team lead by Guan, the work time was shortened from 8 ~ 10 months to within one day!

译文：得益于林明领导的工程团队所发明的新方法和管延安领导的一线安装团队的努力，工作时间从 8 ~ 10 个月缩短到了 1 天!

说明：这是一个简单句。thanks to 引出原因状语，其中 the method 与 hard work 是并列宾语；invented by... 与 of front line installation team... 是后置定语，分别修饰 method 和 hard work。句子主干是 the work time was shortened from 8 ~ 10 months to within one day。

6. He can manually grind the surface on which the window glass will be mounted with a very high precision which coplanarity tolerance is only 0.02 millimeters.

译文：经他手工研磨的平面精度很高，平面度误差仅为 0.02 毫米，该平面可用于视窗玻璃的安装。

说明：这是一个复合句。he 是句子的主语，can grind 是谓语，the surface 是宾语。on which 引导定语从句修饰 the surface，后面的 which 引导定语从句修饰 precision。

Exercises

Ⅰ. Answer the following questions according to the passage.

1. In which year China's government put forward the concept of "spirit of craftsman"?
2. What are the widely accepted characteristics of craftsman spirit?
3. Why has China's government put forward the concept of "spirit of craftsman"?
4. Can you mention the trade categories in which these national craftsmen engaged?
5. According to the text, which welding technologies are normally used in ship building industry?

Ⅱ. Match the following phrases in column A with column B.

ColumnA	Column B
1. spirit of craftsman	a. 坚持
2. persistence	b. 精度
3. precision	c. 钳工
4. carbon-dioxide arc welding	d. 磨削
5. locksmith	e. 敢于担当的精神
6. grind	f. 公差
7. can-do spirit	g. 工匠精神
8. tolerance	h. 二氧化碳保护焊

Ⅲ. Complete the following sentences with the proper form of the words in brackets.

1. Spirit of craftsman is a human attribute relating to knowledge and skill at ________ (perform) a task.
2. The reliability of electronic devices is greatly ________ (affect) by the quality of the workmanship.
3. Workers who ________ (accustom to) practicing high standards of workmanship were first recruited to work on production lines in factories.
4. Two thirds of Swiss students start vocational education after ________ (complete) middle school.
5. Swiss middle school students start to ________ (study) vocational careers and have experience of being an apprentice in the second year.

Ⅳ. Translate the following sentences into Chinese.

1. Many successful entrepreneurs have been found to have the spirit of the craftsman which includes determination, patience and the desire to achieve perfection.
2. The apprenticeship system has a big following in Switzerland.
3. Germany has a name for its spirit of the craftsman and superior products.
4. Welder and locksmith are common positions in modern factories.

Practical English:

英文面试技巧

英文面试是进入跨国公司或合资企业的必经之路，因此在英文面试中很好地阐述专业技能和充分表现自己的综合能力尤其重要，下面介绍一些面试技巧，以便帮助大家在英文面试中脱颖而出，获得理想的工作。

1. 面试前的充分准备

（1）Know the company（了解公司） Your knowledge of the prospective employer will contribute to the positive image you want to create. Research the company before the interview.

（2）Know the job（了解所应聘的职位） Learn everything you can get about the job you're interviewing for and make clear how your previous experience and training qualify you for this position.

（3）Know yourself（了解你自己） Review your resume before the interview to have it fresh in your mind, because it will be fresh in the mind of the person who interviews you. Better yet, have it in front of you on the table.

（4）Prepare questions of your own（准备好自己要问的问题） Employers are as interested in your questions as they are in your answers. And they'll react favorably if you ask intelligent questions about the position, the company and the industry.

（5）Be ready for any eventuality（准备好随时可能发生的情况）

2. 面试过程中的要素

Keep the following in mind:

（1）Make a Good First Impression（良好的第一印象） The outcome of the interview will depend largely on the impression you make during the first five minutes.

（2）Be punctual（准时） Do whatever it takes to arrive a few minutes early. If necessary, drive to the company the night before and time yourself. Allow extra time for traffic, parking and slow elevators.

（3）Dress right（着装合体） Your clothing should be appropriate for the position you're seeking. Attire must fit well within the office and be immaculate.

（4）Shake well（握手） Show your confidence with a firm handshake. And make eye contact when you shake.

（5）Speak correct body language（利用好肢体语言） Send the right message by standing straight, moving confidently, and sitting slightly forward in your chair.

（6）Be honest（诚实） Tell the interviewer about your work skills, strengths and experience, including any volunteer work you have done.

（7）Be enthusiastic（热情） Show your clear interest in the job you are seeking and in the business. Smile and make frequent eye contact. Listen attentively and take notes.

（8）Find common ground（寻求共同点） Pictures, books, plants, etc., in the office can be conversation starters.

(9) Listening skills（倾听的技巧） Listen carefully and ask questions to probe deeper into what the interviewer is telling you.

3. 面试结束时的礼貌道别

面试时，面试官会问及很多方面的内容，包括家庭情况，如家人对你的影响等，这些都应事先有所准备。当面试结束时，礼貌地道别是很必要的，这会给面试官留下深刻的印象。

Example

An Interview for a Mechanical Engineer

（机械工程师面试范例 I：Interviewer；A：Applicant）

I：I was told by Mr. Yu that you arrived in Guangzhou this morning. You had a nice trip, didn't you?

A：Yes, I did.

I：You graduated from Henan Poly technic Institute?

A：Yes.

I：What department did you study in?

A：Department of Mechanical Engineering.

I：And what did you specialize in?

A：I specialized in engine designs.

I：Where are you working now?

A：I'm working at China FAW Group Corporation, the biggest of its kind in China.

I：As far as I know, your factory is well-known for its trucks of Jiefang brand. But do you manufacture any other vehicles?

A：Yes. We also produce buses and minibuses. And we have made advanced limousines of Red Flag brand.

I：What is your responsibility there?

A：I'm responsible for designing engines.

I：Can you tell me something about your achievements?

A：Of course. Years ago, I designed a kind of new diesel engine for trucks of Jiefang brand. Such a new diesel engine has decreased diesel fuel consumption by 15%. The year before last I designed a more powerful gasoline engine for limousines. Such a new gasoline engine has greatly increased the speed of the limousines made by our factory.

I：Wonderful. You've really made some contributions to China's automobile manufacturing. You must have received some honors or rewards, right?

A：Yes. I received a reward for outstanding contributions from the government of PRC in 2001. And I received a first-class reward for advanced designs from the People's Government of Jilin Province the year before last. In addition, I have been chosen as a "model worker" in the factory over the past five years in succession.

I：OK. Now please tell me what you know about this company.

A: Your company is a Sino-Japanese joint venture which produces Honda cars. This kind of car runs fast with less gasoline consumption. So your products sell well both at home and abroad. Demand always exceeds supply.

I: Do you have any plan for the new job if we hire you as a senior mechanical design engineer with this company?

A: If I get the privilege of working at this company, I want to further improve the engine so as to raise Honda's competitiveness with such cars as Lincoln, Benz, and Crown.

I: Like knows like. That is just what we want to do. It is for this reason that we are anxious to seek an enterprising and creative mechanical design engineer. I have interviewed quite a number of job seekers. But none of them are satisfactory to me. Today I've found a suitable man like you at last.

A: Thank you, sir.

I: How long can we expect you to work here?

A: If I feel I'm making progress in the work, I'll stay until the age limit.

I: Very good. We'll provide you with a yearly salary of 160, 000 yuan. And we'll give you a 30-day paid vacation a year so that you can have a happy reunion with your family in Changchun. Do you have any particular conditions you'd expect the company to take into consideration?

A: Nothing particular. But may I ask for an apartment?

I: That's out of question. We'll supply you with an apartment of two bedrooms and a living room. When can you start to work here?

A: In one month. I must go back to Changchun to hand over my work and to go through necessary procedures.

I: We'll look forward to your coming back. I wish you a nice trip, Mr. Yao.

A: Thank you, sir. See you next month.

I: Good-bye.

Reading: WorldSkills Competition

What Is WorldSkills Competition and How to Compete

The WorldSkills *Competition* occurs every two years and is the biggest *vocational* education and skills *excellence* event in the world that truly reflects global industry. The Competitors represent the best of their peers and are selected from skill competitions in WorldSkills Member countries and regions. A Competitor at the WorldSkills Competition must not be older than 22 in the year of the Competition. Competitors in Information Network Cabling, Mechatronics, Manufacturing Team Challenge, and Aircraft *Maintenance*, must not be older than 25 years in the year of the Competition. They *demonstrate* technical abilities both *individually* and *collectively* to execute specific tasks for which they study and/or perform in their workplace.[1]

One of the main *legacies* of the WorldSkills Competitions is to give visibility and importance to professional education, as one of the true tools of *socioeconomic transformation*.[2] The competition also provides leaders in industry, government and education with the opportunity to exchange information and best practices regarding industry and professional education. New ideas and processes inspire school-aged youth to *dedicate* themselves to technical and technological careers and towards a better future.

WorldSkills Competitions and competitors embrace many common values: commitment, persistence, and a competitive spirit coupled with a *generous* sharing of success - whether it is medal winning or not. *Stakeholders* in the skills movement are varied across sectors; and they include youth, educators, industry professionals, and business and economic leaders. Tap into the many resources available to support building excellence and building a future.

Skills Competition Resources

To continue to *hone* your skills, there are prize-based contests and scholarship challenges. Learn about exciting student challenges in global health and eco-driven vehicles. Design a winning website; compete with student *trendsetters* in fashion; get a taste of the future competing with international *pastry* students. These varied resources will broaden your awareness of *compelling*, exciting and meaningful challenges that will help you continue to build the skill and will to change the world. [3]

What Is A Skill?

For 63 years and counting, the WorldSkills Competition has acted as a mirror to the evolution of trades across the world. Throughout these years many competition skills have disappeared, many have merged, and many new ones have *materialized*.

What Is A Skill? sets out to examine if these new skills are really representative of new trades, or just a passing phase? [4] How do you actually define a trade? This project will take a closer look at how individual skills can be defined, and how the trades these skills represent evolve over time. It will analyse the changing needs of multiple industry workplaces worldwide, and the training that must be provided to fill the needs of these industries in the future.

The Happening Event and Coming Events

The 44th WorldSkills Competition has closed in *Abu Dhabi* of United Arab *Emirates* in October, 2017. China rank the first for golden metals points 15 and total medal points 109, the following next is Korea for total medal points 88 and Switzerland for total medal points 81.

The WorldSkills flag has started its long journey to *Kazan*, where the 45th Worldskills competition will be held in the year of 2019.

China was selected On 20, October 2017 to host the 46th WorldSkills Competition in Shanghai in 2021—an event it hopes to use for building a stronger team of skilled workers nationwide and to enhance international exchanges in vocational skills. The Chinese leadership attached great importance to its bid to host the competition, as the event will greatly promote the development of highly skilled talent in China; Shanghai will use the opportunity to draft *preferential* policies to encourage the spirit of *craftsmanship* and improve vocational education to cultivate skilled talent.

New Words and Phrases

competition [ˌkɒmpəˈtɪʃn] n. 竞争；比赛，竞赛

socioeconomic [səusɪəuˌiːkə'nɔmɪk] adj. 社会经济学的
transformation [ˌtrænsfə'meɪʃn] n. 变化；<核>转换；<语>转换
pastry ['peɪstri] n. 糕点；油酥糕点；油酥面皮
trendsetters ['trendsetə(r)] n. （在服装式样等方面）创新风的人
materialize [mə'tɪəriəlaɪz] vi. 具体化，实质化，实现 vt. 使成真，实现
vocational [və(ʊ)'keɪʃ(ə)n(ə)l] adj. 职业的，行业的
excellence ['eks(ə)l(ə)ns] n. 优秀；美德；长处
legacy ['legəsi] n. 遗产；遗赠；传统
demonstrate ['demənstreɪt] vt. 证明；展示；论证 vi. 示威
individually [ɪndɪ'vɪdjʊ(ə)lɪ] adv. 个别地，单独地
collectively [kə'lektivli] adv. 共同地，全体地，集体地
dedicate ['dedɪkeɪt] vt. 奉献，献身
maintenance ['meɪntənəns] n. 维持，保持；保养，保管；维护；维修
stakeholder ['steɪkhəuldə (r)] n. 股东；利益相关者
hone [həun] vt. 用磨刀石磨；磨孔放大 n. 磨刀石
generous ['dʒenərəs] adj. 慷慨的，大方的；肥沃的；浓厚的
compelling [kəm'pelɪŋ] adj. 引人入胜的；非常强烈的；不可抗拒的
emirate ['emɪərət] n. 酋长国
craftsmanship ['krɑːftsmənʃɪp] n. 技术，技艺；工力
preferential [ˌprefə'renʃl] adj. 优先的；优先选择的；特惠的
Abu Dhabi ['æbuː'ðæbiː] [地名] 阿布扎比（阿拉伯联合酋长国首都）
Kazan [kə'zaːn] n. 喀山（伏尔加河中游城市）
WorldSkills Competition 世界技能竞赛
eco - drive 生态驱动

Notes

1. They demonstrate technical abilities both individually and collectively to execute specific tasks for which they study and/or perform in their workplace.

译文：他们在完成他们所研究的以及在工作场地所执行的特定任务时，表现出了个体的和集体的技术能力。

说明：这是一个复合句。主句在前，they demonstrate technical abilities 是主谓宾结构；both individually and collectively to execute specific tasks 是方式状语从句；for which 引导定语从句修饰 tasks。

2. One of the main legacies of the WorldSkills Competitions is to give visibility and importance to professional education, as one of the true tools of socioeconomic transformation.

译文：使专业教育作为社会经济转型的真正工具而变得引人注目，是世界技能大赛的一项主要贡献。

说明：这是一个比较复杂的简单句。主语是 one of the main legacies；is 是系动词，其后接不定式短语作表语：to give visibility and importance to professional education；as one of the true tools of socioeconomic transformation 是句子的方式状语。

3. These varied resources will broaden your awareness of compelling, exciting and meaningful challenges that will help you continue to build the skill and will to change the world.

译文：这些多样的资源将拓宽你对这些引人入胜的、令人兴奋的和意义非凡的挑战的认识，这将有助于你继续培养技能，激发改变世界的愿望。

说明：这是一个复合句。主句是 These varied resources will broaden your awareness of compelling, exciting and meaningful challenges。these varied resources 为主语，will broaden 为谓语，your awareness 为宾语；其后的 of compelling, exciting and meaningful challenges 是后置定语，修饰 awareness，that 引导定语从句也修饰 awareness。

4. ***What Is A Skill***? sets out to examine if these new skills are really representative of new trades, or just a passing phase?

译文：*What Is A Skill*（负责）考察这些新技能是否真正代表新的行业，或者只代表一个过渡阶段?

说明：这是一个复合句。***What Is A Skill***? 在句中指的是负责定义“什么叫作技能”的一个国际组织，故整体作为一个名词成分。sets out to examine 是复合谓语，if 引导的从句在复合句中充当宾语。

Exercises

Ⅰ. Decide whether the following statements are True (T) or False (F) according to the passage.

1. As sports Olympics, the WorldSkills Competition which is regarded as the skill Olympics occurs every four years.

2. A Competitor at the WorldSkills Competition must not be older than 25 in the year of the competition.

3. The skill showed in WorldSkills Competition will exist forever whatever how it evolves.

4. China won the biggest medal points in the 44th WorldSkills Competition and was selected to host the 46th competition in Shanghai.

Ⅱ. Translate the following sentences into English.

1. 中国的消费者对“德国制造”产品的印象往往很好。

2. 中国产品进入国外市场的步伐近年来稳步加快。

3. 一个国家的品牌至少包括三个要素：经济形象、政治形象和技术形象。

Chapter V　The Communication Skills Training for Careers

职场交际技能训练

Dialogue 1　Pick Up New Customers in Airport

Secretary: Excuse me, are you Mr. Anderson from the P&G Company?

Customer: Yeah, call me Anderson please. Thank you for meeting me at the airport.

Secretary: It's my pleasure. Welcome to China. My name is Wang Lin. This is my business card.

Customer: Thank you, Miss Wang. This is mine.

Secretary: Would you please let me get your luggage and go to the hotel now? By the way, you may call me by my English name, Linda.

Customer: Thank you, Linda. It's very kind of you.

Secretary: OK, Anderson. You're welcome.

参考译文　机场迎接客户

秘书：打扰一下，请问您是来自宝洁公司的安德森先生吗？

客户：是的，请直接叫我安德森吧。非常感谢你来机场接我。

秘书：我的荣幸，欢迎您来到中国。我叫王琳，这是我的名片。

客户：谢谢你，王小姐。这是我的名片。

秘书：请把您的行李给我，我们现在去宾馆吧。顺便说一下，您可以叫我的英文名，琳达。

客户：谢谢你，琳达。你真周到。

秘书：不客气，安德森先生。

Dialogue 2　Email Writing

Mary: Sally, could you please teach me how to write email with Outlook software?

Sally: Ok, it's my pleasure.

Mary: The first question is what is the meaning of CC?

Sally: CC is the abbreviation of "carbon copy." When you would like to share the information of the mail to a person who is not the receiver, you can write his or her email addresses in the blank of CC.

Mary: Then what is the meaning of BCC?

Sally: BCC is the abbreviation of "blind carbon copy." When you would like to share the information to people who are not receiver, and don't like the receiver and people who has been "CC" ed to know that you have shared the information to others, you can write the email address in the blank of BCC.

Mary: Ok, another question is that some people forward emails of others to me with only three letters "FYI". What does it mean?

Sally: It is the abbreviation of "for your information", that means it is for your reference or check.

Mary: The last question, how can I automatically attach my personal informations such as name, title, mobile phone number, company address, and so on when I write email every time?

Sally: Oh... m... m, you need to take several minutes to set up the signature. First click the command "signature" in the menu, then wirte down you personal informations as you just mentioned and save the informations after you input them. It is better to have our company's logo there. At the end of the signature, don't forget to write down the complimentary close to others, it can be "best regards" "yours sincerely" "sincerely yours", and etc.

Mary: You are so nice, thank you very much!

Sally: You're welcome.

参考译文　写电子邮件

玛丽：莎莉，你能教我怎样用 Outlook 来写邮件吗？

莎莉：没问题。

玛丽：第一个问题是，CC 是什么意思？

莎莉：CC 是"carbon copy"的缩写。当你想把邮件里边的信息分享给收件人以外的其他人时，你可以把他们的邮箱地址写在 CC 后边的空格里。

玛丽：那 BCC 是什么意思呢？

莎莉：BCC 是"blind carbon copy"的缩写。当你想把邮件信息分享给收件人以外的其他人，且不希望收件人和被抄送的人知道你已经分享了信息给他人时，你可以将他们的邮箱地址写到 BCC 后面的空格里。

玛丽：好的。还有一个问题是有的人把别人的邮件推送给我时，只写三个字母"FYI"，这是什么意思呀？

莎莉：那是"for your information"的缩写，就是供你参考或查阅的意思。

玛丽：最后一个问题，当我每次写邮件时，怎么才能自动添加我的个人信息呢？比如姓名、职位、手机号码、公司地址等信息。

莎莉：哦…… 那你要花费几分钟的时间来设置签名。首先点击菜单中的"签名"命令，然后输入你提到的个人信息并保存。在签名的最后，别忘了写下敬语，通常可以是"best regards"，"yours sincerely""sincerely yours"等。

玛丽：你真是太好了，太感谢了。

莎莉：不客气。

Dialogue 3　AM Machines

Tony: Louis, how is it recent days?

Louis: Quite busy. Now, we are attending a business fair in Beijing and our company will introduce a new model of AM Machine.

Tony: AM Machine? What does it stand for?

Louis: It stands for the Additive Manufacturing.

Tony: I am still not very clear to the concept.

Louis: Do you know 3D printing?

Tony: Yeah, I understand the way a little.

Louis: There are many ways of additive manufacturing. 3D printing is just one way of them, some people even use the words "3D printing" to replace additive manufacture when they talk some simple AM methods.

Tony: Got it.

Louis: Do you plan to order one set AM machine?

Tony: That's depending on the budget. My task this time is to follow up the development trend of new technology. If the machine is not very expensive, I will buy one so that we can produce some prototype within short time.

Louis: Please tell me when you need help at any time.

Tony: Sure. Thank you very much, bye!

Louis: Bye!

参考译文　增材制造设备

托尼：路易斯，最近怎样？

路易斯：非常忙。最近我们正在北京参加一个商务交易会，我们公司准备引进一台新的AM设备。

托尼：AM设备？指的是什么？

路易斯：AM就是增材制造的意思。

托尼：我还是有点不明白这个概念。

路易斯：你知道3D打印吗？

托尼：知道一些。

路易斯：现在有很多增材制造的方法。3D打印是其中之一，有的人谈到比较简单的增材方法时甚至用"3D打印"这个词替代增材制造的概念。

托尼：我明白了。

路易斯：你打算买一套增材制造设备吗？

托尼：这取决于预算。我这次的任务是跟踪新技术的发展趋势。如果这个设备不是很贵，我就买一套，这样我们就能在较短的时间内生产样品了。

路易斯：你有需要时随时联系我。

托尼：一定。非常感谢！再见。

路易斯：再见。

Dialogue 4　New Job Orientation

Bobby: Hello, Robert.

Robert: Hello, Bobby. What are you going to do?

Bobby: I am just going around the building so as to get familiar with the layout of our company as soon as possible. You know, I am a new comer here. Do you have time to show me that or introduce to me here now?

Robert: Ok, let us talk here. Now, we are in the lobby, in front of us is the receptionist. Behind the receptionist table is an entrance to go upstairs, R&D department and quality department lie in the second floor. In the third floor, there are financial department, purchase & logistic department, sales department. While the HR department and GM office locate in the fourth floor.

Bobby: OK. Then where is the workshop?

Robert: Go into the door beside the receptionist, there is the workshop. There are around 300 people every shift.

Bobby: Ok, there are so many people every day. Then where is the canteen?

Robert: The canteen locates at the north-east corner of the workshop, and there is also an entertainment room where you can play billiards or table tennis after off duty.

Bobby: That's great! Thank you very much, Robert!

Robert: You're welcome.

参考译文 熟悉新的工作环境

鲍比：你好，罗伯特！

罗伯特：你好，鲍比！你要去做什么？

鲍比：我想各处转一转以尽快熟悉公司的布局，你知道的，我是新员工。你有时间带我熟悉一下或者在这里介绍一下公司的布局吗？

罗伯特：好的，那我们就在这说吧。我们现在在大厅里，前面是接待处。从接待处桌子后面的入口可以上楼，研发部和质量部在二楼，财务部、采购物流部、销售部在三楼；人事部和总经理办公室在四楼。

鲍比：好的。那么生产车间在哪里呀？

罗伯特：从接待处旁边的门进去就是车间了。那里每班有300人左右。

鲍比：哦。每天有这么多人呀。那食堂在哪里呀？

罗伯特：食堂在车间的东北角。那里还有个娱乐室，你下班后可以玩台球或者打乒乓球。

鲍比：那太好了，非常感谢！罗伯特。

罗伯特：别客气。

Dialogue 5 New Buzzwords Online

Mary: Hi, Shirley, do you have time to talk with me on some new buzzwords? I meet these words online frequently or hear some people talk about them, but I don't know the meanings.

Shirley: Ok, please go ahead.

Mary: The first one is AI.

Shirley: It stands for "artificial intelligence." You know, some people like the convenience of a car but they don't like to drive themselves, not everybody can afford the cost of hiring a driver, so there comes the automatic drive, this is one kind of AI technology. Another example is the electronic translator, since to master a foreign language is not easy, so some companies began to develop translating software.

Mary: OK. I often heard B2B, P2P, B2C, and C2C... I am totally confused. What do they stand for?

Shirley: Actually, they are some abbreviations. "2" means "to", since the number "2" has the same English pronunciation with pronunciation "to", it was used to replace "to" so as to have a simple expression. "B" is the abbreviation of "business", "P" is the abbreviation of "Peer", "C" is the abbreviation of "customer". Now I think you will not confuse any longer.

Mary: Sure. Then what is UMV?

Shirley: UMV is the abbreviation of "unmanned vehicle." It is referred to vehicles which have automatic driving function.

Mary: Really? Thank you very much for your kind help and patience.

Shirley: You're welcome.

参考译文　网络热词

玛丽：雪莉，你有时间和我聊聊一些新的热词吗？我经常在网上看到它们或者听人们谈论它们，但我不知道它们的意思。

雪莉：可以，你说吧。

玛丽：第一个热词是AI。

雪莉：它是“人工智能”的意思。你知道的，有的人喜欢汽车带来的便利却不喜欢开车，也并不是每个人都雇得起司机，所以就出现了自动驾驶技术，这就是一种人工智能技术。另一个例子是电子翻译，掌握一门外语是不容易的，所以有些公司就开始开发翻译软件。

玛丽：好。我经常听到B2B、P2P、B2C和C2C等热词，我完全不清楚它们的意思。

雪莉：实际上，它们是一些简写词。“2”代表“to”，“2”的英文发音与“to”一样，它就用来代替“to”以使表达变得简洁。“B”是“business（商业）”的简写，“P”是“peer（对等者）”的简写，“C”则是“customer（顾客）”的简写。现在你应该不再困惑了。

玛丽：明白了。那什么是UMV？

雪莉：UMV指的是无人驾驶汽车，它指具有自动驾驶功能的汽车。

玛丽：真的吗？那非常感谢你的耐心帮助。

雪莉：不客气。

Dialogue 6　Finding a Job Online

Emily: Hi, Bruce. I have stayed at home for 3 months, can you share your experience to me on getting a job or getting some recruitment informations?

Bruce: You can ask help from you relatives and friends first.

Emily: M... m, already, but that doesn't work.

Bruce: I am a little bit sad hearing that. Then you can register your information on some recruitment websites where you can find a lot of recruitment information.

Emily: So, could you please recommend some recruitment websites to me?

Bruce: You can register your information on zhaopin. com or 51job. com first, these two websites are for both freshman and experienced people. A lot of companies release their recruitment information there frequently. You can send you resume to them when you find the suitable position.

Emily: Will they reply me whatever they think I am the best candidate or not?

Bruce: Not surely. The normal practice of China job market is that the HR will not promise you to get a reply when they decide to refuse you application since there are so many applicants everyday. But they will inform you when they will arrange an interview with you.

Emily: Are there any other frequently visited recruitment websites?

Bruce: To my understanding, linkedin. com and liepin. com are two websites mainly for experienced people. Normally, some headhunters visit these websites everyday. They will contact the people if they found the people is a suitable candidate for their vacancies.

Emily: Mm..., how do you think if I register my information on linkedin. com or liepin. com?

Bruce: Frank to say, it seems not very necessary since you don't have a lot experience at the moment.

Emily: Ok, thanks for your kind help.

Bruce: You're welcome.

参考译文 网上应聘

艾米丽：布鲁斯，我已经无事在家三个月了，你能分享一下你找工作或者获取招聘信息的经验吗？

布鲁斯：你可以先找亲戚或者朋友问问。

艾米丽：哎，已经试过了，不管用的。

布鲁斯：听到这个消息，我有些难过。那你可以在招聘网站上登记一下你的信息，在那里可以找到大量的招聘信息。

艾米丽：那你可以推荐一些招聘网站给我吗？

布鲁斯：那你先在智联网或者前程无忧网站上注册。这两个网站对新人和有工作经验者都适用。很多公司经常在这两个网站发布招聘信息，看到合适的职位时，你可以发简历给他们。

艾米丽：不管他们认为我是否合适，是不是都会答复我？

布鲁斯：不一定。中国就业市场中的惯例是：人事拒绝你时不承诺给你通知，因为每天的申请人太多了。但当他们决定面试你时肯定会通知你的。

艾米丽：还有其他的常用招聘网站吗？

布鲁斯：据我所知，领英网和猎聘网是两个主要针对有经验人士的网站。一般情况下，很多猎头每天都会浏览这两个网站。当他们发现有人适合空缺的岗位时，他们就会联系这些候选人。

艾米丽：那……你觉得我在这两个网站上注册如何？

布鲁斯：坦白地说，意义不大，因为你目前的工作经验不多。

艾米丽：谢谢你的帮忙。

布鲁斯：别客气。

Dialogue 7 Living in China

Frank: Hello! Peter. How is it recent days after you come to China?

Peter: Fantastic! The people's life here has beyond my imagination.

Frank: Really? Then what's impressed you the most?

Peter: The so called new four great inventions of China!

Frank: It is my first time to hear that, and then what are they?

Peter: They refer to online payment, share economy, high-speed train, online shopping.

Frank: mm..., please explain them more clearly.

Peter: OK. For online payment, the most popular ways are Alipay and Wechat, you can pay nearly everywhere without taking money with you, and a mobile phone is enough! For online shopping, nearly every kind of goods can be purchased online without going to physical shopping mall. While for share economy, with the help of app, you can rent a bike even a car, and give it back at everywhere, it is so convenient!

Frank: Then how about the high-speed train?

Peter: It is so fantastic! The travel hour from Beijing to Shanghai by train is within five hours, and it is also as punctual as a subway. I went to Shanghai last week, you can even transfer from high-speed train to airplane in the same building!

Frank: I went to Shanghai by train last month. The traffic hub is fantastic. Do you know there are two civil airports in Shanghai?

Peter: Really?

Frank: Sure. There will be two civil airports in Beijing in the coming days, too. The biggest airport in the world is under construction in Beijing now, and it will be put into use within two years.

Peter: China is developing very fast. I am very glad to have the opportunity to work here.

Frank: M...m, it is time for meeting, I have to go, and then see you later.

Peter: See you.

参考译文 生活在中国

弗兰克：你好，彼得！你来中国后过得怎样？

彼得：好极了！这里的生活超出了我的想象。

弗兰克：是吗？那你印象最深刻的是什么？

彼得：就是人们所说的中国“新四大发明”！

弗兰克：这是我第一次听说，它们都指什么？

彼得：它们指的是在线支付、共享经济、高速列车和网络购物。

弗兰克：那…… 你再详细说说。

彼得：好的。关于在线支付，最常用的是支付宝和微信，你只需带着手机，几乎在哪里都可以支付，根本不需要带钱的。至于网购，几乎所有商品都可以在网上买到而无需到实体店去了。关于共享经济，通过手机应用程序，你可以租辆自行车甚至汽车，然后在任意地点归还就可以，真是太方便了！

弗兰克：那高速列车呢？

彼得：它真是太棒了！从北京到上海还不到五个小时，而且像地铁一样准时。我上周去了上海，甚至可以在同一个建筑里从高铁转乘飞机！

弗兰克：我上个月乘高铁去的上海。上海的交通枢纽是很棒。你知道上海有两个民用机场吗？

彼得：真的吗？

兰克：是的。北京也要有两个民用机场了。而且北京正在建设世界上最大的机场，它将在两年内建成并投入使用。

彼得：中国发展得太快了。我很高兴能有在这里工作的机会。

弗兰克：哦，开会时间到了，我必须走了，再见。

彼得：再见。

Dialogue 8 Introduction of New Products

Salesman: Good afternoon, sir. Can I help you?

Customer: Yes, I'm looking for a new air-conditioner.

Salesman: Air-conditioner? It is in the Houseware Section. By the way, which brand do you prefer?

Customer: Well, I prefer New Aux— the famous brand in China.

Salesman: Certainly. You're making a good choice. The New Aux has made a lot of design improvements in its new products.

Customer: What comes with the new improvements?

Salesman: The New Aux has low sound level and a classical shape and it can be multi-functional in different seasons. Besides, it is cheaper than the previous version.

Customer: Great! One more thing, could you tell me which section the digital TV is in?

Salesman: OK. They are in the Section C next to the Aisle B.

Customer: Thank you! Could you please recommend one for me?

Salesman: Yes, I think the Hisense-New Blueprint is a good choice. You can have a discount of it at present.

Customer: Thank you very much.

Salesman: You're welcome, this way, please.

参考译文 新产品介绍

销售员：先生，下午好，有什么能帮您的？

顾客：我想买一台新空调。

销售员：空调？空调在家电区。顺便问一句，您想买哪个品牌的空调？

顾客：哦，我倾向新奥克斯——它在中国是著名品牌。

销售员：的确，您做了一个明智的选择。新奥克斯在新产品设计上做了很多的改进。

顾客：你能介绍一下它有哪些改进吗？

销售员：新奥克斯空调采用经典的外观设计，噪声小，其多功能适用于不同季节。除此之外，新产品的售价也比之前的产品低。

顾客：太好了！再问你一下，数字电视在哪个销售区呢？

销售员：好的，电视在C区，紧邻走廊B。

顾客：谢谢，你能给我推荐一款吗？

销售员：可以，我想海信—新蓝图就是一个不错的选择。现在买的话还有折扣。

顾客：非常感谢！

销售员：不客气，这边请。

Appendixes

Appendix A 参考译文与习题答案

第1章 机械与模具制造

第1单元 工 程 材 料

在产品的设计和加工过程中，对材料和加工工序进行充分了解是非常必要的。材料在物理性能、机加工性能、成形方法以及可能的使用寿命等方面有很大不同。设计师应该在选择最适合产品的经济性材料和加工工序时考虑到上述因素。

工程材料有两种基本类型：金属材料和非金属材料。非金属材料细分为有机材料和无机材料。因为非金属材料与纯金属、合金的种类一样繁多，所以在选择合适的材料时需要做大量的研究工作。

工业用材料很少是在自然界中以元素的形式存在的。自然金属化合物如氧化物、硫化物和碳酸盐在能够进行深加工前必须经过分离或提纯。分离后它们会获得一定的原子结构，这种结构在常温下很长时间内是稳定的。在金属加工中，铁是最重要的自然元素。虽然纯铁的工业用途很少，但当铁与其他元素相结合变成合金后，就成了最重要的工程金属材料。非铁金属材料包括铜、锡、锌、镍、镁、铝、铅等，在我们的经济中占有重要地位，而且每一种材料都有其特殊的性能和用途。

制造环节需要有能达到经济性和精确性等生产要求的工具和设备。要达到经济性要求，就要选择合适的设备和工序，这样可以得到理想的产品，还要有最佳的操作、高效的人工执行和设备支持。生产批量的大小会影响对上述条件的选择。通常对于确定的产量会有一种最为适合的机器。对于小批量生产，通用机器如车床、钻床和铣床是最好的选择，因为它们适应性强，生产费用低，维护简便，能灵活适应条件的改变。而对于大批量标准件的生产则要采用专用机器。对应于某一工序或操作的专用机器在一个半熟练操作工的操作下会工作得又快又好且成本低，比如用于加工活塞或者气缸端面的磨光机。

许多专用机器和工具与标准型机器不同，因为专用机器融合了操作人员的一些技能。一个简单的螺栓既可以用车床加工，也可以用自动化的螺栓生产机器制造出来。车工不仅必须明白如何加工螺栓，而且还需要有足够的技能来操作机器。在自动机床上，工具的操作顺序和运动是由凸轮和制动器来控制的，并且每一零件的生产过程都与前一个过程相同。机器或自动化生产中的这种“技能迁移”对操作者的技能熟练程度要求有所降低，但在监督管理和维护方面要求更高。通常情况下，使机器完全自动化并不经济，因为机器生产管理费用太高。

对给定的产品，选择最佳的机器和工艺需要具有关于生产方法的知识。需要考虑的因素有产品的数量、产品的质量和生产该产品所用设备的性能（优势和局限）等。大多数零件可以有数种加工方法，但通常只有一种方法是最经济的。

Key to Exercises of the Text

Ⅰ. 1. facts 2. Nonmetallic 3. pure, alloys 4. special-purpose, standard

Ⅱ. 1. Materials differ widely in physical properties, machinability characteristics, methods of forming, and service life.

2. The natural compounds of metals, such as oxides, sulfides, or carbonates, must undergo a separating or refining operation before they can be further processed.

3. On the automatic machine the sequence of operations and movements of tools are controlled by cams and stops, and each item produced is identical with the previous one.

阅读材料　钢铁材料和非铁材料

金属在生活中随处可见。金属被划为两大类：钢铁金属和非铁金属（表1-1和表1-2）。这两类金属应用的广泛程度基本相同。

表1-1　部分钢铁金属及其性能

名称	成分	性能	用途
低碳钢	$w_C=0.1\%\sim0.3\%$ $w_{Fe}=99.9\%\sim99.7\%$	韧性好，抗拉强度高，易表面硬化，且易锈	车间中常见的金属，用于一般金属制品和工程中
高碳钢	$w_C=0.6\%\sim1.4\%$ $w_{Fe}=99.4\%\sim98.6\%$	韧性好，可以进行硬化及回火处理	切削工具，比如钻头
不锈钢	铁、镍和铬	韧性好，抗生锈，抗污	餐具，医疗器械
铸铁	$w_C=2\%\sim6\%$ $w_{Fe}=98\%\sim94\%$	强度好但很脆，具有很高的抗压强度	铸件、井盖、发动机
锻铁	几乎含100%铁	纤维状的，韧性好，具有良好的延展性，抗生锈能力强	装饰门和栏杆，现在已经很少使用了

表1-2　部分非铁金属及其性能

名称	颜色	成分	性能	用途
铜	红棕色	不是合金	延展性好，可以敲打成形，导电及导热性好	电线，管材，水壶，碗，管状物
黄铜	黄色	含铜-锌混合物，常见的锌元素比例为65%～35%	坚硬，铸造及机械加工性能好，表面暗，导电	电子配件，装饰用品
铝	浅灰色	$w_{Al}=95\%$ $w_{Cu}=4\%$ $w_{Mn}=1\%$	延展性好，柔软，可锻压，机械加工性能好，非常轻	窗框，飞机，厨具

（续）

名称	颜色	成　分	性　能	用　途
银	灰白色	主要是银，与铜合成标准纯银	延展性好，可锻性好，焊接性能好，抗腐蚀	首饰，钎料，装饰
铅	蓝灰色	不是合金	柔软，重，延展性好，压力下会变形	钎料，管材，电池，屋顶材料

Key to Exercises of the Reading

1．用于制作切削刀具的钢铁金属　2．用于制作电气设备零件的非铁金属　3．用于制作刀叉餐具和医疗器械的钢铁金属　4．用于制作电线的非铁金属　5．用于制作检查井盖的金属　6．用于制作飞机的非铁金属

第2单元　机械零件

无论多么简单的机床，都是由单个组件，即通称为机械零件或部件组成的。因此，如果把机床完全拆开，就可以得到像螺母、螺栓、弹簧、齿轮、凸轮及轴等简单零件——所有机器的基础元件的集合。因此，机械零件就是用来设计可以执行某一特定功能，且能与其他零件配合的简单零件。某些特定的零件必须成对地工作，如螺母与螺栓、键与轴，在其他情况下，一组零件组成一套装配件，例如：轴承，联轴器及离合器。

机械零件中最常用的是齿轮，它实际上是由轮子和杆件组成的带有齿的轮子。在轴套或轴上旋转的齿轮驱动其他齿轮做加速或减速运动，这取决于主动齿轮的齿数。

其他基本机械零件都由轮和杆演化而来。轮子必须装在轴上才可以转动。轮子由联轴器固定在轴上，轴必须安装在轴承里，由带或链条与第二根轴相连，并带动第二根轴转动。支承结构可由螺栓、铆钉连接，或通过焊接固定在一起。这些机械零件的正确使用，取决于是否了解作用于结构上的力及所用材料的强度等相关方面的知识。

单个机械零件的可靠性是估计整台机器预期使用寿命的基本因素。

许多机械零件是完全标准化的。普通结构和机械部件的最佳尺寸可以通过测试或实际经验来确定。采用标准化，可以获得实际使用上的一致性和最终的经济性。然而，并不是所有的零件都是标准的。在自动化工业中，仅仅只有紧固件、轴承、轴套、传动链及传动带是标准的。曲轴及连杆是非标准的。

Key to Exercises of the Text

Ⅰ．略

Ⅱ．1．F　2．T　3．T　4．T　5．T　6．F

Ⅲ．1．最常用的机械零件是齿轮，它实际上是由轮子和杆组成的有齿的轮子。

2．齿轮的硬度决定了它的耐磨能力。

3．制造工程师们集中精力研制标准化的零件。

4．这些零件是在大批量、高规格和低成本的条件下生产的。

阅读材料

亨利·福特

亨利·福特于1863年出生于底特律附近的一个农场。他喜欢拆卸机器，也喜欢把各种

零件组装起来。他总是梦想着能有让工作变得轻松的机器。

1896年，第一辆福特汽车建造完成。它就像是一口箱子被放在了四个硕大的自行车轮子上。这辆车有一台双气缸发动机，但没有制动装置。要倒车的话，就得下车用力推。他驾驶着那辆速度有限、噪声很大的四轮车，穿行在空荡荡的街道上。实验成功了！事实上，这辆车正常运行了好几年。不过，亨利很快就对它失去了兴致。他已经在计划着要发明一种更大、更好的汽车了。

1898年，他的新车造好了。当时，机动车已经开始引起广泛关注。几年后，亨利自己开办了一个汽车制造厂，招聘了自己的员工。1903年，他卖出了第一部自己制造的轿车。到了1904年，A型福特轿车销量极好，公司的业务也因此蒸蒸日上。

亨利·福特追求的是面向普通大众的轿车。他想要做的是大批量生产一模一样的轿车，如此一来，一辆车上的零部件就可以适用于所有其他的车子，这样可以降低车的价格。在亨利·福特看来，轿车价格便宜些，会促使更多的人购买。有车的人多了，又会促进道路建设的完善，而改善了的道路状况又会促使更多的人买轿车，从而使轿车的价格更为便宜。第一辆T型福特车于1908年驶下了生产线。这种车型体积小、样子难看，但易于驾驶，就连小孩子都能开。

在世界各地共售出1500万辆T型福特车。与此同时，T型轿车行销全球。到1919年，福特工厂月产轿车达8.6万辆。到1922年11月，这个数字增至24万辆。同时亨利也开始并购其他产业。

1927年，新的A型轿车问世。人们蜂拥而至，一睹其风采。两年之后，亨利·福特在他68岁的年纪——这个年纪的大多数人正安享辛勤工作换来的安逸之时，福特又一次一展身手。他造出了大车身的V8福特车。

有人说亨利是历史上最富有的人，有人说他是那个年代世界上最富有的人。但正是因为亨利·福特，人们的出行才变得如此便捷又经济。

Key to Exercises of the Reading

Ⅰ. 1. F　2. F　3. T　4. T

Ⅱ. 1. He grows flowers as well as vegetables.

2. I have arranged that one of my staff will meet you at the airport.

3. Have you figured out how much the holiday will cost?

4. I hope his dream of becoming a pianist will come true.

第3单元　机　　床

大多数的机械加工通常是在五种基本类型的机床上进行的：

钻床；

车床；

牛头刨床或龙门刨床；

铣床；

磨床。

钻　　孔

钻孔是用一种称为钻头的旋转刀具进行的。大多数金属钻孔用的是麻花钻。用于钻孔的机器叫做钻床。铰孔和攻螺纹也归类于钻孔。铰孔是在一个已钻好的孔的基础上再切削少量的金属；攻螺纹是在孔里加工出螺纹的过程，以便螺钉或螺栓能旋合进去。

车削和镗削

车床通常被称为“机床之父”。在车削操作中，车床用一个单刃刀具在旋转的工件上进行金属切削。车削用来切削不同的圆柱形，如轴、齿轮毛坯、滑轮、丝杠等。镗削用来扩大、精加工和准确确定孔位。

铣　削

铣削用一个称为铣刀的旋转多刃刀具进行金属切削。铣刀被制成不同的式样和尺寸。有些铣刀只有两个刃口，而有些铣刀有多达30个或更多的刃口。根据所用的铣刀形状，铣削可加工平面或斜面、槽口、缝、齿轮的齿和其他的轮廓。

刨　削

刨削用单刃刀具加工平面。在刨削操作中，当工件自动朝刀具方向进给时，牛头刨床上的刀具做往复或前后移动。在刨削操作中，龙门刨床上的工件被固定在一个沿着刀具做往复运动的工作台上。每一行程刀具自动向工件进给一点。

磨　削

磨削利用磨粒来进行切削。磨削可根据不同的目的分为精磨和粗磨。精磨应用于接近于精密公差和表面粗糙度很高的磨削，粗磨用于精度要求不高工件的多余金属的切除。

Key to Exercises of the Text

Ⅰ. 略

Ⅱ. 1. F　2. F　3. F　4. F

Ⅲ. 1. 麻花钻　2. 切削刃　3. 表面粗糙度

4. 根据所用的铣刀形状，铣削可用来加工平面或斜面、槽口、缝、轮齿及其他的轮廓。

5. 车削用来切削不同的圆柱形，如轴、齿轮毛坯、滑轮、丝杠等。

阅读材料　坐标镗床

坐标镗床既类似于钻床又类似于立式铣床。其主轴箱类似于钻床，而工作台则与立式铣床极为相似。坐标镗床用来加工尺寸要求非常精确的钻模、夹具和冲模。由于工件能可靠地固定在做纵向、横向运动的工作台上，又由于主轴箱的制造精度很高，所以在这种机床上进行孔的定位和钻削要比其他任何机床更精确。

坐标镗床的坐标测量系统用来确定工作台的纵向和横向运动。最简单的系统之一是以两根导螺杆为基础的。另一个为许多坐标镗床所采用的系统是以精密端部测量杆为基础的。有些坐标镗床装有电气或光学测量系统。

Key to Exercises of the Reading

Ⅰ. 略

Ⅱ. 1. longitudinal and transverse table movements　2. lead screw

3. measuring rod　4. optical system

第4单元　金属的热处理和热加工

我们可以通过各种方式改变钢的性能。首先，含碳量低的钢通常比含碳量高于1.5%的高碳钢的硬度低。其次，我们可以将钢加热到临界温度以上，然后再以不同的速度将其冷却。在达到临界温度时，金属的分子结构开始发生改变。退火就是把钢加热到其临界温度以上，然后再让其缓慢冷却的过程。这就使得金属的硬度降低，从而更容易进行机械加工。退火还有另一个优点，那就是它可以消除金属中的内应力。金属在锻打加工或快速冷却时极易出现内应力。快速冷却的金属外层收缩比内部收缩速度快，这将引起不均匀收缩，增加其出现畸变和破裂的可能。因此冷却速度慢的金属比冷却速度快的金属产生的内应力少。

另一方面，我们可以通过快速冷却提高金属的硬度。我们将金属加热到临界温度以上，然后在水或其他液体中淬火。快速冷却使钢在临界温度产生的结构得以固定，使得材料的硬度更大。但是这样淬硬的钢条比普通钢材更容易脆裂。因而我们将其重新加热到临界温度以下，并将其缓慢冷却，这种处理称为回火。回火有利于消除金属的内应力并使其脆性降低。当制造的工具需要高硬度的钢时，我们可以用回火钢。高碳钢比回火钢硬，但加工也更困难。

上述热处理可以安排在各种成形工序中。我们可以通过使用轧钢机轧制金属以获得钢条和钢板。其中冷轧的轧制压力比热轧的轧制压力大得多，但冷轧可以使操作工人生产的钢板精密度更高，均匀性更强，且表面精度更高。其他成形工艺还包括拉丝、模铸和锻造。

金属的机械加工是借助某些机械工具对冷或热状态的金属进行成形加工。这既不包括去除金属材料的切削或磨削加工成形，也不包括需借助模具的金属铸造成形。在机械加工的过程中，金属通过锻造、弯曲、挤压、拉拔或剪切达到其最终形状。这些工序可以进行金属热加工，也可以进行冷加工。虽然室温通常用于钢铁的冷加工，但是冷加工有时也在介于室温和再结晶温度之间的温度下进行。金属热加工是在再结晶温度以上或淬火温度区进行的。钢铁的再结晶温度在650～700℃之间，但大多数钢铁的热加工温度都高于其再结晶温度。在达到最低的再结晶温度之前，机械加工的金属没有硬化的趋势。某些金属，如铅、锡的再结晶温度很低，在室温下就可以进行热加工，但大多数工业用金属都需要加热。合金的成分对适于加工的温度范围有极大的影响，通常可以扩大再结晶的温度范围。并且先前的冷加工也可以增大其再结晶的温度范围。

Key to Exercises of the Text

Ⅰ. 略

Ⅱ. 1. molecular structure of the metal　　2. unequal contractions

3. at the critical temperature　　4. various shaping operations

5. mechanical working of metal　　6. cold or hot working of steel

7. We can make steel harder by rapid cooling.

8. The metal is shaped by pressure-acutely forging, bending, squeezing, drawing, or shearing it to its final shape.

阅读材料　　钎焊和焊接

将金属连接在一起的方法很多，选用哪种方法要取决于金属的类别和所要求的接缝强度。

1. 软钎焊与硬钎焊

软钎焊能使薄的钢件、铜件或黄铜件获得合乎要求的接缝。但软钎焊的接缝强度比硬钎焊、铆接或焊接的接缝强度低得多，硬钎焊、铆接或焊接等这些接合方法通常用于高强度的永久接合。软钎焊是靠施加熔融状态的第三种金属连接两块金属的焊接方法。钎料由锡、铅组成，而加入铋和镉是为了降低钎料的熔点。钎焊的重要工序之一是将需要焊接的接缝表面清理干净，这一工序可采用某种酸性洗液。虽然通过清理操作可除去氧化物，但清洗完毕之后，会立即产生新的氧化膜，从而阻碍了钎料与金属件表面的连接。助焊剂可用来去除氧化物和防止需要焊接的金属表面的氧化，以使钎料能自由地流动并与金属件连接。在钎焊大多数黑色金属和有色金属时氯化锌是最好的助焊剂，钎焊铝时，可用硬脂酸和凡士林作为助焊剂。钎焊用的纯铜烙铁是连接在一钢条上的纯铜，钢条上有手柄。纯铜烙铁被做成不同的长度、形状和重量。钎焊的质量在很大程度上取决于纯铜烙铁的形状和尺寸。只有在接缝表面吸收足够热量后，使钎料在一段时间内保持熔融状态，这样才能将两部分焊牢。

在有些情况下，用高熔点的硬锌钎料来焊接金属是必要的，这种钎焊称为硬钎焊。

2. 焊接

焊接是借助于热量，在金属软化后通过加压、锤锻或使之熔合的手段将两块金属连接在一起的一种工艺方法。机器、汽车、飞机、船舶、桥梁及各种建筑物的许多构件都是焊接的。

氧乙炔焊（气焊）用氧气和乙炔气混合物燃烧的火焰加热两块金属。氧气和乙炔气分别贮存在两个钢制的贮气瓶内，由贮气瓶流入焊炬，这两种气体在焊炬内混合后通向火焰。氧炔焰是常规用途中最热的火焰，其温度约为 6300℉（3482℃），氧炔焰可用于切割铁和钢。电焊即电弧焊，是用电能对施焊的两件金属加热，其热量是工程应用中所能获得的最高热量，约为 7232℉（4000℃），因而焊接端被熔合成焊缝。点焊是借助电能以焊点形式将两块金属焊接在一起的焊接工艺。人们采用点焊机来完成这一工作。锻焊是将两块金属的焊接端在炉中加热软化，然后将它们锤击在一起的焊接工艺。

第 5 单元 模具简介

模具是工业生产的基础技术装备。规模化工业产品是在为其特别设计与制作的模具中成型的。模具是零件生产过程的核心，因为其型腔赋予了产品形状。模具有很多种，如铸造模具与锻造模具、拉伸模具、注塑模具、挤出模具、压缩成型模具等。

下面简单介绍一些模具成型工艺及所用到的相关模具。

压缩成型

压缩成型是模具成型工艺中最为简单的工艺方法，也是小批量大型工件的理想成型方式。对于数量较低的生产需求来说，制作一个压缩成型模具比制作一个注塑模具更经济。压缩模具常用于产品原型制作，所制作的试样用于组合件间的配合测试并最终成型为组件的一部分。这样一来，在制作适用于大批量生产的注塑模具之前，可在设计上做进一步的变更和完善。压缩成型最适合公差要求不严格的设计。

注塑成型

注塑成型是模具成型工艺中最为复杂的工艺方法。因为注塑模具更为复杂的设计，其费用比铸造模具或压缩模具高。尽管模具的费用很高，但产品的生产周期比其他工艺要短，且产品的生产成本低，尤其是在成型过程实现全自动的条件下。注塑成型非常适合成型形状结构精巧的零部件，这是因为其模具可维持很高压力（高达 29 000 psi），从而将材料填充至模具型腔的每一角落。

铸造成型

铸造成型方法有两种：开放式铸造成型和压力铸造成型。采用开放式铸造成型时，将液体混合物倒入敞开的模具型腔中使其固化。采用压力铸造成型时，将液体混合物倒入敞开的模具型腔中，然后将盖子盖好，对型腔加压。压力铸造用于较为复杂的零件和泡沫材料的成型。

原则上，压力铸造成型和注塑成型相同，只是材料类别不同而已。事实上，铸造成型与注塑成型可生产相同几何外形的零件。在许多情况下，由于零件成本的降低，注塑成型可以替代铸造成型。但是，对于结构件，尤其是那些厚壁零件，铸造成型往往是更好的选择。

挤出成型

虽然挤出模具相当简单，但是挤压成型工艺要求在生产准备、制造和最终处理的过程中投入大量的精力，以确保产品的一致性。压力是由带有通孔腔的模板来施加的，模板上的通孔腔已经加工出所需的形状。进给速率、温度及压力的变化都需要好好控制。

大多数挤压模具外形就是一个简单的圆形钢件，上面设有所要挤出线状产品的截面形状的孔，并留有余量以便于复合物的收缩和膨胀。挤出模具是最为简单的模具。

Key to Exercises of the Text

Ⅰ. 略

Ⅱ. 1. 注塑模具　2. 挤压模具　3. 压缩模具　4. 进给速率　5. cycle time　6. in principle

Ⅲ. 1. Automated　2. mixture　3. delicately　4. variations

Ⅳ. 1. Now they lead in the manufacture of weapons of mass destruction.

2. We accepted the item in principle.

3. There is no substitute to hard work and dedication.

4. This project involves a lot of complex technical problems.

阅读材料　模具材料

根据各种不同工艺方法的工艺参数和不同产品的生产运行时间（比如考虑要加工的成品数量），模具必须满足很多不同的要求。因此，模具可采用各种材料制作，包括像纸模、石膏这样的奇异材料。然而，由于大部分加工过程要求高压并经常伴有高温，到目前为止，金属仍然是最重要的模具材料，其中钢是最主要的金属材料。有趣的是，在许多情况下，选择哪种模具材料不仅要考虑材料的特性和最佳的性价比，还要考虑模具的制造方法，因为它们会影响整体设计。

一个典型的例子，铸造金属模具和机械加工模具的冷却系统区别很大。另外，生产技术也有影响。例如，经常可以了解到，为了制造简捷，时常借助于计算机辅助设计（CAD）和计算机集成制造（CIM）这样的最新技术，将实体毛坯加工成原型模具。与以前利用试样的制造方法相比，CAD 和 CIM 的应用代表了当今更为经济的制造方法，不仅因为厂内现有的生产能力可以得到利用，而且考虑到若采用其他工艺技术，订单必然依赖厂外供应商。

总之，尽管高级材料也常被使用，但原则上，模具制造通常使用标准材料。例如，像陶瓷这样的新型高性能材料几乎不用。这与相关事实有关，模具并不需要陶瓷在高温下仍能保持较好稳定性的特性；但陶瓷的缺点，如抗拉强度低、导热性差等，都使得模具制造中烧结材料的使用比较有限。与多孔透气的烧结金属材料相比，用等压热成型的粉末冶金方法做出的现代材料和零件的抗拉强度、导热性的性能要好一些。

在许多不同的成型方法中，人们都必须将型腔中的气体排出，使用多孔金属材料来解决上述问题的呼声很高。专门制造的排气装置的优势与其潜在的问题一样明显，特别是在熔融物料前沿相遇的地方，比如熔接痕处。一方面，要防止这类材料表面纹理出现在成品上；另一方面，须防止微孔很快被残渣堵塞。在这种情况下，因采用这种材料而带来的全新的模具设计和制造工艺的可能性也是很有意思的。使用烧结金属时的排气、冷却和顶出工序步骤可以用图 5-1 做很好的展示。

用侧浇口成型模具来成型带有复杂底部形状的杯状制品时，为保证将底部区域的气体顺利排出，在技术上需要付出很大的努力。采用多孔材料做型芯和型腔半模，就不再需要其他的排气措施。而且因为是在型腔的整体表面排气，所以材料填充型腔的速度更快，原则上，也因此可以自由选择浇注口的位置。

从狭长的型芯中散发所需要散去的大量热量是非常困难的。在这种情况下，通过型芯中的微孔系统来分流冷却空气，从而可加快冷却和缩短生产周期。这个方法带来的另一附加优点是提高了模具表面的温度均匀性。

可以结合其他工具（顶出器、脱模圈），或不用其他工具而直接在型芯半模中引入高压气体来顶出产品。这样就可减少或消除产品表面的顶出痕迹，且在顶出过程中无须克服真空问题。

Key to Exercises of the Reading

1. exotic　　2. predominant　　3. related　　4. performance

第 6 单元　模具设计

压铸法是一种用于批量生产金属制品和零件的工艺技术。在此过程中，模具设计是最重要的步骤之一，因为模具的形状与属性会直接影响到最终的产品。压铸工艺使用高压将熔化了的金属注入模腔，该过程的完成要求模腔的规格必须精准。

模具设计的重要性

模具设计会影响通过压铸工艺生产出来的产品的形状、结构、质量及其均匀度。不合适的规格会导致工具或原材料受到腐蚀，还会造成产品质量低劣，而有效的设计会提高效率，节省生产时间。

影响模具设计质量的因素

要为一个生产项目确定合适的规格，在进行模具设计时，需要考虑很多因素。主要包括以下因素：①起模斜度；②倒圆角；③分型线；④凸起；⑤筋条；⑥孔和开窗；⑦符号；⑧壁厚。

起模斜度

起模斜度指的是模芯可被拔锥的角度。要使铸件顺利地从模具中脱出，需要一个精确的斜度，然而，斜度值不是一成不变的，会随着侧壁角度的不同而变化，诸如所用熔融合金的类型、内壁的形状、模具的深度等特性也会影响铸件的脱出和斜度。模具的几何形状也会影响到斜度。一般来说，由于有收缩的风险，设计上没有斜度的孔需要拔锥。同样，内壁也会收缩，因此比外壁需要更大的斜度。

倒圆角

圆角是一种凹形的连接形式，用来光滑过渡有棱角的表面。尖锐的转角会妨碍压铸的过程，许多模具都通过倒圆角来产生圆边，从而降低出现生产差错的风险。除有分型线的模具外，模具可在任意部位倒圆角。

分型线

分型线，或者分型面，将模具的不同部位连接起来。如果分型线的位置不准确，或由于工作压力而变形，原材料可能会从模具工件之间的间隙渗出，会引起成型不均匀及接缝溢出。

凸台

凸台是压铸模中的一种凸起结构，在模具设计中作为产品安装的支点或者托脚。制造商通常会在凸台的内部添加孔，以确保模压制品有均匀的壁厚。高度较大的凸台不易被金属填充，所以借助于倒圆角及添加筋条来缓解这个问题是有必要的。

筋条

对于壁厚不足，无法满足特定用途的产品，可用压铸筋条来提高材料强度。选择合理位置安放筋条可减少应力开裂及壁厚不均的几率；同时，也有利于减轻产品重量，改善填充能力。

孔和开窗

压铸型里设置孔或开窗会使成型模具的脱出更容易，并能产生实用的拔锥。孔和开窗还具有如溢流、闪络、横浇口的一些特点，这些对于阻止孔内不必要的压铸及不良原材料绕孔流动，都是必要的。

符号

制造商设计压铸产品模具时通常会设计上品牌名称及产品标志。尽管这些符号的添加一般不会使压铸过程更复杂，但会增加生产成本。特别是凸起的商标或符号意味着每一个生产的工件都需要增加熔融金属的含量。反之，凹陷的符号会使用较少的原材料，从而节省费用。

Key to Exercises of the Text

Ⅰ. 略

Ⅱ. 1. A　　2. B　　3. A　　4. A　　5. B　　6. A

Ⅲ. 1. I believe that each of us can contribute to the future of the world.

2. Nowadays, a great deal can be done to alleviate back pain.

3. The document requires substantial changes.

4. I like all fruits with the exception of grape.

阅读材料参考译文及答案略。

第7单元　注射成型和注塑机

注射成型是最常用的塑料工件的加工制造方式。用注射成型的方式制造的产品多种多样，包括各种尺寸、各种复杂程度、各种用途的产品，不一而足。注射成型过程中需要用到注塑机（图 7-1）、塑料原料和模具。塑料在注塑机里熔化后被注入模腔，并在模腔里冷却、凝固成最终的产品。

注射成型可用于生产各种用途的薄壁塑件，其中最常见的是塑料外壳。塑料外壳是一种薄壁包围物，内部通常需要设置很多筋条和凸台。这些塑料外壳用于很多产品，包括家用电器、电子消费品、电动工具和汽车仪表板。其他常见的薄壁产品还包括不同种类的开口容器，如水桶。注射成型还用来生产一些日常的生活用品，如牙刷或小的塑料玩具。

工艺过程

注射成型的工艺过程非常短暂，一般在 2 秒至 2 分钟之间，由以下四个阶段组成：

1）夹紧：在材料注入模腔之前，模腔的两半模必须先通过锁模装置紧紧关闭。模腔的两半模紧附在注塑机上，其中一个半模可滑动。液压驱动的锁模装置将模具的两半模合拢，并在注入材料的时候，施加足够的力以保持模腔紧闭。关闭及锁紧模腔所需的时间取决于所用的机器——大的机器（夹紧力更大）需要的时间更长。这个时间可以根据机器的干燥周期来估算。

2）注入：塑料原料通常以小颗粒的形式被进给注入注塑机，并通过注射装置推入模腔。在此过程中，材料受热受压而熔化。熔化后的塑料很快被注入模腔，压力的积聚使原材料成型并保压。注入的材料的数量称作注塑量。由于熔化的原材料在模腔里的流动具有复杂性和变化性，注塑的时间难以精准计算。但是，注塑时间可以通过注塑量、注塑压力和功率来估算。

3）冷却：模腔里熔化的塑料一接触到模腔的表面，就开始冷却。随着塑料冷却，它会凝固成目标工件的形状。然而，在冷却的过程中，工件可能会出现某种程度的收缩。在注入阶段，原材料的成型使得补充的原材料流入模腔，从而减少了可见的收缩量。冷却的时间到了之后，才能打开模腔。冷却所需的时间可通过塑料原材料的一些热力学性质以及工件的最大壁厚估算出来。

4）弹出：经过足够的时间，冷却了的工件会通过弹出系统从模腔脱出，脱出系统附着在模腔的后半部。当打开模腔时，有一个机构会将工件从模腔里推出。由于在冷却过程中，工件收缩并黏附在模腔里，所以需要外力来脱出工件。为了方便工件的脱出，可在注入原材料之前，向模腔的表面喷脱模剂。打开模腔并脱出工件所需的时间可以根据机器的干燥周期来估算，并应包括工件从模腔落下所需的时间。一旦工件脱出，模腔即可紧闭，准备下一轮注入。

在注射成型周期结束之后，通常还需要一些后加工处理。冷却期间，模腔材料流道中的原材料会凝固并附着在工件（图 7-2）上。这些多余的材料，连同一些毛边，必须修剪掉，常用的方法是使用刀具进行修剪。对于有些种类的材料，如热塑材料，因修剪而产生的废料可放进塑料磨粉机中再循环利用。塑料磨粉机也叫研磨机或粉碎机，可把废料研磨成小颗粒。由于重磨后的材料存在一定程度的降解，重磨后的材料必须以适当的重磨比例与原材料混合，才能在注射成型加工过程中实现再利用。

设备

注塑机有很多组件，也存在不同结构的注塑机，包括卧式机和立式机。然而，不管如何设计，所有的注塑机都由电源、注射系统、模具组件、锁模装置来完成工艺周期的四个阶段。

材料

多种材料可用于注射成型加工工艺。大多数聚合物，包括所有的热塑材料、部分热固材料和人造橡胶都可以使用。这些材料用于注射成型加工工艺时，其最初的形式通常是小颗粒或呈精细的粉末状态。在加工过程中，还可添加色素以控制最终产品的颜色。为制造注射成型件而进行的原材料的选择，不仅仅基于最终产品的特性。每种材料都有其不同的性质，这些会影响到最终产品的强度与功能，这些性质也会决定加工这些材料所需的参数。

Key to Exercises of the Text

Ⅰ. 1. F　2. T　3. F　4. F

Ⅱ. 1. 家用电器　2. 电子消费品　3. 工艺周期

4. automotive dashboards　5. mould release agent　6. mould assembly

Ⅲ. 1. elapse　2. estimated　3. facilitate　4. solidified

Ⅳ. 1. One must have a correct estimate of oneself.

2. The influence we exert is powerful.

3. The island offers such a wide variety of scenery.

4. Champagne sprayed from the bottle.

阅读材料　注塑机组成部件

对于热塑性塑料，注塑机将粒状原材料通过熔融、注射、保压和冷却这样的循环过程转化为最终的模塑制品。典型的注塑机由注射机构、模具系统、液压系统、控制机构、锁模机构组成。

注射机构

注射机构包括料斗、往复式螺杆、料筒和注射用喷嘴。该部分在喂料、压缩、排气、熔融、注射和保压阶段起到包容和运送物料的作用。

（1）料斗　热塑性物料以小颗粒状供给成型，注塑机料斗要盛装这些粒料，物料以重力输送形式通过料斗颈部进入料筒和螺杆。

（2）料筒　注塑机料筒承载往复式塑化螺杆并由电加热圈加热。

（3）往复式螺杆　往复式螺杆可用来压缩、熔融和输送物料，由三段功能区组成：加料段、压缩段（过渡段）和计量段。

当螺杆外径保持恒定时，自加料段到计量段开始处，往复式螺杆螺纹区段的槽深逐渐变浅，这段螺杆将物料压向料筒内壁并产生黏性（剪切）热。

（4）喷嘴　喷嘴连接料筒与模具上的主流道衬套，并且在料筒和模具之间形成密封。喷嘴温度应设为物料的熔融温度或稍低，这取决于材料供应商的推荐值。

模具系统

模具系统包括拉杆、动模板、定模板以及开设有型腔、主流道、分流道、顶出杆和冷却水道的模具基板。模具其实是一个热交换系统，热塑性熔体在其中固化成为型腔给定尺寸和形状的制品。

液压系统

注模机上的液压系统可提供开、合模的动力和锁模力，旋转和驱动往复式螺杆、推动顶杆及带动型芯移动。这需要一系列的液压元件来提供动力，包括泵、阀、液压马达、管接头、管道和液压油箱等。

控制机构

控制机构可保证设备连续重复工作，能监测和控制成型参数，包括温度、压力、注射速度、螺杆转速和位置、液压位置。过程控制对于制品最终质量和过程的经济性有直接影响。过程控制系统可以只有简单的继电器开/关控制功能，也可以是非常复杂的基于微处理器的闭环控制系统。

锁模机构

锁模机构用于开启和闭合模具，支承和移动模具的组成部件，产生足够的推力以防止模具（意外）开启。锁模力可通过机械（肘杆）式锁紧机构、液压锁紧机构或这两种基本形式的组合锁紧机构来实现。

Key to Exercises of the Reading

1. 小孩子应该学会有效地沟通。
2. 此外，就是要在各种制造方法方面具有坚实的理论基础知识和实践经验。
3. 很多外观精美的产品，如由彩色塑料或其他特殊材料制成的产品，都十分畅销。

第 2 章 计算机数字控制（CNC）

第 8 单元 数 控 机 床

早期机床通常由工人操作并由他们决定机床速度、进给量、切削深度等变量。随着科学技术的发展，一个新术语——数字控制（NC）诞生了。数字控制（NC）就是利用穿孔纸带或储存的程序来控制机床。美国电子工业协会对数控的定义为“通过在点位直接插入数字数据来控制系统动作，此系统至少必须能够自动解译这些数据中的一部分信息”。

过去，人们使用的机床结构简单，以便降低成本。随着劳动成本日益增长，人们研制出性能更好，并配有电控设备的机床，这样企业可以生产更多、更好、具有价格优势的产品，以便和海外企业的产品竞争。

数控技术已经应用于从简单到复杂的各种类型的机床。最常见的机床有：单轴钻床、车床、铣床、车削中心及加工中心。

1. 单轴钻床

单轴钻床是最简单的数控机床之一。多数数控钻床可在三个坐标轴上编程：

1） X 轴控制工作台左右运动。

2） Y 轴控制工作台靠近或离开立柱。

3） Z 轴控制主轴上下运动，进行钻孔。

2. 车床

普通车床是生产效率最高的机床之一，它是加工回转体零件非常有效的工具（图 8-1）。大部分数控车床可在两个坐标轴上编程：

1）X 轴控制刀具横向运动（切入或切出）。

2）Z 轴控制机床滑板靠近或离开主轴箱。

3. 铣床

铣床一直是工业生产中加工方式最多的机床之一（图 8-2），像铣削、成形加工、齿轮加工、钻削、镗削、铰孔等只是可在铣床上进行的一小部分加工方式。数控铣床可在三个坐标轴上编程：

1）X 轴控制工作台左右运动。

2）Y 轴控制工作台靠近或离开立柱。

3）Z 轴控制升降台或主轴垂直（上下）运动。

4. 车削中心

曾有研究表明，所有金属材料的切削操作大约 40% 是在车床上进行的。据此，早在 20 世纪 60 年代中期，人们就研制出了车削中心。这种数控机床比普通车床具有更高的加工精度和生产效率。一般数控车削中心仅在两个坐标轴上工作：

1）X 轴控制刀具转塔横向运动。

2）Z 轴控制刀具转塔纵向运动（靠近或离开主轴箱）。

5. 加工中心

加工中心是在 20 世纪 60 年代发展起来的。有了加工中心，工人们就不必把零件从一台机床转移到另一台机床来完成各种加工。由于工件经过一次装夹能进行更多种加工操作，所以大大提高了生产效率。加工中心主要有两类：卧式加工中心和立式加工中心。

（1）卧式加工中心在三个坐标轴上工作：

1）X 轴控制工作台左右运动。

2）Y 轴控制主轴垂直运动（上下）。

3）Z 轴控制主轴水平运动（切入或切出）。

（2）立式加工中心在三个坐标轴上工作：

1）X 轴控制工作台左右运动。

2）Y 轴控制工作台靠近或离开立柱。

3）Z 轴控制主轴垂直运动（上下）。

Key to Exercises of the Text

Ⅰ. 略

Ⅱ. 1. single-spindle drilling machine　　2. the engine/horizontal lathe

3. machining center　　4. production rate　　5. gear cutting

6. vertical movement　　7. a punched tape　　8. the Electronic Industries Association

Ⅲ. 1. NC is being used on all types of machine tools, from the simplest to the most complex.

2. The engine/horizontal lathe is one of the most productive machine tools.

3. Studies showed that about 40 percent of all metal cutting operations were performed on lathes.

4. The horizontal spindle-machining center operates on three axes.

阅读材料参考译文及答案略。

第9单元 CNC机床的组成

CNC机床系统由六个基础部分组成，即工件加工程序、机床控制装置、测量系统、伺服控制系统、实际的CNC机床（车床、钻床、铣床等）以及自适应控制系统。在实际编写零件的数控加工程序前，了解CNC机床的每个组成部分是十分重要的。

工件加工程序

工件加工程序是机床必须遵循的一组详细指令。每条指令都规定了工件在笛卡儿坐标系中（X，Y，Y）的位置、机床动作（工件或刀具运动）、加工参数以及开启/停止函数。工件加工程序员应该非常熟悉机床、加工工艺、工艺变量的影响以及CNC机床控制的局限性。工件加工程序可手动编写，也可用像APT（自动编程工具）这样的计算机辅助程序语言来编写。

机床控制装置（MCU）

机床控制装置（MCU）是一台微型计算机，其中存有加工程序，并能够控制机床执行程序指令来完成各种动作。MCU主要由数据处理装置（DPU）和回路控制装置（CLU）两部分组成。DPU软件包含控制系统软件、计算算法、将工件加工程序转换成MCU可用格式的编译软件、实现刀具流畅运动的插补算法及工件加工程序编辑等功能。DPU处理工件程序中的数据并将结果输送到CLU，CLU则启动装在机床丝杠上的驱动器并接收各轴实际位置与速度的反馈信号。CLU由以下电路组成：位置与速度控制回路、减速和冲击缓冲电路、功能控制（如主轴的启停控制）电路。

在CNC系统中，DPU的功能往往是由CNC计算机内的程序控制完成的，而CLU的主要工作部分通常是在最复杂的CNC系统内进行的。

测量系统

在CNC系统中，测量系统是指机床把工件从参考点移动到目标点所采用的方法程序。目标点是指钻孔、铣槽或进行其他加工操作的某一具体位置。CNC机床采用绝对测量系统和增量测量系统两种形式。绝对（坐标）测量系统使用固定参考点（原点），所有位置数据都以此点为基准。换言之，不论工件移动到哪里，其位置数据必须是与最初的固定参考点相关的距离。增量（Δ）测量系统采用浮动坐标系统，因此每当工件移动后，机床会建立一个新原点或参考点。需要注意的是，在此系统中，每个新位置的X、Y坐标都由前一个位置进行计算。这种系统的缺陷之一是，出现的任何错误若没有被发现并纠正，会在整个加工过程中反复出现。

伺服控制系统

CNC伺服机构是使工作台或滑台沿坐标轴精确运动的装置。不同于传统机床需要手动操纵旋转曲柄或手轮使机床运动，CNC机床是在CNC控制下通过伺服电动机实现运动并由工件程序指导加工过程的。一般来说，几乎所有CNC机床中，运动方式（快速、直线、圆弧）、加工轴移动、移动量、移动速率（进给率）都可以由程序控制。

滑台既可以电动驱动，也可以液压驱动。目前已经有很多液压系统可用，这些系统经过充分验证，效率很高、响应快速。不过到目前为止，最常用的动力源还是电动机。常用电动机有三种，即步进电动机、非步进电动机、磁性直线驱动电动机。步进电动机是一种在电脉冲激励下，以有限步长按序旋转的特殊电动机。在相等功率的条件下，交流伺服电动机比直流电动机的体积要大一些，价格也高。但交流伺服电动机便于保养维护，这是其更受欢迎的一个原因。磁性直线驱动系统是最新的机床滑台驱动系统，仅靠大功率直线电动机产生的磁力便足以驱动机床运动。

采用开环系统的CNC机床，没有反馈信号来保证加工轴移动了所要求的距离。也就是说，如果接收的输入信号要求某个特定工作台移动1.000英寸，伺服装置一般会驱动工作台移动1.000英寸，但没有办法将工作台的实际移动量与输入信号进行比较，唯一能够保障工作台移动1.000英寸的因素就只能是所用伺服系统的可靠性了。当然，开环系统比闭环系统价格便宜一些。

闭环系统将实际输出（工作台移动1.000英寸）与输入信号进行比较，并对所得任何偏差进行补偿。反馈装置将工作台实际移动量与输入信号进行对比。闭环系统所用的反馈装置有传感器、电子或磁性标尺，以及同步装置。闭环系统的采用大大提高了CNC机床的可靠性。

控制器中执行的每一条CNC指令（通常通过程序来实现）会指令驱动电动机转动一系列精确的次数，驱动电动机的转动又使得滚珠丝杠转动。滚珠丝杠进而驱动直线轴运动。滚珠丝杠尾端的反馈部件使控制器能够确认指令要求的转数已经完成。

实际的CNC机床

实际使用的机床有以下几种：车床、铣床、三坐标测量机等。与传统机床不同，CNC机床具有较高的强度和刚度、较好的稳定性和抗振性以及较小的间隙。

自适应控制系统

对某一项加工操作来说，术语“自适应控制”是指能够检测某些过程变量并利用检测结果调节转速或进给量的控制系统。自适应控制加工系统已经使用的过程变量有主轴变形量或受力、主轴转矩、切削温度、振动幅度及驱动功率。换句话说，几乎所有可检测的金属切削变量，在实验性自适应控制系统中都被尝试使用过。研究自适应加工系统的动机在于想要更有效地控制加工过程。衡量加工性能的主要标准有两项：金属切削率和单位体积的切削成本。

自适应控制在机械加工中的主要优点如下：生产效率提高、刀具寿命延长、工件保护程度高、操作者干预少和编程更加便捷。

Key to Exercises of the Text

Ⅰ. 1. machine control unit (MCU)　2. feedback signal　3. milling machine
4. 自适应控制系统　5. 伺服控制系统　6. 开环系统

Ⅱ. 1. slide　2. components　3. execute　4. command　5. magnetic

Ⅲ. 1. T　2. F　3. F　4. F　5. T

阅读材料　CNC机床安全防范措施

为减少安全隐患、保持设备功能的正常运行，落实下列安全防范措施是十分重要的。

1. 操作

为了最大程度地降低事故风险，请遵循以下安全防范措施：

1）穿安全鞋、佩戴防护眼镜、戴安全帽。

2）穿工作服。衣服不应宽松，特别是衬衫袖口应该扣紧或紧贴在手臂上。

3）禁止戴手套操作机器。

4）在机器运转时，禁止触碰工件或主轴。

5）为机器操作环境提供充足的光线照明，并且尽量保持环境的清洁。

6）禁止使用高压空气在控制器附近吹尘或者清理切屑。

7）操作区的地面要坚固。

8）安装机器的地面应该足够坚固。

9）如果感到不舒服，请勿操作机器。

10）禁止把工具放在工作台上或工作板上。

11）请在操作前检查机器的状况，确保机器的最佳性能。

12）每天打开机器后，机器需要进行预热以确保主轴的使用寿命。所需的最短时间为30分钟，主轴转速设定为3000转/分钟。

13）安装工件时请停止机器的所有功能。

14）在工作过程中禁止打开防护门。

2. 机器

在操作机器之前，操作者需详细地理解该手册。

1）操作者或维修人员应当注意所有的警示标志。禁止撕毁或移除机器上的任何标志。

2）非调整、修理需要，所有的门都应关闭，保持不必要的物品远离机床。

3）进行调整、修理或维护操作时，确保使用合适的工具。

4）不要为了改变机器的行程而移动或改变行程开关的位置。

5）机器发生任何问题时，请立即停止操作并按下紧急停车按钮。

6）日常工作应注意以下事项：

在操作过程中，禁止将身体的任何部位放在可移动的部件上，如主轴、换刀器、工作台和切屑输送机。

禁止徒手清洁工具和/或工作台上的切屑。所有清洁工作都必须在机器完全停止后才能进行。

调整切削液、空气和流体喷嘴应在机器完全停止后进行。

7）正常情况下，应按照下列步骤停止日常工作：

按下紧急停止按钮→关掉电源→清洁工作台→在主轴和工作台上喷防锈油。

3. 电气

检查和维护期间，应采取以下预防措施：

1）在任何情况下，禁止用力敲击控制器。

2）只使用制造商指定的电缆。只使用合适长度的电缆。如在地面铺设电缆，应提供适当的电缆保护措施。

3）只允许制造商和授权代理人修改控制器的设置参数。

4）禁止改变控制器的设置值和控制按钮。

5）禁止超载使用插座和导线。

6）检查和维修电器元件前，应断开控制器和主电源，并将其锁定在“OFF”位置。

7）禁止任何潮湿的工具触碰电器元件。

8）只使用经制造商认可的熔丝；禁止使用高容量的熔丝或者以铜线代替熔丝。

9）非经过培训的合格人员禁止打开电气柜门进行维修。

Key to Exercises of the Reading

1. They were held together to deal with emergency.
2. They had sufficient food to sustain life.
3. This approach reduces development and maintenance costs.

第10单元　CNC编程

在NC（数字控制）机床中，刀具由编码系统控制，从而使加工过程中人工监管达到最小化、加工操作的可重复性也很好。CNC（计算机数字控制）机床与NC机床的操作系统相同，不同的是CNC机床的刀具是由计算机来监控的。

操作手动机床的一些原则同样适用于NC机床或CNC机床的编程。主要区别在于手动机床通过转动曲柄使滑台移到指定点，而NC机床或CNC机床则是将这些尺寸信息事先储存在机床控制装置的存储器中，每次程序运行时控制器将控制机床移动到指定位置。

要运行VF系列立式加工中心，就需要设计和编写工件程序，并将其输入到控制器的存储器中。工件程序通常采用离线方式编写，也就是说，在脱离CNC机床的设备中进行编程，编好后保存并传输到CNC控制器中。最常用的工件程序传输方式是通过RS-232接口传输。HAAS VF系列立式加工中心配有RS-232接口，该接口与目前大多数计算机和CNC机床都是兼容的。

为了操作CNC机床并对其进行编程，必须对机床切削加工和数学知识有基本的理解和掌握。对控制面板以及键盘、开关、显示器等元件的布置非常熟悉也是十分重要的。

完全掌握CNC机床之前，必须首先了解制造厂如何处理即将在CNC机床上加工的工件。下面列出的条目是编制CNC程序时相当常见的任务逻辑次序，这些条目的逻辑次序只是一个推荐的顺序，供您进行进一步的评估。

获取或绘制零件图。

选择加工机床。

确定加工工序。

选取加工刀具。

对编程坐标进行数学运算。

根据刀具和工件材料计算转速和进给量。

编写加工程序。

准备机床设置表和刀具清单。

把加工程序发送到机床。

校验加工程序。

如无需更改，运行加工程序。

多年来，手动编程（不使用计算机）一直是编制工件程序时最常用的方法。最新的CNC控制器通过使用固定或重复的加工周期、变量编程、刀具运动图形仿真、标准数学输入以及其他简捷功能，使手动编程比以往方便了很多。为进一步提高CNC机床编程的效率与精度，人们开发了多种使用计算机编制工件程序的方法。这种计算机辅助 CNC 编程多年来已经得到了广泛的应用。

Key to Exercises of the Text

Ⅰ. 略

Ⅱ. 1. verify 2. display 3. monitor 4. simulation 5. incompatible

Ⅲ. 1. control console 2. program coordinate 3. manual programming

4. We'd better present the facts and reason things out instead of quarreling.

5. The technicians have developed a variety of methods that use a computer to prepare part programs.

阅读材料参考译文及答案略。

第11单元 计算机辅助设计

计算机对设计师而言非常实用，因为它简化了零件设计中经常包含的冗长且乏味的计算过程。1963 年，美国麻省理工学院演示了一个叫作 Sketchpad（画图板）的计算机系统，它能够生成图形信息并将其在阴极射线管（CRT）显示屏上展示。该系统很快成为人们所熟知的 CAD，它使设计师或工程师根据铅笔草图或者模型便可绘制最终的工程图样，如果达不到预期功能要求，还可在显示屏上即时修改。此外，从三视图投影的角度，设计师能把视图转变成三维图形；使用合适的软件，还可模拟零件的实际使用情况。这样，设计师可在显示屏上重新设计零件图、规划如何使用零件，并能够在短时间内进行连续的设计变更。

CAD 的组成

CAD 类似于电视系统，它接收计算机发来的电子信号，并在 CRT 显示屏上生成图像。多数 CAD 系统使用台式计算机，并与主机或服务器相联；同时配以键盘、光笔，或电子平板、绘图仪等附件，操作者可以绘制任何图形或生成要求的任意视角。

有了草图，操作者通常使用光笔或电子平板就能在 CRT 显示屏上绘制适当比例的零件图，并储存于计算机内存之中。如果需要改进，设计师可用光笔、机电光标或电子平板在 CRT 显示屏上直接创作和修改零件图或线条。

完成设计后，工程师或设计师可以检验零件是否达到预期性能。若需要改进，工程师或设计师可迅速、方便地对图样或设计的任意部分进行修改，而无需重绘原图。一旦确认设计结果正确无误，便可用绘图仪直接打印最终的零件图样。

CAD 的优点

CAD 给工业生产带来很多优势，可使工作更加高效而精确。下面列出了 CAD 的一些常见优点：使画图者的工作效率更高、缩短绘图时间、便于修改图样、提高绘图与设计精度、细节规划更好、图样质量更好、提高零件标准化程度、装配工艺更合理且可以减少报废量。

Key to Exercises of the Text

Ⅰ. 1. T 2. T 3. F 4. F 5. F

Ⅱ. 1. successive 2. tedious 3. transform 4. standardization 5. simplified

Ⅲ. 1. So, what should a real interview consist of?

2. The minister of foreign affairs has redesigned a new foreign policy.

3. Generally, it means not putting all of your eggs in one basket.

4. The advantage of this method is its simplicity.

阅读材料　　计算机辅助制造

计算机辅助制造（CAM）可定义为使用计算机系统来规划、管理和控制制造工厂的运作，这个过程中与工厂生产资源的对接可通过直接或间接的计算机接口来实现。CAM 首次于 1971 年用于车身设计和加工。

CAM 的功能主要集中在四个方面：数字控制、工艺设计、机器人技术和工厂管理。

CAM 通过 CAD 编程使生产物理模型成为可能。CAM 可以在一个软件包内进行现实版器件的创建。CAM 可将 CAD 软件数据直接转换为一组生产指令。

工人长时间工作会感到疲惫，导致犯错误。执行生产功能的机器不会感到劳累。因此，出现错误的概率被大大降低了。

结合 CAD，CAM 使制造商得以通过人工操作的最小化来降低商品生产的成本。

所有这些技术发展使得运营成本降低，终端产品价格下降，而制造商的利润增加。

Key to Exercises of the Reading

1. This matter is hard to define.

2. I must convert sorrow into strength.

3. This method can be used in conjunction with other methods.

4. When you do something with confidence, it will result in success.

第 12 单元　柔性制造系统

柔性制造系统（FMS）是 CNC 机床、机器人、工件传输装置组成的集成系统，该系统能对粗坯件或铸件进行所有必要的加工、工件处理与检验，形成最终工件或组件成品。它完全是一个基于软件的全自动无人生产线/装配线。柔性制造系统包含四个主要部分：CNC 数控机床、坐标测量机、工件处理与装配机器人、工件/刀具传输装置（图 12 – 1）。

柔性制造系统的主要元素是 CNC 加工中心或车削中心。只要配置适当的支持系统，自动换刀功能可使这些机床在无人监督的条件下运行。CNC 机床具有内置刀具监测系统，可监测并更换磨损刀具。柔性制造系统运行的主要障碍不在加工中心，而在于机床支持系统，例如工件的装卸与传输系统等。

柔性制造系统的检验功能由坐标测量机实现。坐标测量机在程序控制下向工件各处移动，这一点与 CNC 机床十分相似。不同的是，坐标测量机不用旋转主轴与切削刀具，而是通过电子测头来测量工件的各个特征。测量结果将与编入程序的可接受极限值进行比较。

柔性制造系统常用机器人给机床装卸工件。由于机器人是程序控制装置，缺乏判断能力，所以传输装置需使用专用夹具定位工件，以便机器人能正确处理工件。使用专门设计的机床夹具与夹紧机构，以确保机床上工件位置及夹紧正确。通过工件传输装置、机器人、CNC 机床协调工作，所有工件处理必须按特定顺序完成。未来的机器人可能具有某种人工智能，这使它们对工件定位有一定的判断能力，且能采取必要的纠错动作。

柔性制造系统的第四个重要组成部分是工件/刀具传输装置。这些装置在各机床之间移送工件，也可在机器上来回移动刀库，从而保证每台 CNC 机床上锋利切削刀具的正常供给。现代柔性制造系统主要使用四种传输装置：自动导向车（AGV）、导线导向车、气垫车和硬件导向车。

自动导向车依靠自身携带的传感器和/或程序来确定其运行路径，没有硬件使其与柔性制造系统相连。自动导向车的优点是人们能够通过编程改变其运行路径，每次改变路线无需重新布置导轨或导线；其相应的缺点是，由于缺乏硬件连接，自动导向车是最难以发挥作用的工件传输装置。

导线导向车使用一根埋在地下的导线来确定其路径。车上装有传感器，用于检测导线位置。导线导向车的主要优点是，不同于自动导向车，导线导向车具有使用导线的能力，但不需要如悬空导轨或地面导轨这样的硬件系统。导线导向车的缺点是，如果要求改变行车路线，则需要安装新的地下导线。

气垫车由外部硬件装置（如悬空导线）导向，但它是在气垫上滑行，而不是在导轨上行驶。使用气垫车时，一定要注意在柔性制造系统内部设置切屑排除与控制功能。气垫车行驶路线上如有切屑，则气垫车无法运行。这种传输装置通常用于直线行车路径运输。

硬件导向车是传输装置中最可靠但柔性最差的装置，其传输路径由地面导轨或悬空导轨控制。硬件导向车的优点是可靠、易于和系统其余部分协调。其主要缺点当然是每当改变传输路径或想试验一条新路径时，都要架设新的悬空导轨或铺设新的地面导轨。根据生产线上不同部分的需求，一个大型柔性制造系统可能使用多种类型的传输工具。

Key to Exercises of the Text

Ⅰ. 1. F　2. T　3. T　4. F　5. T

Ⅱ. 1. flexible machining system　　4. 坐标测量机

2. automatic guided vehicle　　5. 悬空导轨

3. production line　　6. 硬件导向车

Ⅲ. 1. adequate　2. obstacle　3. acceptable　4. delivery　5. capability

阅读材料参考译文及答案略。

第 3 章　电子信息技术

第13单元　交　流　电

不同于方向保持不变的直流电，交流电是指大小和方向都会做周期性变化的电流。

1. 交流电

交流电流总是伴随着交流电压存在（或由交流电压产生）。在英语中，缩略词 AC 可以通用，有时既可表示交流电压，也可表示交流电流。

阶跃函数和脉冲函数可用于分析电路首次通电或出现突发性、不规则的输入变化时电路的响应，这称为暂态分析。但如果想知道电路对一个有规律的或重复的输入信号的响应——

稳态分析，到目前为止最有用的函数是正弦函数。

正弦波是交流电最常见的波形，有时我们把正弦交流电简称为交流电。交流电电压 v 可以用时间的数学函数公式表示：

$$v(t) = V_{peak} \sin(\omega t)$$

图 13-1 所示为一个周期（360°）的正弦波波形，虚线表示的是有效值，约为峰值电压的 0.707 倍。其中：V_{peak} 是峰值电压（单位：伏特）；ω 是角频率（单位：每秒弧度）；t 是完成一次周期性变化所需的时间，称为周期（单位：秒）。角频率 ω 与物理频率 f 即每秒振荡的次数（单位：赫兹）有关，有 $\omega = 2\pi f$。

关于正弦波，我们有如下结论：

1）如果一个线性时不变电路的输入是一个正弦信号，则它的响应是一个同频率的正弦信号。

2）一个正弦稳态响应的幅值和相位角可以通过实正弦波或复正弦波求解。

3）如果一个正弦电路的输出响应比输入信号先达到峰值，则称此电路是超前网络，反之，称之为滞后网络。

4）采用相量和阻抗的概念，对同一频域的正弦电路可以采用类似于分析电阻电路的方法，即用基尔霍夫电流定律、电压定律，节点电流分析法，网孔分析法和回路分析法等的相量（复数）形式进行分析。

虽然发电机和许多其他物理现象自然产生的都是正弦波形交流电，但交流电并不只有这一种形式。在电路中也常产生交流电的其他波形，图 13-2 给出了一些波形示例和它们的名称。

这几个简单的波形经常用到，因此有指定的名称。即使在希望出现“纯”正弦波、方波、三角波和锯齿波电压/电流的电路中，现实生活中的实际波形也是理想波形的变形形式。一般来说，将任何接近正弦波的波形称为正弦波，其他统称为非正弦波。

2. 交流电功率

之前在直流电路中曾定义功率是电压和电流的乘积，在电压和电流都是常数的条件下，瞬时功率与平均功率相等。在交流电路中，电压和电流都是正弦量，瞬时功率仍然是电压和电流的乘积，但随时间变化，并不等于平均功率。

在交流电路中，储存能量的元件，如电感和电容（只能储存或释放能量），可能引起能量流方向的周期性反转。能量流中真正（从电源到负载）做单向传输的那一部分，其大小等于交流波形在一个完整周期中的平均功率，称为有效功率（也可称为有功功率）。而由于存储元件（储能时）引起的那部分能量流，在每个周期（放能时）要流回电源的，称为无效功率（中文称无功功率）。

有功功率、无功功率和视在功率的关系可以用矢量的方法来表示。有功功率可表示成水平矢量，无功功率可表示成垂直矢量（图 13-3），则视在功率矢量就是连接有功功率和无功功率矢量构成的直角三角形的斜边。这种表示形式称为功率三角形。根据勾股定理，有功功率、无功功率和视在功率之间的关系为：

$$\text{视在功率}^2 = \text{有功功率}^2 + \text{无功功率}^2$$

有功功率与视在功率的比值称为功率因数，其值通常在 0 ~ 1 之间。

Key to Exercises of the Text

Ⅰ. 1. peak voltage 2. linear, time-invariant circuit 3. angular frequency 4. root mean square (RMS) 5. reactive power 6. apparent power 7. transient analysis

Ⅱ. 1. T 2. F 3. F 4. F

阅读材料 三相电路

含有不同相位的正弦电压源的电路称为多相系统。这个概念的重要性在于多数发电机和配电网的功能都是用多相系统实现的。最常用的多相系统是三相对称系统，它的瞬时输出功率为常数，因此可以减小旋转式发电设备发电时的振动。

频域中Y（星形）连接的三相电源连接如图 13-4 所示，a，b，c 三端称为相线（火线）端，n 端称为中线端。如果相电压 $v_{an}(t)$，$v_{bn}(t)$，$v_{cn}(t)$（或相量 $\dot{v}_{an}$，$\dot{v}_{bn}$，$\dot{v}_{cn}$）的振幅相等，即 $v_{an}=v_{bn}=v_{cn}$；且三相电压之和等于 0，即 $v_{an}(t)+v_{bn}(t)+v_{cn}(t)=0$（或相量式 $\dot{v}_{an}+\dot{v}_{bn}+\dot{v}_{cn}=0$），则称这个电源为三相对称电源。

设每个正弦电压的幅电度为 V，如果我们任取 v_{an} 的初相位为 0°，即 $v_{an}(t)=V\sin(\omega t+0°)$（或 $\dot{v}_{an}=V\angle 0°$），则一个平衡的电源可能源于以下两种情况：

第 1 种情况	第 2 种情况
$v_{an}(t)=V\sin(\omega t+0°)$	$v_{an}(t)=V\sin(\omega t+0°)$
$v_{bn}(t)=V\sin(\omega t-120°)$	$v_{cn}(t)=V\sin(\omega t-120°)$
$v_{cn}(t)=V\sin(\omega t-240°)$	$v_{bn}(t)=V\sin(\omega t-240°)$

对于第 1 种情况，$v_{an}(t)$超前 $v_{bn}(t)$120°，而 $v_{bn}(t)$超前 $v_{cn}(t)$120°，这称为正相序或 abc 相序。同样，第 2 种情况称为负相序或 acb 相序。显然，负相序可以通过简单地变换端子标称转换为正相序，因此我们只需研究正相序即可。

现在把平衡的三相电源连接到平衡的星形连接的三相负载上，如图 13-5 所示。相线（火线）端之间的电压称为线电压，a 和 A、b 和 B，c 和 C 之间的电流称为线电流。对Y连接三相平衡负载，有：$V_{line}=\sqrt{3}V_p$ 和 $I_{line}=I_p$。

如果线路阻抗都相等，有效负载是三相对称的，则中线电流等于 0，因此可以移除中线。

与三相对称 Y 负载连接相比，负载的 Δ（三角形）接法应用更多，将一个Y连接三相电源连接到一个三角形对称连接的三相负载上，如图 13-6 所示，可以看到每个负载都直接接到两个相线之间，因此在三角形接法中加上或除去一个负载比星形接法中更容易。对 Δ 连接三相对称负载，有：$V_{line}=V_p$ 和 $I_{line}=\sqrt{3}I_p$。

无论一个相位负载是星形连接还是三角形连接，对于线电压、线电流和负载的阻抗角，我们可以用同一个公式表示三相平衡负载吸收的总功率

$$P=\sqrt{3}V_{line}I_{line}\cos\varphi=3V_pI_p\cos\varphi$$

Key to Exercises of the Reading

1. 三相平衡系统 2. 正相序 3. 负相序 4. 线电压 5. 线电流

第 14 单元 电子元件

大量电子元件都有其相应的表示符号。能够识别更为常见的元件并掌握它们的实际用途

是很重要的。下面画出了一些电子元件，值得注意的是，一种类型的元件往往可以用几种不同的符号表示。

电阻

电阻阻碍电流的流动。比如，一个电阻可以与一个发光二极管（LED）串联来限制流过LED 的电流。图 14-1 所示为电阻实物图及其电气图形符号。

电阻可以连接在任一回路中。它们不会因焊接高温而损坏。

电容

电容存储电荷。由于电容可使交流信号轻易通过，而阻止直流信号，因此它们经常应用于滤波电路中。图 14-2 所示为电容实物图及其电气图形符号。

电感

电感为无源电子元件，它以磁场的形式存储能量。电感就是一个将导线缠绕了若干匝数的线圈，通常会把它缠绕在像铁芯这样的磁性材料上。图 14-3 所示为电感实物图及其电气图形符号。

二极管

二极管只允许电流单向流通。其电气图形符号中的箭头表示电流能够流通的方向。二极管是真空管的电子版，实际上早期的二极管就称为真空管。图 14-4 所示为二极管实物图及其电气图形符号。

晶体管

标准晶体管分为两类，NPN 型和 PNP 型，它们在电路图中的符号不同。字母（N 和 P）表示制造晶体管的半导体材料不同。图 14-5 所示为晶体管实物图及其电气图形符号。

集成电路（芯片）

集成电路通常称为 IC 或芯片。它们将复杂电路固化在小型半导体（硅）芯片上。该芯片被封装在一个塑料固定物上，其引脚间隔距离为 0.1 英寸（2.54mm），这样的栅格将适合带形板和电路试验板的孔距。在封装里面用很纤细的导线连接芯片引脚。图 14-6 所示为集成电路实物图及其电气图形符号。

发光二极管（LEDs）

有电流通过时，LEDs 会发光。

LEDs 必须连接到正确的回路中，电路图中可用“a”或者“+”表示正极，用“k”或者“-”表示负极（注意，的确是用“k”而非“c”来表示阴极）。负极是较短的引脚，并且在 LEDs 的圆形体上可能有窄小的平面端面。图 14-7 所示为 LED 实物图及其电气图形符号。

其他电子元件

图 14-8 所示为其他电子元件的实物图和电气图形符号。包括可变电阻器、麦克风、扬声器、熔丝、信号灯/灯丝、电动机、带磁芯的电感器和开关。

Key to Exercises of the Text

Ⅰ. 1. T　2. F　3. T　4. T　5. T

Ⅱ. 1. d　2. e　3. a　4. c　5. b

Ⅲ. 1. Resistors　2. filter　3. version　4. plastic　5. only one

Ⅳ. 1. It is essential/necessary that you can recognize the more common electronic components and understand what they actually do.

2. Resistors are not damaged by heat when soldering.

3. Metals do not melt until heated to a definite temperature.

4. The electronic components include the electron tube, transistor, integrated circuit and so on.

阅读材料参考译文及答案略。

第15单元　晶体管电压放大器

放大器对很多种电子设备来说是必不可少的，如收音机、示波器和录音机等。通常小的交流电压需要放大。一个结型晶体管连成共射模式，在其集电极回路中接入合适的电阻，也称负载，即可构成电压放大器。

将一个小的交流输入电压 U_i加在基极 - 发射极电路上，基极产生的小的电流变化会使负载的集电极电流产生较大的变化。负载将交变电流转变成交变电压，即输出交流电压 U_o。（U_o远比 U_i大）。

晶体管电压放大器电路示例如图15-1所示。了解电压的放大过程，首先考虑没有输入电压的情况，即 $U_i=0$，称为静态。

晶体管在放大状态工作时，其发射结必须为正向偏置。为保证发射结正偏，一个简单的方法是接入一个电阻 R_B，称为基极偏置电阻。一个稳定（直流）的基极电流 I_B从电源正极流出，经 R_B流入基极，再经发射极流回电源负极。一旦放大器最佳状态下的 I_B值确定，就可以计算出 R_B的值。

如果 V_{CC}为电源电压，V_{BE}为基极 - 发射结电压（NPN 硅管的电压常为 +0.6V），则对于基极 - 发射极回路，因为直流电路电压为各项的加和，我们可以写成

$$V_{CC}=I_BR_B+V_{BE} \tag{15-1}$$

I_B引发了一个更大的集电极电流 I_C，I_C则在负载 R_L上产生电压降 I_CR_L。若 V_{CE}为集电极 - 发射极电压，则对集电极 - 发射极回路，有

$$V_{CC}=I_CR_L+V_{CE} \tag{15-2}$$

当输入交流电压 U_i且为正值时，基极 - 发射极电压略有增加（如从 +0.60V 增加到 +0.61V）。当 U_i转为负值时，基极 - 发射极电压略有下降（如从 +0.60V 下降到 +0.59V）。这使得一个小的交流电流叠加在静态基极电流 I_B上，其结果是使直流电压产生变化。

当基极电流增大，集电极电流大幅增大，由式（15-2）可知，将相应引起集电极 - 发射极电压的大幅下降（因为 V_{CC}固定）。同理，基极电流减小将引起集电极 - 发射极电压大幅增加。实际上，U_i毫伏数量级的上下波动，就会导致负载 R_L上的电压上升或下降几个伏特，集电极 - 发射极电压也做相应幅度的变化。

集电极 - 发射极电压可以认为是一个叠加在稳定直流电压之上的交流电压，比如叠加在静态电压 V_{CE}上。只有交流部分是被需要的，电容器 C 隔绝了直流部分，但是允许交流电通过，比如输出电压 U_o。

Key to Exercises of the Text

Ⅰ. 1. F　2. F　3. F　4. T　5. T

Ⅱ. 1. a　2. e　3. c　4. b　5. f　6. d

Ⅲ. 1. A trained dog can act as a guide to the blind.

2. Does electricity convert easily into other forms of power?

3. In this way they can better apply theory to practice.

4. People regard wind power as clean energy.

阅读材料　集成电路

集成电路是一些相互连接的电路元件的组合，如晶体管、二极管、电容器和电阻器等，它是用半导体材料制成的较小的电子器材。第一块集成电路是在20世纪50年代由德州仪器公司的Jack Kilby和仙童半导体公司的Robert Noyce合作开发的。

这些构成集成电路的电气连接元件称为集成元件。如果集成电路只包含一种类型的元件，便称之为组件或一套元件。

集成电路在很多设备上都有应用，包括微处理器、音频和视频设备以及汽车等。集成电路通常根据其包含的晶体管和其他电子元件的数量来归类，包括：

- SSI（小规模集成电路）：每个芯片中有至多100个电子元件。
- MSI（中规模集成电路）：每个芯片中有100～3000个电子元件。
- LSI（大规模集成电路）：每个芯片中有3000～100000个电子元件。
- VLSI（很大规模集成电路）：每个芯片中有100000～1000000个电子元件。
- ULSI（超大规模集成电路）：每个芯片中有100万个以上的电子元件。

随着一个集成电路集成更多数量晶体管的能力的提高，对专用集成电路的需求已更加普遍，至少对大批量的应用来说如此。硅芯片技术的发展使得集成电路设计者可以在一个芯片上集成多于几百万个的晶体管，现在甚至可以在单一芯片上集成一个完整的中等复杂的系统。

集成电路的发明是电子工业的一项重要革命。凭借这一技术，完全可以实现尺寸的显著减小和重量减轻。更重要的是，能够实现高可靠性、良好的工作性能、低成本和低功耗。集成电路已广泛应用于电子工业。

Key to Exercises of The Reading

Ⅰ. 略

Ⅱ. 1. Their success is the result of a fortuitous combination of circumstances.

2. They'll have to make up time lost during the strike.

3. He carefully began to classify the results of his examinations.

第16单元　因特网简介

因特网是各个独立的网络通过中间网络设备连接到一起形成的一个功能独立的大型网络集合。它将世界上数百万的计算机连接在一起，网络内的任何计算机都能全天候地向其他计算机传送信息。这些计算机可能是属于家庭的、学校的、大学的以及政府部门的或大小企业的。它们可以是任何类型的，例如单独的私人计算机或者是一个公司或学校网络的工作站。互联网经常被描述为“网络的网络”，因为所有较小的网络结合到一起构成了叫作互联网的巨型网络。所有连接到互联网的计算机都是平等的，唯一不同的是连接速度，这取决于你的互联网服务供应商和调制解调器。

网络发展历史

最早的网络是使用主机和终端的分时复用网络。此类网络主要由 IBM 的系统网络体系（SNA）和数字网络体系结构实现。

局域网（LANs）随着个人电脑的发展得到了发展。局域网可以使相对较小的地理区域内的多个用户交换文件和信息，以及连接文件服务器和打印机等共享资源。

广域网（WANs）是将地理位置上分散的局域网用户互连在一起形成的互通网络，主要使用的一些连接技术包括 T1、T3、ATM、ISDN、ADSL、帧中继、无线链路等。每天都会有连接分散局域网的新技术出现。

当今，高速局域网和互联网被广泛地使用，主要因为它们可提供并支持高速率的传输和高带宽的应用，例如多媒体和视频会议。

TCP/IP

互联网中所有的计算机都使用传输控制协议/互联网协议群进行通信，简称 TCP/IP。TCP/IP 是互联网中传输数字数据的通信协议。TCP/IP 是由美国国防部（DOD）的一个研究项目开发的，用于连接不同网络供应商设计的不同网络，从而形成一个网络的网络（即互联网）。互联网中的计算机采用客户端/服务器架构，这意味着由远端的服务器为本地的客户端机器提供文件和服务。客户机可以安装软件以利用最新的接入技术连接入网。

网络服务

互联网用户可以获取广泛的服务，如电子邮件、文件传输、大量信息资源、加入兴趣群组、交互式协作、多媒体显示、实时广播、突发重大新闻、网络购物等。

互联网代表的是全球性的信息和知识资源，也提供了一个商务合作的平台，这就导致了大量信息服务供应商的出现。不幸的是，在互联网上也存在一些不健康的网络服务提供者，例如那些从事儿童色情交易和极端行为宣传的。也许我们过于简单化了，但网上确实有好的信息和坏的信息。信息也可以用另外一种方式分类，即开放的信息和内部的信息，而后者是需要安全保护措施的。

互联网中的计算机可以使用下述一个或全部的网络服务：

1）电子邮件。你可以收发电子邮件。

2）远程登录。你可以远程登录到另外一台计算机，就像你在那里一样直接使用它。

3）文件传输协议。使你的计算机可以迅速地从远端检索完整的复杂文件资源，然后浏览或保存到自己的计算机里。

4）地鼠检索程序。早期的一个只能获取文本的互联网文档接入方式。地鼠程序已经完全包含在万维网中，但你仍可以在网页中找到链接的地鼠程序文件。

作为最大和增长最快的互联网服务，万维网（WWW 或者 Web）包含上面提到的所有互联网服务甚至更多。通过万维网，你可以检索文档，观看图片、动画、视频，收听音频文件，语音通信。

当你使用网景或微软公司的互联网浏览器，或其他的浏览器登录到因特网上，那你就正在查看万维网上的文档。当前万维网的运行基础是 HTML 编程语言。超文本可使网页包含链接，这些链接是网页的某些区域、按钮或者图像，你可以用鼠标点击它们，从而检索到其他的文档资料。这种使用超文本链接功能的“点击能力”是万维网最独特和具有革命性的特征。

Key to Exercises of the Text

Ⅰ. 1. Local – Area Networks（LANs） 2. high – bandwidth applications 3. Wide – Area Networks（WANs）
4. Electronic mail（E – mail） 5. real – time broadcasting 6. interactive collaboration
7. global resources 8. FTP or File Transfer Protocol 9. Transmission Control Protocol

Ⅱ. 1. 广泛的服务 2. 互联网服务供应商
3. 系统网络体系结构 4. 数字网络体系结构
5. 交换文件和信息 6. 高速局域网和互联网
7. 客户端/服务器架构 8. 相对较小的地理区域内
8. 多媒体和视频会议 10. 获取互联网文档的方法

Ⅲ. 1. A typical PDA can function as a mobile phone, fax sender, and personal organizer.
2. Infrared is widely used in industry and medical science.
3. These proposals represent a realistic starting point for negotiation.
4. After reflow, carefully remove substrate with components and allow to cool.

阅读材料 比尔·盖茨

威廉（比尔）·亨利·盖茨是微软公司主席、首席软件设计师，微软是全球个人及商务计算机领域软件、服务和互联网技术的领军企业。微软公司现有超过11.8万名员工分布在世界上60多个国家，截至2015年6月底的财政年度总收入为936亿美元。

盖茨1955年10月28日出生，与他的两个姐姐一起在西雅图长大。盖茨曾就读于一所公立小学和私立湖滨中学。在那里，盖茨对软件产生了兴趣，13岁时便开始编制计算机程序。

1973年，盖茨作为新学生入读哈佛大学，住在史蒂夫·鲍尔默的楼下，后者现在是微软公司的首席执行官。在哈佛就读期间，盖茨为第一台微机—MITS Altair 开发了 BASIC 编程语言的一个版本。

大学三年级时，盖茨从哈佛退学，专心于微软公司的业务，这个公司是盖茨在1975年与他的儿时好友保罗·艾伦一起创建的。他们坚信计算机将是每张办公桌上和每个家庭里非常有用的工具，基于这一信念，他们着手开发个人计算机软件。盖茨对个人计算机的远见卓识是微软公司和软件行业成功的关键。

1999年，盖茨写了一本名为《未来时速》的书，书中介绍了计算机技术怎样以全新的方式从根本上解决商务问题。此书有25种语言的版本，在60多个国家均有销售。《未来时速》得到评论界的广泛赞誉，被《纽约时报》《今日美国》《华尔街日报》和亚马逊网站列入最畅销书目录。盖茨在1995年出版的另一本书《未来之路》，曾连续七周高居《纽约时报》畅销书排行榜榜首。

除了对计算机和软件充满热爱以外，盖茨对生物技术也有兴趣。他是数家生物技术公司的投资人，还创立了致力于开发世界上最大的视觉信息资源的Corbis公司。此外，盖茨与移动电话的先驱克雷格·麦考共同投资了 Teledesic 公司，这家公司正在实施一个雄心勃勃的计划：使用数百颗低轨卫星为全世界的用户提供双向的宽带远程通信服务。

Key to Exercises of the Reading

1. B　　2. C　　3. B　　4. D　　5. A

第 17 单元　4G 网络技术

4G 是目前正在开发的具有高速宽带移动能力的第四代无线通信技术。它的特点是可实现更高速的数据传输和更好的声音质量。虽然 ITU（国际电信联盟）尚未给出确切的定义，行业已经确认了以下技术为 4G 技术：

1）WiMAX（全球微波互联接入）。

2）3GPP LTE（第三代合作伙伴长期演进项目）。

3）UMB（超移动宽带）。

4）Flash - OFDM（快速低时延接入无缝切换的正交频分复用）。

4G 技术正在被开发以满足 QoS（服务质量）和网络流量优先化的速率要求，从而保证良好的服务质量。这些机制对于使用大带宽的应用调节是必需的。此类应用如无线宽带上网、MMS（多媒体信息服务）、视频聊天、移动电视、HDTV（高清电视）、DVB（数字视频广播）、实时音频、高速数据传输。

ITU 为 WiMAX 和 LTE 设定的目标是：当用户相对于基站高速移动时，数据传输速率达 100Mb/s；位置固定时，速率达到 1Gb/s。

大量的 4G 兼容设备随着行业的发展出现了。不仅局限于 4G 手机或笔记本电脑，在数十种不同的移动设备，如摄像机、游戏装置、自动售货机和冰箱等设备上也有应用。

现在的趋势是给每一个可以提供和纳入嵌入式 4G 模块的便携式装置提供无线互联网接入。4G 技术不仅可以提供互联网宽带连接，而且具有高级别的安全性，有利于像自动售货机和计费装置等包含金融交易的设备安全运行。

上述 4G 技术具有以下共同的主要特征：

1）多输入多输出：通过包含多天线和多用户多入多出的空间处理手段达到超高的频谱效率。

2）频域均衡（SC - FDE），如下行链路中的多载波调制（正交频分复用 OFDM）或单载波频。

上行链路中的频域均衡：在没有复杂均衡的条件下利用频率选择性信道特性。

3）Turbo 原理的纠错码：尽量减小接收端需要的信噪比（SNR）。

4）基于无线信道的调度：采用时变信道。

5）链路适配：自适应调制和纠错码。

6）利用移动 IP 的移动性。

7）基于 IP 架构的家庭基站（连接到固定网络宽带基础设施的家庭节点）。

Key to Exercises of the Text

Ⅰ. 略

Ⅱ. 1. T　2. F　3. F　4. F

Ⅲ. 1. Currently, the Euro - American and Japanese cartoons have already developed into a rather mature stage.

2. Workers should ensure mould coating in good quality and on time delivery.

3. How does a private pilot get access to the airways?

4. Several different names have emerged to characterize different parts of the spectrum.

阅读材料　交互式网络电视 (IPTV)

交互式网络电视（IPTV）系统中的电视服务使用以互联网协议族为基础的分组交换网络（如互联网）来传递信号，而不是通过传统的地面、卫星信号和有线电视格式传送的。

IPTV 业务主要可分为三大类：

1）电视直播，可以（也可以不）与正在播出的相关电视节目进行互动；

2）时移电视：电视回看（回放几小时或几天以前播出的电视节目），电视重播（从头开始重新播放当前电视节目）；

3）视频点播（VOD）：浏览视频节目单，这个节目单与当前正在播放的电视节目无关。

IPTV 与互联网电视的不同主要体现在它持续的标准化制定进程（比如欧洲电信标准化组织）和基于用户的通信网络优先部署方案，这些方案中的频道能够通过机顶盒或其他客户端设备高速连接至终端用户处所。

IPTV 被定义为能够为用户安全可靠传送娱乐视频和相关服务的一种服务。这些服务可以包括像电视直播、视频点播（VOD）和互动电视（ITV）类的节目。这些服务都是通过与入口无关的采用 IP 协议的分组交换网络来传输音频、视频和控制信号的。与公共互联网上的视频传播相比，IPTV 的部署使网络安全和性能都受到严格管理，从而保证了卓越的娱乐体验，也因此为内容提供商、广告商和客户提供了非常有吸引力的商业环境。

基于互联网协议的平台有显著的优势，可将电视与其他基于 IP 的服务（如高速互联网接入和 VoIP）进行集成。

交换 IP 网络还可以提供更多内容和功能的传送。在典型的电视或卫星网络中，视频广播技术可使所有内容不断流向每个客户，客户可在机顶盒上进行内容切换。客户可以从电信、有线或卫星公司提供的内容中进行各种选择，这些内容被这些公司灌入“管子”里并被发送到客户住所。交换 IP 网络的工作方式则不同，内容保留在网络中，只有客户选择的内容被发送到客户的家中。这样就释放了带宽，而客户的选择不受“管道”大小的限制。这也意味着客户的隐私可能会受到比使用传统电视或卫星网络更大程度的损害。它还可以提供一种侵入手段，或至少可以破坏私有网络（如拒绝服务）。

IPTV 技术将视频点播（VOD）引入电视，允许用户浏览在线节目或电影目录，可观看预告片，然后选择已经录制好的节目。被选定的节目几乎可以即刻在客户电视或 PC 上进行播放。

从技术上讲，当客户选择电影时，在客户的解码器（机顶盒或 PC）与流媒体服务器之间将设置点到点的单播连接。RTSP（实时流协议）保证了特技播放功能（暂停、慢动作、前进/倒退等）所需的信号传输。

Key to Exercises of the Reading

Ⅰ. 1. F 2. F 3. T 4. F

Ⅱ. 1. The interface is very simple and users can define the mails they received as junk mails.

2. I have orders to deliver it to Mr. James personally.

3. As a basketball player, his height gives him significant advantage.

第18单元 物 联 网

物联网（IOT）也称传感网，是指将各种信息传感设备与互联网结合起来而形成的一个巨大网络，这个网络可使所有的物品与网络连接，以便于识别和管理。因其具有全面感知、可靠传递和智能处理的特点，它被认为是继计算机、互联网和移动通信网络之后的又一次信息产业浪潮。

轻触一下电脑或者手机的按钮，即使在千里之外，你也能了解到某件物品的状况或者某个人的活动情况。发送一个短信，你就能打开风扇；如果你的住宅被非法侵入，你还会收到自动电话报警。这些已不只是好莱坞科幻大片中才有的场景了，物联网使这些场景正逐步出现在我们的生活中。

这些能够得以实现是因为物联网里有一种叫作射频识别（RFID）的可存储物体信息的关键技术。RFID系统由两部分组成（图18-1）：天线收发器（通常集成在一个读入器里）和应答器（标签）。天线发出无线电信号激活标签并读出和写入数据。被激活后，标签发射返回数据到天线。标签传输的数据可能提供识别或定位信息，或者是有关产品标签的具体信息，如价格、颜色和购买日期等。低频RFID系统（30～500kHz）的传输距离较短（一般小于6英尺）。高频RFID系统（850～950MHz和2.4～2.5GHz）的传输距离更长（超过90英尺）。通常，频率越高，系统的价格越昂贵。

比如在手机里嵌入RFID－SIM卡，手机内的“信息传感设备”就能与移动网络相连，这种手机不仅可以确认使用者的身份信息，还能缴纳水电燃气费、彩票投注、飞机票订购等多种支付服务。

只要在物体中嵌入特定射频标签，传感器和其他与互联网连接的设备就能形成一个庞大的网络系统。通过这个网络，即使远在千里之外，人们也能轻松获知和掌控物体的信息（如图18-2所示）。

更具体地说，让我们想象这样一个世界，围绕着我们的大量事物是“自主”的，因为它们有：

名字：一个具有独特代码的标签。

存储器：存储那些不能立即从网上获得的信息。

通信方式：移动通信和高能效通信，如果可能的话。

传感器：为了与环境互动。

习得行为或自主行为：根据主人给出的目标，按特定逻辑行动。

当然，像地球上其他事物一样，这些都必须以电子信息的形式存在于一个网络当中。

一些专家预测，10年内物联网就可能非常普及并发展成为上万亿规模的高科技市场。然后在个人健康、交通控制、环境保护、公共安全、工业监测及老年护理等几乎所有领域发挥作用。也有专家表示，只需3~5年时间，物联网就会改变人们的生活方式。

物联网有很大的发展前景，但这些系统被广泛接受之前，必须解决商业、政策和技术上的挑战。早期应用者需证明，这种新的传感器驱动的商业模式将创造卓越的价值。行业组织和政府监管机构应该研究数据隐私和数据安全的规则，特别是针对那些可接触到消费者敏感信息的使用情况。在技术方面，传感器和执行机构的成本必须下降至某个水平才能实现广泛的应用。网络技术和相应的标准，必须改进到支持数据在传感器、计算机和执行机构间自由流动。收集和分析数据的软件，以及图形显示技术，必须提高至海量数据可以被人们接收的水平。

Key to Exercises of the Text

Ⅰ. 略

Ⅱ. 1. h　2. g　3. a　4. d　5. c　6. b　7. f　8. e

Ⅲ. 1. A healthy diet should consist of wholefood.

2. The teacher created his own slides, then embedded in his classroom blog.

3. The boy first showed promise as an athlete in grade school.

阅读材料　互联型汽车如何改善驾驶体验

为什么开车让人感到厌烦？

马切伊·克兰兹，思科互联工业集团副总裁兼总经理，给出了世界上尚未互联的汽车的严峻的统计结果：数据显示11%~13%的通勤时间都浪费在城市交通拥堵上，共计900亿小时（似乎一半的时间浪费在旧金山到圣何塞的101号高速公路上）。约7%~12%的城市拥堵是由司机寻找停车位造成的（貌似都发生在旧金山的北海滩附近）。10%~17%的城市燃料浪费在没有交叉车流却需要等待交通信号灯放行上。80%的交通事故（630万起）是由司机精神不集中造成的。

互联型汽车如何改善驾驶体验

道路拥堵，意味着需提高交通管理和优化道路网络。对于停车问题，互联型汽车可以连接应用程序找出最近的、最经济的可用停车位并提供给汽车导航系统。汽车还可以智能调整行车速度，以提高燃油效率。

例如，智能交通信号灯能探测出是否一个方向上有10辆汽车在等待，但在另一个方向上只有一辆，然后调整交通灯的时间来保持交通畅顺。克兰兹说，他们可以沿直线方向建立交通信号的“绿波”，以保持车道内车流的高效流动。

还有一个想法是，如果一辆车能够感知其他车辆、交通信号灯和其他道路基础设施的情况，那他们就可以更有效地做出调整。举个例子来说，如果你的车知道前面的车是要转弯还是制动，那么它就可以在实际察觉到前车动作之前做出反应。

思科估计这可能会减少7.5%的交通拥堵时间和4%的车辆燃料费用浪费，并且可降低汽车维修成本和保险成本。克兰兹说，这种好处对于车队管理来说尤为明显。例如，在一个拥有10000辆运输卡车的公司中使用互联技术来安排预防性维护是非常有价值的。

在防止意外事故发生方面，互联型汽车可以在离前面的车辆太近时对司机做出警告。克兰兹说，“如果司机未做出回应，在某些情况下汽车会自行制动以避免事故的发生。”

物联网应用的例子不胜枚举。到2015年，全球将有60亿物体被连接到互联网。尽管看起来物联网好像是一个非常复杂的概念，但物联网比连接到因特网的对象更简单、更令人惊叹。令人兴奋的是，这些互联的物体会学习并适应用户的行为。物联网可能已经影响到了你的生活，如果还没有，那它的影响会很快到来。

Key to Exercises of the Reading

Ⅰ. 略

Ⅱ. 1. F　2. T　3. T　4. F

第4章　应用技术

第19单元　可编程序逻辑控制器（PLC）

可编程序逻辑控制器，PLC，或者称可编程序控制器，是一种小型计算机，可用于实际流程的自动化生产，比如控制工厂的装配线机械等。可编程序控制器通常使用微处理器进行控制。其程序可控制复杂的加工工序，并由工程师编写。程序通常存储在电池备份存储器和（或）电可擦可编程序只读存储器中。

PLC的基本结构如图19-1所示。

从图19-1中可见，PLC有4个主要的部件：程序存储器，数据存储器，输出设备和输入设备。程序存储器用来存储逻辑控制序列指令。开关状态、联锁状态、数据的初始值等其他工作数据存储在数据存储器中。

与其他计算机比较，PLC最大的区别是有专用的输入/输出装置。这些装置将PLC与传感器和执行机构相连接。PLC能读取限位开关、温度指示器和复杂定位系统的位置信息。有些甚至使用了机器视觉。执行机构方面，PLC能驱动任何一种电动机、气压或液压缸、薄膜装置、电磁继电器或螺线管。输入/输出装置可能安装在简单的PLC内部，或者PLC可通过外部I/O模块与接入PLC的专用计算机网络相连。用可编程序控制器替代要使用成百上千个继电器和凸轮计时器的早期自动控制系统更加实惠。一台PLC通常可以通过编程来替代数千个继电器。可编程序控制器已在自动化的制造工业中使用，通过更新软件取代硬连线控制面板的重新连线。

可编程序控制器经过数年的发展，其功能包括典型继电器控制，复杂运动控制，过程控制，分布控制系统和复杂网络技术。

最早的PLC采用简单的梯形图来描述其逻辑程序，梯形图源自于电气连线图。电气技师们使用梯形逻辑图很容易找出原理图中的电路问题。这样大大减少了技术人员的忧虑。

如今，PLC的功能已经非常可靠，但是可编程计算机仍然有其发展空间。依照国际电工委员会的61131-3标准，现在PLC可以使用结构化的编程语言和逻辑初等变换来进行编程。

某些可编程序控制器可使用一种称为顺序功能图表的图形编程符号。

然而，值得注意的是，PLC不再是非常昂贵的产品（通常数千美元），而成了一种普通的产品。如今，功能完善的PLC也只要数百美元。使机床实现自动化还有其他的方法，比如传统的基于微控制器的设计，但是这两者之间有一些差别：PLC包含所有能直接操作大功率负载的输出功能，而微控制器还需要电子工程师设计专用的电源和电源模块等。同时，微控制器设计的方式不具备 PLC 现场可编程的灵活性。那也正是 PLC 在生产线上得以运用的原因之一。PLC 控制系统是一些典型的高度用户化的系统，所以与一次请一个设计人员进行专门的设计比起来，一台 PLC 的成本要低一些。另一方面，在大批量生产中，因为其构件成本低廉，采用的用户化系统能很快收回成本。

Key to Exercises of the Text

Ⅰ. 略

Ⅱ. 1. F　　2. F　　3. T　　4. T　　5. T

Ⅲ. 1. The PLC usually uses a microprocessor.

2. PLCs read limit switches, temperature indicators and the positions of complex positioning systems.

3. The earliest PLCs expressed all decision making logic in simple ladder logic.

4. PLCs contain everything needed to handle high power loads right out of the box.

阅读材料　　PLC 编程

最初 PLC 编程基于与继电器逻辑布线图相同的技术（梯形图），这样电工、技术员和工程师都不需要学习计算机编程。但是这种方法被普遍接受了，成为今天 PLC 编程的最通用的技术。图 19-2 所示为梯形逻辑图示例，为了说明这个图，我们可以想象左边的竖线为电源线，称为火线，右边的是零线。图中有两条通路，每一条通路上都有输入或输入的组合（两条竖线），以及输出（圆形符号）。当通过适当组合使输入开启或关闭时，电流就可以从火线流入，通过输入，控制输出，最终到达零线。输入可以来自传感器或开关。输出可以是 PLC 外接的一些可控制开闭的器件，如灯或电动机。顶层通路中的连接一个常开、一个常闭，如果输入 A 合上（则 A 动作，其常开触点成为闭合），B 断开（则 B 不动作，其常闭触点仍为闭合），则电流会流过输出并使输出有效。而任何其他输入的组合（另 3 种组合为：A 合上，B 合上；A 断开，B 合上；A 断开，B 断开）都会使输出 X 断开（第一条通路不通）。

图 19-2 中的第 2 条通路比较复杂，有多个可能的输入组合会使输出 Y 接通。在通路左侧，如果 C 断开、D 接通，电流可以流通；如果 E 和 F 都接通，电流也可以（或与上条通路同时）流通，这时电流流过半条通路，如果还有 G 或 H 接通，则电流就流到输出端 Y 了。

PLC 编程还有其他方法，其中最早的技术之一是符号指令。这些指令可以直接从梯形图中导出，通过一条简单的编程引脚输给 PLC。图 19-3 所示为符号指令的一个例子，在这个例子中，从顶部到底部，每次读取一条指令。第 0 行是对输入端 00001 的指令 LDN（输入加载并取反）。这条指令会检查输入端 00001，如果断开就输给 PLC 一个 1（或真），如果接通就输给 PLC 一个 0（或假）。下一条用 LD（输入加载）语句来检测输入端 00002，如果输入端断开就记为 0，如果输入接通则记为 1。AND 语句是把上两条语句的值相“与”，如果都为真则结果为 1，否则结果为 0。对 00003 和 00004 输入端的处理过程相同，第 5 行 AND 指令把后两个 LD 指令的输出值相“与”会得到一个输出值。OR 指令对上述 AND 指令所保留的两个值进行“或”运算，如果其中任何一个为 1，输出为 1，否则结果为 0。最后的指令是 ST（存储输出），会把最终的输出值存储起来，如果是 1 则输出端通电，如果是 0 则输出端断开。

图 19-3 所示的梯形逻辑图与代码程序等效，即使你用梯形逻辑图编程，也要把它转换成代码形式输给 PLC，PLC 才能执行。

Key to Exercises of the Reading

Ⅰ. 略

Ⅱ. 1. 这样电工、技术员和工程师都不需要学习计算机编程。

2. 输出可以是 PLC 外接的一些可控制开闭的器件，如灯或电动机。

3. 最早的技术之一是符号指令。

4. 这些指令可以直接从梯形图中导出，通过一条简单的编程引脚输给 PLC。

第 20 单元　自动控制系统

近些年来，控制系统在现代文明和科技的发展和进步中变得日益重要起来。无论多复杂的系统，都是由输入（目标）、控制系统及输出（结果）组成的。实际上，我们日常的活动都会受到某些控制系统的影响。控制系统有两个主要的分支：①开环系统和②闭环系统。

开环系统

开环系统也称为非反馈系统，它是两个系统中较简单的。在开环系统中，没有办法保证实际速度自动地接近期望速度。实际速度可能跟期望速度差得很远，这是因为实际速度会受风速及/或路面情况的影响，如上山或下山等。

闭环系统

闭环系统也称为反馈系统。它设有一套机制来保证实际速度自动接近期望速度。

反馈回路

反馈回路是设计控制系统时常用且有效的方法。反馈回路会参考系统输出，使系统能够调节其性能来实现期望的输出响应。

反馈控制系统主要有两种类型：负反馈控制系统和正反馈控制系统。在正反馈控制系统中，设定值和输出值相加。在负反馈控制系统中，设定值和输出值相减。通常，负反馈系统比正反馈系统更稳定。负反馈也使系统对分量值和输入值随机波动的响应更小。

控制函数的定义方式很多。需明确的是，系统偏差是设定值与实际输出值的差值。当系统输出值与设定值匹配时，偏差为零。设定值与输出值的差值过大会导致较大的偏差。例如，如果期望速率是每小时 50 英里，而实际速率是每小时 60 英里，偏差值就是每小时负 10 英里，那么汽车就应减速。这个数字揭示的规律体现了巡航控制系统的控制函数工作原理。

当谈及控制系统时，需要记住的是，工程师通常已经有一些给定参数的系统，如执行机构、传感器、电动机和其他装置等，并且他们要对这些系统的性能进行调节。很多情况下，不可能打开系统（也就是“被控对象”）从内部对其进行调节：必须从系统外部做一些改变，以使系统按期望的行动做出反应。这个目标是通过在系统中加入控制器、补偿器和反馈结构来实现的。

Key to Exercises of the Text

Ⅰ. 1. T　2. F　3. F　4. F

Ⅱ. 1. 基本部件　2. 闭环系统　3. 负反馈　4. 控制函数

5. positive feedback　6. component value　7. parameter set　8. open-loop system

Ⅲ. 1. Her legs have now ceased to function.

2. Birds in outside cages are immune to airborne bacteria.

3. The air humidifier is of fine performance, it can adjust itself automatically according to the temperature and moisture inside.

4. High-end speed sensor and curve feedback system are set in the Formula 1.

阅读材料　控制系统组件

众所周知，一个闭环控制系统包括三个基本组件。分别是：

1. 误差检测器。该装置接收低功率输入信号和具有不同物理性质的输出信号，并将它们转换成同一个常见物理量来求差，求差后输出一个低功率的具有正确物理性质的误差信号来驱动控制器。误差检测器通常包含传感器，用来将一种物理形式的信号转换为其他形式。

2. 控制器。控制器是一个放大器，接收低功率误差信号，以及外部电源的供电。接着，一个大小可控的功率（具有正确的物理性质）便提供给了输出元件。

3. 输出元件。它将从控制器接收的具有正确物理性质的功率信号传给负载。

齿轮箱、"补偿"设备等其他设备经常出现在控制系统中，但它们通常被认为是形成其他部件的一部分。

Key to Exercises of the Reading

Ⅰ. 1. three　2. an amplifier　3. transducers　4. controller

Ⅱ. 1. 众所周知，一个闭环控制系统包括三个基本组件。

2. 它将从控制器接收的具有正确物理性质的功率信号传给负载。

3. 诚实是她成功的一个重要因素。

第21单元　基本型机器人

概述

机器人（robot）这个词来源于捷克语 robota，意思是工作。韦氏字典将机器人定义为"一种可执行通常由人类完成的功能的自动装置"。美国机器人协会对工业机器人做出了更加精确的描述："机器人是一个可进行重复编程的多功能操纵器，用于移动材料、零件、工具或执行特种作业。"简而言之，机器人是一个可编程的具有外部传感器的通用操纵器，可执行各种装配作业。

根据这个定义，机器人必须智能化，这通常由与其控制和传感系统有关的计算机算法实现。机器人是通用的计算机控制的操纵器，它由几个刚性杆件组成，这些杆件由关节连接成开放运动链。关节通常是可旋转的（转动副）或线性移动的（移动副）（图21-1）。转动副关节就像一个铰链，可使两个杆件进行相对转动。移动副关节可使两个杆件进行线性相对运动。常规上用 R 来表示转动副关节，用 P 来表示移动副关节。关节如图所示。每个关节代表 l_i 和 l_{i+1} 两个杆件之间的相互连接。同样，如果关节使得 l_i 和 l_{i+1} 杆件相互连接的话，那么转动副关节的旋转轴或移动副关节的滑动轴由 Z_i 表示。转动副关节变量由 θ_i 表示，移动

副关节变量由 d 表示，这些变量代表了相邻杆件之间的相对位移。

腕关节和终端执行器

从机械角度而言，机器人由手臂和腕关节组件以及相关器械组成，可触达位于其加工区域内的工件。加工区域指的是机器人的可操作范围，在此范围内，机器人的手臂可将腕关节组件送到任意位置。通常，手臂组件的移动有 3 个自由度。（各自由度方向）运动的组合将腕关节定位在工件上。腕关节组件通常可实现 3 个旋转运动。根据物体的外形，这些运动的组合可定向移动工具以进行小心拾取。最后的 3 个旋转动作常常被称为投掷、侧摆和转动。因此，对于一个 6 关节的机器人而言，手臂组件是定位机构，而腕关节组件是定向机构。图 21-2 和图 21-3 所示的 Unimation PUMA 型机器人的手臂结构体现了上述概念。

Key to Exercises of the Text

Ⅰ. 1. F　2. T　3. T　4. F

Ⅱ. 1. sensing system　2. relative motion　3. degree of freedom　4. work volume

5. orientation mechanism　6. position mechanism　7. common normal relative motion

Ⅲ. 1. The houses in my hometown were chiefly composed of wood.

2. Coal of all kinds originated from the decay of plants.

3. These proposals represent a realistic starting point for negotiation.

4. Mother warned the children not to associate with bad companions.

阅读材料　机器人的应用

1995 年，大约有 70 万个机器人在工业化国家得以应用。其中日本使用了 50 多万个，在西欧约有 12 万，在美国约有 6 万。许多机器人应用于危险或人类不愿意做的工作任务中。在医学实验室里，机器人可处理有潜在危险的材料，如血液或尿液样本。在其他情况下，机器人可用于执行重复性的、一成不变的任务，而人类做这些任务的效率会随着时间的推移下降。机器人可以每天 24 小时执行这些重复的、高精度的业务而不疲劳。

机器人的一个主要应用是汽车行业。通用汽车公司使用了大约 16，000 个机器人执行例如定位焊、涂装、机装载、零件传送和组装等任务。组装是增长最快的工业机器人应用之一。它比焊接或涂装要求的精确度更高，并且依赖于低成本的传感器系统和强大的廉价计算机即可。机器人被用于电子装配，它们会把芯片装载在电路板上。

一些对人类来说很危险的环境中的活动，如寻找沉船、清理核废料、探索水下矿藏和勘探活火山，非常适合机器人。同样，机器人可以探索遥远的行星。美国宇航局的伽利略号无人太空探测器曾在 1996 年前往木星，并完成许多任务，如确定了木星大气的化学成分。

机器人被用来帮助外科医生安装假肢，超高精度机器人可以帮助外科医生进行精细的眼部手术。机器人可在医院里来回穿行，运送医疗用品、药品、食品托盘或者其他各种东西到护理站。一旦完成任务，它又回到充电站，并等待下一个任务。

Head There 公司已推出了远程监控机器人，可以通过互联网远程控制其位置移动。该机器人使得用户能够听、看、说，并实现远距离定位。从某种意义上说，这种机器人可充当用户的替身。

在初、高中教育和公司的机电一体化技术培训中，组装机器人正变得越来越流行。历史上被用于教育的机器人包括Turtle机器人（和编程语言紧密联系）和Heathkit HERO系列机器人。

Key to Exercises of the Reading

1. B　2. B　3. A　4. B　5. A

第22单元　增材制造

增材制造（AM）是描述3D物体生成技术的一个恰当的名称，这种技术通过一层一层地添加材料来生成物体，这些材料可以是塑料、金属、水泥，甚至有一天可能会是人体组织。

AM技术的共性是使用计算机、3D造型软件（计算机辅助设计或称为CAD）、机器设备和分层材料。在CAD框架生成后，AM设备从CAD文件中读出数据，然后生成材料层或者在基材上连续添加液体层、粉末层、片状材料层或者其他材料层，以一层一层叠加的形式来制造3D物体。

AM这个术语的含义包含很多技术分类，像3D打印、快速成型（RP）、直接数字制造（DDM）、层叠制造和增材制造等。

AM技术的应用非常广泛。早期AM技术中快速成型的应用侧重于预生产可视化模型。最近以来，AM正用于制造飞机、牙科修复物、医疗植入物、汽车、甚至时尚产品的终端产品。

虽然一层一层增加材料的方法是简单的，但是AM技术的许多应用具有各不相同的复杂程度以满足多种需要，包括设计中可视化工具的使用，为顾客和专业人员制作高度定制产品的手段，如制作工业模具，生产小批量零件，以及将来某一天生产人体器官的手段。

在AM技术的发明地——麻省理工学院，有大量的项目来支持一系列的前瞻性应用，从多结构混凝土到可以制造机器的机器；而**Contour Crafting**（轮廓工艺）可支持制造人们可以生活和工作在其中的建筑结构。

一些人将AM看作基础性减材制造（如去除材料的钻削）和更小程度的成形制造（如锻造）的补充。无论如何，AM可以向消费者和专业人员提供参与创建、定制和修理产品的可能，并且在该过程中改进当前的生产技术。

无论简单还是复杂，AM确实令人惊叹，层层叠加的方式对这个工艺进行了最好呈现，无论材料是塑料、金属、混凝土或者有一天是人体组织。

增材制造的例子

SLA（立体光固化成型法）

这是一个利用激光技术来一层一层地固化光敏树脂（曝光时会改变性质的聚合物）的高端技术。

构建在树脂池中进行。激光束直射入树脂池中，通过跟踪模型的横截面形状定位目标层并且固化它。在构建周期内，用于构建操作的平台将不断重新定位，每次降低一个单层厚度的高度。该过程不断重复，直到模型构建完成并呈现出令人满意的外观。可能需要专门的材料来支持模型的其他特性。模型可以进行加工并作为注塑成型、热成型或其他铸造工艺的样品。

FDM（熔融沉积成型）

这种工艺定位于使用热塑性材料（加热时变成液体并在冷却时固化成固体的聚合物），通过转位喷嘴将其注射到平台上。喷嘴跟踪目标层的剖面形状，其中的热塑性塑料已为下一层的固化提前硬化。该过程不断重复，直到模型构建完成并呈现出令人满意的外观。可能需要专门的材料来支持模型的其他特性。与SLA类似，其模型可以进行机械加工或用作样品。

MJM（多重喷射成型）

多重喷射成型与喷墨打印机原理类似，有一个能够来回穿梭移动的（3个维度——x，y，z）打印头，通过数百个小喷嘴一层一层地喷射成型热塑性塑料层。

3DP（三维打印）

该工艺在填充有淀粉或石膏基粉末材料的容器中构建模型。由喷墨打印机机头穿梭移动并施加少量黏合剂来形成目标层。当一层黏合剂施加完成，马上在其上撒一层粉末，然后施加下一层黏合剂。该过程不断重复，直到模型构建完成。由于模型是由散粉构成的，则不需要其他支承。此外，这是唯一一个可以构建彩色模型的工艺。

SLS（选择性激光烧结）

有点像SLA技术，选择性激光烧结（SLS）利用高功率激光来熔化塑料、金属、陶瓷或玻璃的小颗粒材料。在构建周期内，用于构造操作的平台将不断重新定位，每次降低一个单层厚度的高度。该过程不断重复，直到构建或模型完成。与SLA技术不同的是，这个技术过程不需要额外支承，因为构造过程中的未烧结材料可以提供支承。

Key to Exercises of the Text

Ⅰ. 略

Ⅱ. 1. f 2. e 3. a 4. b 5. c 6. d 7. h 8. g

Ⅲ. 1. 现在每个人都知道3D打印这个术语，但很多人谈论到它时实际上指的是多种增材制造工艺之一。

2. 每个增材制造工艺因使用的材料和设备技术不同而各不相同。

3. 在进行3D打印之前首先要准备一个3D数字模型。

4. “3D打印”这个术语的最初含义是指这样一个过程：用喷墨打印机头在粉末基床上一层一层沉积粘接材料。

阅读材料 美国科研人员致力于活体组织替代品的打印

3D打印机的工作原理和平常的桌面打印机非常相似，但和在纸上喷墨不同，3D打印机是将层层活性物质进行堆积从而制造一个3D物体。这项技术已经有将近20年的历史，为牙医、珠宝商以及机械师提供了快捷的工作手段，甚至还为那些希望不用打造模具就可以为人们定制巧克力的生产商实现了梦想。

在21世纪初，科学家和医生们就看到了该技术的潜力——人们可以利用它制作活性组织，甚至是人体器官。他们称这项技术为3D生物打印，而它也成为组织工程学新兴领域的热门分支。

在世界各地的实验室里，化学、生物、医药和工程学领域的专家们正致力于通过多种路径来研究和实现一个大胆的目标：用患者自己的细胞打印出功能齐全的肝、肾或者心脏，也就是打印新的器官。如果他们成功了，器官捐献者名单将成为过去。

生物打印技术要打印出这么复杂的器官可能还要好几年甚至几十年，但是科学家们已经打印出了皮肤和腰椎间盘并将它们植入了生命体内。目前为止，人类还没有接受过这样的移植手术，但未来两到五年内会有一些打印的可替代器官进入人体试验。

科学家称最大的技术挑战并不是制造器官本身，而是复制器官内部错综复杂的血管网络，这些血管起到了滋养器官、为器官提供氧气的作用。

许多生物组织工程师认为如今最好的办法可能是仅打印出一个器官最大的连接性血管，然后给大血管细胞提供充足的时间、空间和理想环境来自生成剩余的血管部分，最终的完整器官便可以被植入体内。

Key to Exercises of the Reading

Ⅰ. 1. F　2. F　3. F　4. T

Ⅱ. 1. Industrial 3D printers have existed since the early 1980s and have been used extensively for rapid prototyping and research purposes.

2. 3D printing has been considered as a method of implanting stem cells capable of generating new tissues and organs in living humans.

3. The STL file format has been the industry standard for transferring information between design programs and additive manufacturing equipment since the mid－1980s.

第23单元　工业4.0简介

工业4.0起源于制造强国德国。然而这个概念性的想法已经被很多其他工业国家广泛接受和采用，这些国家包括欧盟内部的成员国，更远影响到中国、印度和其他亚洲国家。工业4.0这个名字指的是第四次工业革命，前三次工业革命分别通过机械化、电力和IT技术引发。

第四次工业革命，也因此称之为工业4.0，将通过物联网技术和服务互联网与制造环境的集成而实现。前三次工业革命带来的好处已经是既成事实，而我们有机会积极地引导第四次工业革命的发展来改变我们的世界。

工业4.0的远景是，将来的工业企业将建立全球网络来连接他们的机器、工厂和仓储设施从而形成一个信息物理系统（CPS），这个系统将通过分享引发操作的信息智能地连接各个部分并进行相互控制。这些信息物理系统将呈现智能工厂、智能机器、智能仓储设施和智能供应链等各种形式。这将改进工业过程，使生产制造成为一个整体，这种改进将通过工程、材料使用、供应链和产品生命周期管理的提升来实现。上述便是我们所说的横向价值链，对横向价值链的展望是，工业4.0将与横向价值链的每个阶段进行深度集成从而使工业生产过程发生巨大的变化。

实现这个远景的核心是智能工厂，它将改变生产方式，这个改变基于智能机器和智能产品。不仅智能设备等信息物理系统是智能的，被组装的产品也包含嵌入式智能装置，这样的话它们在生产过程中的任何时间都能被分辨和定位。微型的无线射频识别装置标签使产品变得智能，以便能够知道它们是什么、什么时候被生产，更为关键的是，能够知道它们目前的状态和达到期望状态还需要的步骤。

这就需要智能产品知道它们自己的历史和把它们变成完整产品还需要的工序。这些工业生产制造的知识将嵌入到产品中使得它们可提供生产工艺的替代路线。比如，当智能产品知道了它当前的状态和成为完整产品所需要的后续生产过程时，智能产品将能够指挥它所应跟随的生产线的输送带。之后，我们将看看这项工作实际上是怎样运行的。

然而，现在我们需要看看工业4.0远景的另一个关键因素，那就是价值链中纵向制造过程的集成。所期待的远景是：嵌入式横向系统与纵向业务过程（销售、物流、财务和纵向业务过程中的一些其他项）和相关的 IT 系统相集成。它们将使得智能工厂能够对从供应链到服务和产品生命周期管理的整个制造过程进行连续控制。这个运营技术（OT）和信息技术（IT）的融合不是没有问题，我们早在讨论工业互联网时就发现了这个问题。然而，在工业 4.0 系统里，这些独立的问题可以看作是同一类问题。

智能工厂不仅仅与大公司相关，实际上他们的灵活性使他们更适用于中小企业。例如，对横向制造过程和智能产品的控制使得我们能更好地进行决策制定和动态过程控制，在能力和灵活性方面来适应最新的设计变更或者改变生产来满足顾客在产品设计方面的偏好。更进一步说，这个动态过程控制使小批量生产成为可能，而小批量生产还是盈利的且能够适应个性化订单。这些动态业务和工程过程使得创造价值的新方法和创新的商业模型成为可能。

总的来说，工业 4.0 需要信息物理系统集成在制造和物流过程中，同时在制造过程中引进物联网和服务互联网。这将产生新的方法来创造价值、商业模型并为中小企业提供下游服务。

Key to Exercises of the Text

Ⅰ. 1. The Industry 4.0 concept originated from Germany which is one of the strongest industrial countries in manufacture.

2. The first three industry revolutions come about through mechanization, electric technology and information technology.

3. The fourth industry will come about via Internet of Things and the Internet of services becoming integrated with manufacturing environment.

4. The horizontal value chain of industrial process consists of engineering, material usage, supply chains, and product lifecycle management.

5. The vertical business processes or vertical value chain consist of sales, logistics, finance and etc.

Ⅱ. 1. f 2. d 3. a 4. b 5. g 6. e 7. c

Ⅲ. 1. 第一次工业革命通过使用水和蒸汽的力量实现了机械化生产。

2. 第二次工业革命始于亨利·福特在 1913 年引进的组装线，从而使产能大幅提升。

3. 第三次工业革命是在 20 世纪 70 年代将计算机引入到生产现场的结果，这引起了自动组装线的大量出现。

4. 工业 4.0 的展望是“信息物理生产系统”——在这个系统中，布满传感器的“智能产品”会告诉机器如何处理他们。

阅读材料　　中国制造 2025 和德国工业 4.0 的合作机会

随着中国政府对提升中国大陆工业的纲要规划《中国制造 2025》战略的发布，2016 年 3 月通过的“十三五”规划对该战略在下一个五年（2016 - 2020 年）的实施进行了部署。这也引起了大家对中国工业发展方向的兴趣，一些工业观察员已经把这个战略与德国为提高其工业效率设计的工业 4.0 战略相提并论。

值得一提的是，已经有人开始担心这两个战略将使中国和德国之间的工业竞争加剧。尽管如此，两国已在2015年7月签署了加强智能制造技术开发领域合作的谅解备忘录。而且去年十月份在北京的会议上，中国总理李克强与到访的德国总理安格拉·默克尔一致同意扩展战略合作来开发符合中国制造2025和德国工业4.0的新机会。

实际上，两国在工业方面的相对发展和不同的战略发展侧重点，相对于竞争而言显示出了更多的合作机会，包括在工业机器人领域的合作。此外，中国和德国工业在全球供应链中的不同位置也预示着中外合作项目方面的相关参与者有更多机会。

从本质上讲，德国工业4.0倡导的是在生产方法中采用最先进的信息和通信技术来进一步提高工业效率。这一战略的发展要基于德国强大的装备和制造业基础，其相应的IT能力和嵌入式系统、自动化工程领域的专业积累有利于巩固德国在制造工程行业的全球领先地位。

德国工业4.0的目标是：连接现有的嵌入式IT系统生产技术与智能过程，来转变和升级工业价值链和商业模式，从而实现智能生产。这将要求德国在深化集成制造系统等方面加强研发。还需要新的工业和技术标准来连接不同公司和装置的系统，同时数据安全系统也应该升级以防止系统中数据信息的误用和未授权访问。所有这些发展都有望加强德国工业的效率和创新产能，同时节约资源和成本。

至于中国制造2025，其核心是创新和质量，以及引导中国工业从低附加值活动向中、高端制造运营活动转变，而不是一味追求产能的扩张。淘汰低效和过时的产能，帮助企业开展更多的自主设计和自有品牌业务，也是这个战略的目标。这些目标将通过以下举措来推进：建立制造创新中心，加强知识产权保护，建立新的工业标准，促进重点和战略部门的发展。

Key to Exercises of the Reading

Ⅰ. 1. F　2. T　3. T　4. T　5. T

Ⅱ. 1. Industry 4.0 is currently more of a vision than a reality, but it is one with potentially far reaching consequences.

2. Programmable logic will become increasingly important since it will be impossible to anticipate all the environmental changes to which control systems will need to dynamically respond.

3. Whether revolution or evolution, industrial production is about to become more efficient.

第24单元　大国工匠

中国政府在2015年提出了“工匠精神”的概念，作为提升工业水平的措施之一，以应对来自其他国家的激烈竞争。公认的工匠精神的内涵是：应该以完美、精确、专注、耐心和坚持不懈的态度投入到工作中。

为提倡工匠精神，从2015年起，中国政府和组织对很多手艺精巧的工作者授予各种荣誉称号，并在很多场合和媒体中宣传他们的事迹。

下面对其中的一些大国工匠进行简要介绍。

高凤林——制造火箭的“心脏”

高凤林是中国航天科技集团公司的一名高级技师。他在过去的35年里一直从事火箭发动机喷管的焊接工作。长征五号系列运载火箭的发动机喷嘴上有几千个空心管子，每个管子的焊接都要求作业者一丝不苟、技艺高超，一些管子的直径只有0.16mm左右。作为一名精通特种焊接技术的工匠，他曾被诺贝尔奖获得者丁肇中邀请，作为美国宇航局的一位特邀专家，领导反物质探测器项目的一个子项目，因为这个探测器的结构非常复杂，对焊接技师焊接技能的要求也非常高。

经过高凤林焊接的长征系列火箭的发动机多达 130 多个，这些发动机将火箭推送到了太空中，而这一数字占到了长征火箭总数的一半以上。

庞辉勇——钢铁行业的博士工匠

庞辉勇，39 岁（2017 年时），是河北钢铁集团舞钢公司的一名高级工程师。作为一名材料科学博士，他以一个炼钢车间普通技术员的身份开始了自己的职业生涯，并在车间工作了3 年，以熟悉炼钢工艺。他多次亲手实践操作。他专注的表现、出色的专业能力和扎根基层的工作态度，使他被任命为一个项目组长，领导开发替代进口产品的特殊钢材。他的团队开发出 10 余种产品，填补了国内一些钢材的空白。

当超低温钢的开发进入关键阶段的时候，结婚后的第一个春节即将到来，他不得不向新婚妻子及其亲友道歉，因为他不能去她的家乡看望父母了。当新年的钟声响起的时候，新钢种的试制成功了！这是中国第一次生产出这种可用于建造 LNG 船的钢材。

用于核岛建设的特殊钢材的成功制造在他个人历程中值得纪念。经过近 4 个月的 24 次实验，他们成功生产出了这种特殊钢材，价格仅为高达 80，000 元/吨的进口钢材的 25%！从那时起，第三代核电站所需的各种钢材我们都可以自己生产了。在此之前，世界上只有一家法国供应商能提供这种钢材。

张东伟——LNG 船舶焊接的多面手

张东伟是中船集团沪东中华造船有限公司的一名焊工。他因在 LNG 船体焊接领域的高超技术而广为人知。

LNG 船体结构中有一种特种钢材叫作殷瓦钢，这种钢材的焊接条件非常苛刻，对焊工的要求也非常高。船体焊接中最为广泛使用的焊接技术是二氧化碳保护焊和氩弧焊。

张东伟 2015 年时只有 34 岁，而他已经在焊接行业工作了几乎 15 年。他克服了无数个困难，不遗余力地学习各种焊接技术，最终成为这一领域的专家。目前，在多种船体建造和殷瓦钢焊接技术方面，张东伟已经训练和指导了 40 多名工人，其中 30 多人已成为这一领域的熟手并获得船级社颁发的资格证书。

宁允展——铁路工业的无敌钳工

宁允展是原南车集团青岛四方有限公司的一名钳工技师。他的磨削及研磨技术在高速铁路行业非常有名。在一次国外引进高速列车的试制过程中，宁允展开始崭露头角。在这次试制过程中，他发明了转向架定位臂手动打磨的新方法。后来他反复试验，开发了“气动磨削机的手动研磨法”，可用于大批量生产。他生产的零件表面有独特的花纹，零件精度也很高。利用他自己发明的工艺方法，以及十分认真的工作态度，在过去十年里他生产的转向架定位臂不合格数是零。

宁允展不只是一名钳工技师，也是一个精通焊接、电工以及维修技术的多面手。他开发了许多维护、修复的方法和工具，并创造了巨大的价值。他同时也是两项专利的拥有者。

管延安——深海钳工专家

港珠澳大桥是世界上最大的桥梁，大桥建成后，管延安被称为“深海钳工第一人”。

由于历史的原因，在港珠澳大桥修建之前，中国没有隧道沉管安装的关键技术。因为不能接受外国公司仅仅限于技术指导的超高报价，由林明领导的中国工程团队自己开发了安装方法。沉管的对接和舾装都需要非常细致的作业，更需要严格的管理方法和高超的技术。舾装是指在沉管沉入海底之前在船上将若干系统设备进行装配，这是隧道施工中最复杂的工作段。由于一些方法是由中国人自己研发的，所以没有建立规范的作业指南，在每一个工艺定型之前，管延安都会演练数次。很多次，他是冒着生命危险在严酷的环境中进行试验。按照发达国家公司发明的传统作业方法，最后一节沉管的安装是整个桥梁作业中最为复杂的一环，工作时间一般为8~10个月。得益于林明领导的工程团队所发明的新方法和管延安领导的一线安装团队的努力，工作时间从8~10个月缩短到了1天！管延安因其进取精神、团队精神和对新技术的钻研受到高度赞扬。

顾秋亮——潜水器首席钳工技师

顾秋亮是中国船舶重工集团公司的一名高级钳工。因其在深海水下航行器装配方面的高超技术和敢于担当的精神，顾秋亮被授予“大国工匠”的称号，他在造船行业已经工作了近40年。

作为一名高级钳工，顾秋亮为很多国家重点工程的设备制造做出了很大贡献。他在深海潜水器“蛟龙号载人潜水器”的组装过程中表现出高超的技艺，开发了很多方法和工具来保证装配符合最高的质量标准。他可以手工研磨具有极高安装精度的观察窗玻璃的安装工作面，平面度误差仅为0.02毫米。这个观察窗可以承受700个标准大气压的压力。为了提高锉削、磨削技能，他用锉刀和磨头无数次练习以用手一遍又一遍地感觉尺寸的偏差，最后达到了很高的技术水平，也付出了很大的代价：他的指纹被磨平，无法被考勤机识别了！

Key to Exercises of the Text

Ⅰ. 1. China’s government put forward the concept of “spirit of craftsman” in the year of 2015.

2. Be devoted to make things with perfection, precision, concentration, patience and persistence.

3. The goal is to encourage people to work hard so as to upgrade China’s industry.

4. Six of them work as a welder or locksmith, two of them work as an engraver and a Xuan paper “fishing man.”

5. Carbon - dioxide arc welding and argon arc welding are normally used welding technologies in ship building industry.

Ⅱ. 1. g 2. a 3. b 4. h 5. c 6. d 7. e 8. f

Ⅲ. 1. performing 2. affected 3. accustomed to 4. completing 5. study

Ⅳ. 1. 人们发现许多成功的企业家都具有工匠精神，即具有决心、耐心和追求完美的渴望。

2. 学徒制度在瑞士有很多拥护者。

3. 德国以其工匠精神和优质产品闻名于世。

4. 焊工和钳工是现代工厂中的常见岗位。

阅读材料 世界技能大赛

什么是世界技能大赛以及如何参赛

世界技能大赛每两年举行一次，是世界上最大的关于职业教育和技能的卓越赛事，也是全球工业水平的真实体现。参赛者是同龄人中的佼佼者，他们是从世界技能大赛会员国家和地区的技能竞赛参赛者中选出的。世界技能大赛参赛者在参赛当年的年龄不能超过22岁。信息网络布线、机电一体化、制造团队挑战赛和飞机维修方面的参赛者的当年年龄不得超过25岁。他们在完成他们所研究的以及在工作场地所执行的特定任务时，表现出了个体的和集体的技术能力。

使专业教育作为社会经济转型的真正工具而变得引人注目，这是世界技能大赛的一项主要贡献。这项竞赛还为工业、政府和教育界的领导人提供了一个交流信息的机会，关于工业和专业教育的信息和最好的实践做法得到交流。新的思想和工艺激励着学龄青年投身于技术学习和技术事业中，走向更美好的未来。

世界技能大赛和参赛者拥有许多共同的价值观：承诺、坚持和乐于慷慨分享成功的竞赛精神——无论是否获得奖牌。这项技能活动的利益相关者遍布各个领域，包括青年、教育工作者、行业专业人士、商业和经济领域领袖。挖掘现有的诸多可用资源，以支持铸就卓越和建造美好未来。

技能大赛资源

为继续磨练你的技能，技能大赛设计了有奖金和奖学金的比赛和挑战。了解一下全球健康和生态驱动汽车方面令人兴奋的学生挑战赛吧！设计一个获奖网站；与引领潮流的学生竞争；与国际上糕点专业的学生竞争来品尝未来的味道。这些多样的资源将拓宽你对这些引人入胜的、令人兴奋的和意义非凡的挑战的认识，这将有助于你继续培养技能，激发改变世界的愿望。

什么是技能?

63年以来，世界技能大赛反映了世界上各行各业的发展和变革。这些年来，许多竞赛技能已经消失，很多技能合并，也出现了许多新的技能。

What Is A Skill（负责）考察这些新技能是否真正代表新的行业，或者只代表一个过渡阶段？你如何定义一个行业？这个项目将仔细研究个人技能是如何定义的，以及这些技能代表的行业如何随时间的推移而演变。它将分析全球大量行业工作场所不断变化的需求，以及未来为满足这些行业需要而必须提供的培训。

发生的和即将发生的大事件

第44届世界技能大赛于2017年10月在阿拉伯联合酋长国的阿布扎比闭幕。中国以15枚金牌和奖牌总数109的成绩排名第一，韩国以奖牌总数88的成绩名列第二，第三名是奖牌总数为81的瑞士。

世界技能大赛的旗帜已经开始了通往俄罗斯喀山的漫长旅程，第45届世界技能大赛将于2019年在那里举行。

中国在2017年10月20日当选为第46届世界技能大赛的主办国，大赛将于2021年在上海举行。中国希望通过举办这次大赛，为建设全国范围内的更强大的技能工人团队做出贡献，并加强职业技能方面的国际交流。中国领导层对争取这次大赛的主办权高度重视，大赛将大大促进中国高技能人才的发展；上海将利用这个机会制定优惠政策来鼓励工匠精神，改进职业教育以培养技能型人才。

Key to Exercises of the Reading

Ⅰ. 1. F 2. F 3. F 4. T

Ⅱ. 1. China's consumers often have a positive impression of "Made in Germany" products.

2. The pace at which Chinese products are entering foreign markets has grown steadily over the years.

3. A nation's brand comprises at least three elements: its economic image, political image, and technological image.

Appendix B　Glossary（总词汇表）

Δ（delta）–connected load　三角形连接的负载
a matter of　大约；大概
a punched tape　穿孔纸带
abbreviate［əˈbriːvieɪt］v. 缩写，简写成
abound［əˈbaʊnd］vi. 富于；充满
abrasive［əˈbreisiv］adj. 有研磨作用的；粗糙的 n. 研磨料
abrasive particle　磨料颗粒
Abu Dhabi［ˈæbuːˈðæbiː］［地名］阿布扎比（阿拉伯联合酋长国首都）
AC（Alternating Current）　交流电
acceptable［əkˈseptəbl］adj. 可接受的；合意的
access［ˈækses］vt. 使用；存取，访问；接近 n. 进入；使用权
accommodate［əˈkɒmədeɪt］vt. 容纳；使适应；调解 vi. 适应；调解
accomplish［əˈkʌmplɪʃ］vt. 完成；实现；达到；使完美
accomplished with　完成
accuracy［ˈækjurəsi］n. 精确性，准确度，精度
acetylene［əˈsetiliːn］n.［化］乙炔
acid［ˈæsid］n. 酸 adj. 酸的，酸味的
actuator［ˈæktʃʊeɪtə］n.　执行机构；激励者；促动器
adapt to　适合
adaptive control system　自适应控制系统
Additive Manufacturing　增材制造
additive［ˈædɪtɪv］adj.　附加的；［数］加法的 n. 添加剂，添加物
address［əˈdres］vt.　演说；写地址；向……致辞；处理 n. 地址；致辞
adjacent［əˈdʒeɪsnt］adj. 邻近的，毗邻的；（时间上）紧接着的
adjoining［əˈdʒɔɪnɪŋ］adj. 毗邻的，邻近的
adjustment［əˈdʒʌstmənt］n. 调解，调整；调节器
advent［ˈædvənt］n.（事件、时期等的）出现，到来
advocate［ˈædvəkeɪt］vt. 提倡，主张，拥护 n. 提倡者；支持者；律师
aerospace［ˈeərəspeɪs］航空宇宙；［航］航空航天空间
afield［əˈfiːld］adv. 在战场上；去野外；在远处；远离
aggregate［ˈægrɪgət］vt. 使聚集，使积聚；总计达
agnostic［ægˈnɒstɪk］n. 不可知论者 adj. 不可知论（者）的
air cushion vehicle　气垫车
air–permeable material　透气性材料
alert［əˈlɜːt］adj. 警觉的，警惕的，注意的；思维敏捷的；活泼的　n. 警报；警戒状态
algorithm［ˈælgərɪðəm］n. 运算法则；演算法；计算程序
alleviate［əˈliːvieɪt］vt. 减轻；使……缓和
allocate［ˈæləukeit］vt. 分派，分配，指定
allow... to do...　允许……做……
allowance［əˈlaʊəns］n. 津贴；定量，余量；允许
alloy［ˈælɔɪ］n. 合金　vt. 使成合金；使降低，贬损；掺以杂质　vi. 有合金能力
alloyed［ˈəlɔɪd］adj. 合金的；合铸的；熔合的　vt. 将……铸成合金
alternating［ˈɔːltəneitiŋ］adj. 交互的；交替

的 v.（使）交替

aluminium [ˌæljəˈmɪniəm] adj. 铝的 n. 铝

Amazon. com 亚马逊网站

ambitious [æmˈbɪʃəs] adj. 野心勃勃的；热望的；炫耀的

amount to 意味着；共计；发展成；折合

amplifier [ˈæmplifaiə] n. 放大器，扩大器；扩音器

amplitude [ˈæmplɪtjuːd] n. 广阔，充足；[物] 振幅

analog comparators 模拟比较器

analogous [əˈnæləgəs] adj. 类似的，相似的；可比拟的（analogous to/with）；模拟的

anchorage [ˈæŋkərɪdʒ] n. 抛锚，停泊（处）

animation [ˌænɪˈmeɪʃn] n. 活泼，有生气的动画

annealing [əˈniːlɪŋ] n. （低温）退火；焖火；磨炼

anode [ˈænəʊd] n. 阳极（电解）；正极（原电池）

answering machines 电话答录机，录音电话

antenna [ænˈtenə] n. [电讯] 天线；[动] 触角；[昆] 触须

anticipated [ænˈtisipeitid] adj. 预先的；预期的

anticipation [ænˌtɪsɪˈpeɪʃn] n. 希望；预感；预支

Anti-lock brake 防抱死制动系统，汽车反锁制动系统

antimatter probe 反物质探测器

apart [əˈpɑːt] adj. 分离的；与众不同的 adv. 与众不同地；分离着

applause [əˈplɔːz] n. 欢呼，喝彩；鼓掌欢迎

appliance [əˈplaiəns] n. 器具，器械；装置设备；应用

application [ˌæplɪˈkeɪʃn] n. 应用，运用；申请

apprehension [ˌæprɪˈhenʃn] n. 理解；逮捕；恐惧；忧惧

appropriate [əˈprəupriət] adj. 适当的；恰当的；合适的 vt. 占用，拨出

approximately [əˈprɒksɪmətlɪ] adv. 近似地，大约

arbitrarily [ˈɑːbɪtrəlɪ] adv. 任意地；武断地；反复无常地；肆意地

architect [ˈɑːkɪtekt] n. 建筑师；缔造者

architecture [ˈɑːkitektʃə] n. 建筑，建筑学；体系机构

argon arc welding 氩弧焊

arithmetic [əˈriθmətik] n. 算术，算法；运算

arouse [əˈraʊz] vt. 引起；唤醒；鼓励 vi. 激发；醒来

artificial [ˌɑːtɪˈfɪʃl] adj. 人造的，人工的；虚假的

assemble [əˈsemb(ə)l] vt. 集合，聚集；装配；收集 vi. 集合，聚集

assembler [əˈsemblə] n. 汇编程序；汇编机；装配工

assembly [əˈsembli] n. 组装件，装配；集会，集合，汇编，编译

associate [əˈsəʊʃieɪt] v. 联合，结合，参加，连带

assure [əˈʃʊə] vt. 向……保证；使……确信；<英>给……保险

attendance machine 考勤机

attract [əˈtrækt] vt. 吸引；引起 vi. 吸引；有吸引力

attract attention 引起注意

attribute [ˈætrɪbjuːt] vt. 把……归于 n. 属性，标志，象征，特征

audacious [ɔːˈdeɪʃəs] adj. 大胆的；鲁莽的；大胆创新的

authorized [ˈɒːθəraizd] adj. 权威认可的，审定的，经授权的 v. 授权；批准；辩护

automated [ˈɔːtəʊmeɪtɪd] adj. 自动化的

APT (Automated Programming Tool) 自动编程工具软件
automatic control system 自动控制系统
automatic guided vehicle 自动导向车
automatically [ˌɔːtəˈmætɪklɪ] adv. 自动地; 无意识地; 不自觉地; 机械地
automation [ˌɔːtəˈmeiʃən] n. 自动化; 自动操作
automobile [ˈɔːtəməbiːl] n. <美>汽车; 驾驶汽车
automotive [ˌɔːtəˈməʊtɪv] adj. 自动的, 自动车的
automotive dashboard 汽车仪表板
autonomous [ɔːˈtɒnəməs] adj. 自治的; 自主的; 自发的
axis [ˈæksɪs] n. 轴, 轴线; [政] 轴心 (复数 axes [ˈæksiːz])
axis of rotation 旋转轴
back and forth 反复地, 来回地
balanced three phase system 三相平衡系统
bar stock 棒料
base - emitter 基极 - 发射极
batch [bætʃ] n. 一炉, 一批, 一次所制之量
battery - backed memory 电池备份存储器
be identical to 与……相同
be known as... 称为……
behaviour [bɪˈheɪvjə(r)] n. 行为; 举止; (人、动植物、化学药品等的) 表现方式; 态度
bend [bend] v. 使弯曲, 使屈服, 使致力, 使朝向
bending [ˈbendiŋ] n. 弯曲 (度), 扭曲 (度)
biased [ˈbaiəst] adj. 结果偏倚的, 有偏的, 有偏见的
binder [ˈbaɪndə (r)] n. 黏合剂; 包扎物, 包扎工具; 装订工
biotechnology [ˌbaɪəʊtekˈnɒlədʒi] n. [生物] 生物技术; 生物工艺学
bismuth [ˈbizməθ] n. 铋
blood vessel [blʌd ˈvesəl] n. 血管
bluish [ˈbluːiʃ] adj. 带蓝色的; 有点蓝的
bogie [ˈbəʊgɪ] n. [铁路] 转向架; 小车; 妖怪; 可怕的人
bolt [bəʊlt] n. 螺栓; (门、窗等的) 插销
Boolean logic 布尔逻辑
boon [buːn] n. 实惠
boost [buːst] vt. 促进, 提高 vi. 宣扬 n. 提高; 吹捧
borer [ˈbɔːrə] n. [机] 镗床, 镗孔刀具
boring [ˈbɔːrɪŋ] n. 镗孔, 镗削加工 adj. 无聊的, 无趣的, 令人厌烦的
boss [bɒs] n. 老板; 首领; 凸起, 凸台
brake [breɪk] n. 闸, 制动器
brass [brɑːs] n. 黄铜, 黄铜制品
braze [breɪz] vt. 铜焊, 用黄铜镀或制造; 用铜锌合金焊接
breadboard [ˈbredbɔːd] n. 擀面板; 案板; 电路试验板
bring about 引起; 使调头
brittle [ˈbrɪtl] adj. 易碎的, 脆弱的; 易生气的
browser [brauzə(r)] n. 浏览器
burgeoning [ˈbɜːdʒənɪŋ] adj. 迅速成长的 v. 迅速发展; 发 (芽)
bushing [ˈbuʃɪŋ] n. 套管; 轴衬
cabinet [ˈkæbinət] n. 内阁; 柜橱
cadmium [ˈkædmiəm] n. 镉
calculation [ˌkælkjuˈleiʃn] n. 计算; 估计
cam [kæm] n. 凸轮, (偏心轮; 样板, 靠模, 仿形板 vt. 给 (机器) 配置偏心轮
camcorder [kæmˈkɔːdə] n. 摄录像机; 便携式摄像机
can - do spirit 敢于担当的精神
capability [ˌkeipəˈbiləti] n. 才能, 能力; 容量; 性能
capacitance [kəˈpæsitəns] n. [电] 电容;

电流容量；电容器
capacitor [kə'pæsɪtə] n. 电容，电容器
carbonate ['kɑːbəneɪt] n. 碳酸盐 vt. 使充满二氧化碳，使变成碳酸盐
carbon-dioxide arc welding 二氧化碳气体保护焊
carriage ['kærɪdʒ] n.（机床的）滑板；刀架；拖板，客车；运费
case-hardened adj. 表面硬化的；定型的；无情的
cast [kɑːst] vt. 浇注；投，抛；计算；投射（光、影、视线等）
casting & forging dies 铸造模具与锻造模具
casting ['kɑːstɪŋ] n. 投掷；铸造；铸件 v. 投掷，投向，扔掉；铸造
category ['kætəgəri] n. 种类，类别，范畴
cater ['keɪtə(r)] vt. 投合，迎合；满足需要；提供饮食及服务
cathode ['kæθəʊd] n. 阴极（电解）；负极（原电池）
cathode-ray tube（CRT）阴极射线管
cavity ['kævəti] n. 洞，空穴，型腔，蛀洞
CCW（counter clock wise）逆时针方向
cellular ['seljələ] adj. 细胞的；由细胞组成的 n. 移动电话
cement [si'ment] vt. 巩固，加强；（用水泥等）粘合 n. 水泥
ceramic [sɪ'ræmɪk] adj. 陶瓷的；陶器的；n. 陶瓷；陶瓷制品
chamfering ['tʃæmfəriŋ] n. 倒角，切角
chancellor ['tʃɑːnsələ(r)] n.（德、奥等国的）总理；（英）大臣；校长
characteristic [ˌkærəktə'ristik] adj. 典型的，特有的 n. 特征，特性
chip [tʃɪp] n. 芯片；筹码；碎片，薄片 vt. 削，凿 vi. 剥落，碎裂
chloride ['klɔːraid] n. 氯化物
chromium ['krəʊmiəm] n. [化学] 铬（24号元素，符号 Cr）
CIM（Computer Integrated Manufacturing）计算机集成制造
circuit ['səː kit] n. （= electric circuit）电路，线路，通路，回路；电路图
circular motion 圆周运动
clamp [klæmp] n. 夹钳，夹子 vt.（用夹钳）夹住，夹紧
clamping mechanism 夹紧装置
clamping tonnage 锁模力
Classification Society 船级社
classify ['klæsɪfaɪ] vt. 分类，归类；把……列为密件
clearance ['kliərəns] n. 净空，余隙，间隙；清除，排除
client ['klaɪənt] n. [计] 顾客，客户，委托人
clip [klip] vt. 修剪；夹牢；痛打 vi. 修剪 n. 夹子；修剪；回形针
clog [klɒg] v. 堵塞，障碍
closed-loop [kˌləʊzdl'uːp] adj. 闭环的 n. 闭环电路
closed-loop system 闭环系统
clutch [klʌtʃ] n. 离合器，联轴器；夹紧装置
collaboration [kəˌlæbə'reɪʃn] n. 协作；通敌；勾结
collector-emitter 集电极-发射极
collectively [kə'lektivli] adv. 共同地，全体地，集体地
column ['kɒləm] n. 立柱；柱形物；纵队，列；圆柱；专栏
combination [ˌkɒmbɪ'neɪʃn] n. 组合，结合，联合，合并
come about 发生；产生；改变方向
come on the scene 问世；来到
command [kə'mænd] n. 命令 vi. 命令，指挥 vt. 控制；远望
commercial [kə'mɜːʃl] adj. 商业上的，商业的 n. 商业广告
commute [kə'mjuːt] vi. 通勤；代偿

vt. 减刑；交换 n. 通勤来往（的路程）

compatible [kəm'pætəbl] adj. 兼容的

compelling [kəm'pelɪŋ] adj. 引人入胜的；非常强烈的；不可抗拒的

compensation [ˌkɔmpen'seiʃən] n. 补偿；赔偿金；报酬

compensator ['kɒmpenseɪtə] n. 补偿器

competition [ˌkɒmpə'tɪʃn] n. 竞争；比赛，竞赛

complex ['kɒmpleks] adj. 复杂的；合成的；复合的

complexity [kəm'pleksətɪ] n. 复杂性，复杂的事物；复合物

component [kəm'pəʊnənt] n. 元件，部件 adj. 组成的；合成的；构成的；成分的

component value 分量值

compound ['kɒmpaʊnd] n. 化合物 vt. 合成 vi. 和解 adj. 复合的

comprehensive [ˌkɒmprɪ'hensɪv] adj. 综合的；广泛的；有理解力的

compressing moulds/ compression moulds 压缩模具

computer－integrated machining (CIM) 计算机集成制造

concave [kɒn'keɪv] adj. 凹的 n. 凹面，凹线，凹形

concentration [kɒns(ə)n'treɪʃ(ə)n] n. 浓度；集中；浓缩；专心；集合

conceptual [kən'septjuəl] adj. 概念上的

concrete ['kɒŋkriːt] adj. 混凝土的；实在的，具体的 n. 混凝土

configuration [kənˌfɪgə'reɪʃn] n. 结构，布局，形态

congestion [kən'dʒestʃən] n. 拥挤；堵车；阻塞；稠密

conjunction [kən'dʒʌŋkʃn] n. 结合；联合

connectivity [kɒnek'tɪvɪtɪ] n. 连接；连通性

consist of 由……组成，由……构成

consistency [kən'sɪstənsɪ] n. 连贯，一致性；强度；硬度；浓稠度

console [kən'səul] n. 控制台

consolidate [kən'sɒlɪdeɪt] vt. 巩固，使固定；联合 vi. 巩固，加强

constant ['kɒnst(ə)nt] adj. 不变的，恒定的，经常的 n. [数] 常数，恒量

constituent [kən'stɪtjʊənt] n. 成分；选民；委托人 adj. 构成的；选举的

consumer electronics 电子消费品

contain [kən'teɪn] vt. 包含，容纳；克制；牵制

contour ['kɔntuə] n. 轮廓；等高线；周线；概要 vt. 画轮廓；画等高线

Contour Crafting 轮廓工艺

contouring [kən'tuəriŋ] n. 成形加工；轮廓线；做等值线；外形修整

contract [kən'trækt] v. 收缩，缩紧；感染；订约

contribute [kən'trɪbjuːt] vt. 捐助；投稿 vi. 投稿；贡献；是原因之一

control console 控制面板

control function 控制函数

control grid 控制栅（极）

control loops unit (CLU) 回路控制装置

control variable 控制变量

convention [kən'venʃn] n. 会议；国际公约；惯例

convert [kən'vɜːt] vt. 使转变，转换；使改变信仰 vi. 转变；皈依

coolant ['kuːlənt] n. 切削液，冷却剂

cooperative [kəʊ'ɒpərətɪv] adj. 合作的；协助的；共同的

coordinate [kəu'ɔːdineit] vt. 使协调，使调和；整合，协调 n. 坐标 adj. 并列的，坐标的

coordinate measuring machine 坐标测量机

coplanarity [kəuplə'nærəti] n. 平面度

copper ['kɒpə(r)] n. 铜，铜币；铜制

物 adj. 铜制的

copyright［ˈkɒpɪraɪt］n. 版权，著作权 adj. 版权的；受版权保护的 vt. 保护版权

corresponding［ˌkɒrəˈspɒndɪŋ］adj. 相当的；相应的；一致的

corrosion［kəˈrəʊʒn］n. 侵蚀；腐蚀；锈蚀

counter－sink vt. 钻（孔），穿（孔），使（钻头等）插入 n. 埋头钻；凹陷

coupling［ˈkʌplɪŋ］n. 联轴节，联轴器；联结器，联合器

crack［kræk］vt. 破裂，打开；（使……）开裂 n. 裂缝；试图；缝隙；重击

craftsman［ˈkrɑːf(t)smən］n. 工匠；手艺人；技工

craftsmanship［ˈkrɑːftsmənʃɪp］n. 技术，技艺；工力

Craig McCaw 克雷格·麦考

crankshaft［ˈkræŋkʃaːft］n. 曲轴

critical［ˈkrɪtɪkl］adj. 爱挑剔的；决定性的；评论的；关键的；严重的

cross motion 横向运动

cross－section 剖面

crowdsourcing［kraudˈsoisin］n. 众包

crystal［ˈkristəl］n. 水晶；晶体 adj. 水晶的；透明的，清澈的

crystalline［ˈkrɪstəlaɪn］adj. 水晶的；似水晶的；结晶质的 n. 结晶性，结晶度

curiousness［ˈkjuəriəsnis］n. 好学；好奇；不寻常

current［ˈkʌr(ə)nt］n. 电流；趋势；涌流 adj. 现在的；通用的；最近的

cursor［ˈkəːsə］n. 指针，光标

customize［ˈkʌstəmaiz］v. 定制，用户化，按客户具体要求制造

customized［ˈkʌstəˌmaɪzd］adj. 客户指定的，定制的（产品）vt. & vi. 订做，定制

cutlery［ˈkʌtləri］n. 刀剑制造业；餐具，刀叉

cyber［ˈsaɪbə］adj. 计算机（网络）的，信息技术的

cyber－physical system（CPS） 信息物理系统；网宇实体系统

cycle time 工作周期

cylinder［ˈsɪlɪndə（r）］n. 圆筒；气缸；［数］柱面；圆柱状物

cylindrical［səˈlɪndrɪkl］adj. 圆柱形的；圆柱体的，气缸（或滚筒）的

data processing unit（DPU）数据处理装置

DC（Direct Current）直流电

debug［diːˈbʌg］vt. 调试；排除故障 n. 调试器，排错器

dedicate［ˈdedɪkeɪt］vt. 奉献，献身

deem［diːm］v. 认为，相信

define［diˈfain］vt. 精确地解释；界定

defined［dɪˈfaɪnd］adj. 有定义的，明确的；轮廓分明的 v. 给……下定义，解释

defined as 定义为

deflection［diˈflekʃən］n. 偏向，偏差，挠曲

deformed［dɪˈfɔːmd］adj. 变形的，变丑的，破相了的，畸形的

degrade［dɪˈgreɪd］vt. 降低；使降级；降低……身份

degree of freedom 自由度

delicate［ˈdelɪkət］adj. 微妙的；熟练的；易损的

delicately［ˈdelɪkətli］adv. 优美地；微妙地；精致地；谨慎地；巧妙地

delivery［dɪˈlɪvəri］n. 交付；分娩；递送

demonstrate［ˈdemənstreit］vt. 证明；显示，展示 vi. 示威

dental［ˈdent（ə）l］adj. 牙科的；牙齿的，牙的

deposition［ˌdepəˈzɪʃ(ə)n；diː－］n. 沉积物；矿床；革职

derive［dɪˈraɪv］vt. & vi. 得到，导出；源于，来自；（从……中）提取

designation [ˌdezigˈneiʃən] n. 指定；名称；指示；选派

desktop [ˈdesktɒp] n. 桌面；台式机

detect [diˈtekt] vt. 查明，发现；洞察；侦察，侦查

Detroit 底特律

device [dɪˈvaɪs] n. 装置；设备；策略；设计

diagram [ˈdaɪəgræm] n. 图表，图解，示意图，[数] 线图 vt. 用图表示

diameter [daiˈæmitə] n. 直径

diaphragm [ˈdaɪəfræm] n. (机) 隔板；横隔膜；(物) 光圈 adj. 光圈的，隔膜的

dictate [dɪkˈteɪt] vt. 口授；规定；决定，影响 vi. 听写

die casting mould 压铸型

die draft 起模斜度

die [daɪ] vt. & vi. 死亡，熄灭；凋零，枯萎 n. 模具；冲模；压模

die-casting [daɪkɑːstɪŋ] n. 压模法；铸造法

dimension [diˈmenʃən] n. 尺寸；维 (数)；尺度；范围，方面；线度；[复] 面积；容器

diode [ˈdaɪəʊd] n. [电子] 二极管

Direct Digital Manufacturing (DDM) 直接数字制造

dismantle [dɪsˈmæntl] vt. 分解 (机器)，拆开，拆卸

disperse [diˈspɜːs] v. (使) 分散，(使) 散开，疏散

displacement [dɪsˈpleɪsmənt] n. 替代；停职；[化] 置换

display [ˌdisˈplei] n. 显示器，指示器；陈列，展览 vt. 展示；显示

dissipation [ˌdɪsɪˈpeɪʃn] n. (物质、精力逐渐地) 消散，分散

distinct [disˈtiŋkt] adj. 明显的；独特的；清楚的；卓越的，不寻常的；确切的；有区别的

distinguish [dɪˈstɪŋgwɪʃ] v. 辨别，分清；辨别是非

distort [disˈtɔːt] vt. 扭曲；使失真；曲解 vi. 扭曲；变形

distortion [disˈtɔːʃən] n. 变形，扭曲，曲解，失真

distracion [dɪˈstrækʃn] n. 注意力分散；消遣；精神错乱

Distributed Control Systems 分布式控制系统

domain [dəuˈmein] n. 领域；域名；产业；地产；疆土；管辖范围

donor [ˈdəʊnə (r)] n. 捐赠者

downstream [ˌdaʊnˈstriːm] adv. 下游地；顺流而下 adj. 下游的

draw [drɔː] vt. & vi. 拉，画，吸引

drawing [ˈdrɔːiŋ] n. 冲压成形；绘画；制图；图画；图样

drawing dies 拉丝模具；拉伸模具

dream of / about 梦见；渴望

drill [drɪl] vt. & vi. 钻 (孔)；打 (眼)；操练；训练 n. 操练；钻头；军事训练

drill press 钻床

drive motor 驱动电动机

ductile [ˈdʌktail, ˈdʌktil] adj. 易教导的；易延展的，柔软

dynamic [daɪˈnæmɪk] adj. 动态的；动力学的；有活力的 n. 动态；动力

eco-driven 生态驱动

edge – gated cavity 侧浇口式型腔

EEPROM (Electrically Erasable Programmable Read – Only Memory) 电可擦写可编程只读存储器

eject [iˈdʒekt] v. 喷射；放逐，驱逐

ejector [ɪˈdʒektə] n. 顶出器

ejector mark 顶出痕迹

elapse [ɪˈlæps] v. 逝去，过去 n. (光阴) 逝去

elastomer [ɪˈlæstəmə(r)] n. 弹性体；合成

橡胶

electrical pulse 电脉冲

electromechanical [iˌlektrəumiˈkænikəl] adj. 电动机械的，机电的

electron beam 电子束

electronic [ɪˌlekˈtrɒnɪk] adj. 电子的；电子操纵的；用电子设备生产的

electronic component 电子元件

element [ˈelimənt] n. 元素，成分，要素；原理；自然环境

eliminate [ɪˈlɪmɪneɪt] vt. 排除，消除；淘汰；除掉；<口>干掉

embed [ɪmˈbed] vt. 使嵌入，使插入；使深深留在脑中

embody [imˈbɔdi] vt. 表现，使具体化；包括，包含

embrace [ɪmˈbreɪs] v. 拥抱；包括；接受

emergency [iˈməːdʒənsi] n. 紧急情况；突发事件 adj. 紧急的，应急的

eminently [ˈeminəntli] adv. 突出地；显著地

emirate [ˈemɪərət] n. 酋长国

enclosure [ɪnˈkləʊʒə(r)] n. 附件；围墙；围绕

encompass [ɪnˈkʌmpəs; en-] vt. 包含，包围，环绕；完成

end-effector [endɪˈfektə] n. 终端执行器

end-to-end 端对端；首尾相连

end-user premises 终端用户处所

energize [ˈenədʒaiz] vt. 激励；使活跃；供给……能量 vi. 活动；用力

energy consumption [物] 能量损耗

energy-efficient 高能效的

envision [ɪnˈvɪʒn] vt. 想象；预想

equation [iˈkweiʒən, -ʃən] n. 方程式，等式；相等；反应式

equipment [ɪˈkwɪpmənt] n. 设备，装备；器材，配件

equivalent [ɪˈkwɪvələnt] adj. 等价的，相等的；同意义的 n. 对等物；[化学] 当量

error [ˈerə(r)] n. 偏差；故障

essay [ˈeseɪ] n. 散文；试图；试验 vt. 尝试；对……做试验

establish [ɪˈstæblɪʃ] vt. 确定，制定；建立，创办，产生；使固定

esthetic(al) [iːsˈθetik] adj. 美（学）的；审美的；雅致的

estimate [ˈestɪmət] n. 估价；估计 v. 估计；估价；评价

etched [ˈetʃid] adj. 被侵蚀的，风化的 v. 蚀刻（etch 的过去分词）

evolution [ˌiːvəˈluːʃn] n. 演变；进化论；进展

evolve [iˈvɒlv] v. 进化，演变；开展，发展，展开

excellence [ˈeks(ə)l(ə)ns] n. 优秀；美德；长处

executable [ˈeksikjuːtəbl] adj. 可执行的；可实行的

execute [ˈeksikjuːt] v. 执行（命令）vt. 处死，处决；履行

executive [ɪgˈzekjətɪv] adj. 行政的；执行的 n. 执行者

exert [ɪgˈzɜːt] vt. 运用；施加

exotic [ɪgˈzɒtɪk] adj. 奇异的，外来的

external [ikˈstəːnl] adj. 外面的，外部的；表面上的；外用的

extruding moulds 挤出模具

fabricate [ˈfæbrɪkeɪt] vt. 制造；伪造；装配

facilitate [fəˈsɪlɪteɪt] vt. 促进，帮助，使……容易

fall into 落入；分成，归入，属于；开始（某事）

fatigue [fəˈtiːg] adj. 疲劳 vt. 使疲劳，使疲乏

feed [fiːd] n. 馈送，供给；进给

feed rate 进给速率

feedback system 反馈系统
femtocell [ˈfemtəʊsel] n. 家庭基站
ferrous [ˈferəs] adj. [化学] 亚铁的; 铁的, 含铁的
fibrous [ˈfaibrəs] adj. 纤维的, 纤维性的, 纤维状的
figure [ˈfɪɡə(r)] n. 图形; 数字; 人物; 体型
filament [ˈfɪləm(ə)nt] n. 灯丝; 细线; 单纤维
fillet [ˈfɪlɪt] n. 圆角 vt. 倒圆角
filter [ˈfɪltə] n. 滤波器, 过滤器; 筛选 vt. 过滤, 渗透 vi. 滤过, 渗入
fix [fɪks] vt. 固定; 准备; 修理; 使牢固 n. 困境; 定位于; 受操纵的事; 应急措施
fixed cycle 固定周期
fixture [ˈfikstʃə] n. 夹具, 固定装置, 设备
Flash ROM program memory 闪存程序存储器
flashover [ˈflæʃˌəʊvə] n. 飞弧; 击穿; 闪络; 跳火
fleet [fliːt] n. 舰队; 船队 adj. 快速的; 敏捷的 vi. 疾驰; 飞逝 vt. 使(时间)飞逝
flexibility [fleksiˈbiliti] n. 弹性, 适应性, 机动性, 挠性
flexible [ˈfleksəbl] adj. 灵活的; 易弯曲的; 柔韧的
flexible machining system 柔性制造系统
flexible manufacturing cell (FMC) 柔性制造单元, 柔性生产单元
flight [flaɪt] n. 飞行, 班机; 几节楼梯; 几圈螺纹 vt. 射击; 使惊飞 vi. 迁徙
floating coordinating system 浮动坐标系统
flux [flʌks] n. 焊剂, 助焊剂
fraction [ˈfrækʃən] n. 小部分, 部分; 稍微; [数] 分数
frequency [ˈfriːkwənsi] n. 频繁性; [数] [物] 频率, 次数; 频率分布
function [ˈfʌŋkʃn] n. 函数 v. 起作用
functional [ˈfʌŋkʃənl] adj. 功能的
functioning [ˈfʌŋkʃənɪŋ] v. 起作用 (function 的现在分词); 正常工作
fundamental [ˌfʌndəˈmentl] adj. 基本的, 根本的, 重要的
fundamentally [ˌfʌndəˈmentəli] adv. 根本地, 从根本上; 基础地
furnace [ˈfəːnis] n. 炉子, 熔炉; [冶] 鼓风炉, 高炉
fuse [fjuːz] n. 熔丝; 导火线 vi. 融合; 熔化 vt. 使融合; 使熔化, 使熔融
Fused Deposition Modelling 熔融沉积成型
gauge [ɡeɪdʒ] v. 检验, 校准 (同 gage) vt. 估计; 测量; 给……定规格 n. 测量的标准
gear [ɡɪə(r)] n. 仪器, 装置; 传动装置, 齿轮
gear blank 齿轮毛坯
gear cutting 齿轮加工
Geiger Counter [核] 盖革计数器 (用于测量放射性)
general public license 通用公共许可证
general - purpose [ˌdʒenrəlˈpɜːpəs] adj. 通用; 通用的
general - purpose microprocessor 通用微处理机
generous [ˈdʒenərəs] adj. 慷慨的, 大方的; 肥沃的; 浓厚的
geometry [dʒiˈɒmətri] n. 几何, 几何学; 几何形状, 几何图形, 几何结构
giant [ˈdʒaɪənt] n. 巨人 adj. 巨大的, 庞大的
Gopher 地鼠程序 (信息检索)
granulator [ˈɡrænjʊleɪtə] n. 碎石机 (成粒器)
graphic [ˈɡræfik] adj. 图形的

grid［grid］n. ［计］网格；格子，栅格；输电网
grim［grɪm］adj. 冷酷的，残忍的；严厉的；阴冷的；可怕的，讨厌的
grind［graɪnd］vt. 磨碎；磨快 vi. 磨碎；折磨 n. 磨；苦工作
grinder［ˈgraində］n. 磨床，研磨机，粉碎机
grinding［graindiŋ］adj. 刺耳的；磨擦的，碾的 v. 磨碎，嚼碎 n. 磨削加工；研磨
groove［gruːv］n. 槽 vt. 开槽于 vi. 形成沟槽
group technology 成组技术
hall［hɔːl］n. 过道；食堂
handoff［ˈhændɒf］n. 切换；传送；手递手传球（美国橄榄球）
handy multimeter 指针式万用表
hardening［ˈhɑːdnɪŋ］v. 淬水；（使）变硬；（使）坚固；（使）变得坚强 n. 硬化，淬火；锻炼
hardware guided vehicles 硬件导向车
hard - wired［hɑrdˈwaiəd］硬接线；硬连线
Harvard［ˈhɑːvəd］n. 哈佛大学；哈佛大学学生
Harvard architecture 哈佛结构（一种将程序指令存储和数据存储分开的存储器结构）
Harvard University 哈佛大学
hazard［ˈhæzəd］n. 危险；冒险的事 vt. 赌运气；冒……的危险，使遭受危险
headstock［ˈhedstɒk］n. 主轴箱；头座；车头
Henry Ford 亨利福特
high - precision［haɪprɪˈsɪʒn］adj. 高精密度；高精确度
high - speed access 高速接入
hinder［ˈhɪndə(r)］v. 阻碍；打扰 adj. 后面的
hinge［hɪndʒ］n. 铰链，折叶；转折点
hint at 暗示
hone［həʊn］vt. 用磨刀石磨；磨孔放大 n. 磨刀石
horizontal［ˌhɒrɪˈzɒntl］adj. 水平的，卧式的；地平线的 n. 水平线；水平面
hot rail 火线
household appliance 家用电器
hub［hʌb］n. （轮）毂；中心；木片
hydraulic［haɪˈdrɔːlɪk］adj. 水力的，液压的，水力学的
hydraulic power 液压动力
hypertext［ɪˈhaɪpətekst］超文本
identification［aiˌdentifiˈkeiʃn］n. 识别，鉴别，辨别，鉴定，核对
IEC 国际电工委员会
illegal［ɪˈliːgl］adj. 非法的；违法的；违反规则的
immerse tube （建筑）沉管
immune［ɪˈmjuːn］adj. 免疫的；免于……的，免除的；不受影响的；无响应的
impedance［imˈpiːdəns］n. ［电］阻抗，全电阻；电阻抗
implant［ɪmˈplɑːnt］vt. 种植；嵌入 vi. 被移植 n. ［医］植入物；植入管
implement［ˈɪmplɪmənt］vt. 实施，执行 n. 工具
in conjunction with 连同；结合；与……协力
in contrast to 与……对比（或对照）；与……相反
in large quantities 大量地；批量地
in line with 符合；与……一致
in principle 原则上
incorporate［ɪnˈkɔːpəreɪt］vt. 组成公司；包含 vi. 包含；吸收；合并 adj. 合并的，一体化的
indexing［ˈɪndeksɪŋ］n. 指数化；［机械学］分度，转位

individual [ˌɪndɪˈvɪdʒuəl] adj. 单独的，个别的，特殊的 n. 个人；个体

individually [ɪndɪˈvɪdjʊ(ə)lɪ] adv. 个别地，单独地

inductance [inˈdʌktəns] n. 电感；感应系数；自感应

inductor [ɪnˈdʌktə] n. 感应器，电感；授职者；感应体；扼流圈

Industrie n. 工业（德语）

industry [ˈɪndəstrɪ] n. 工业；产业；工业界

inferior [ɪnˈfɪəriə(r)] adj. 较低的，次等的，不如的 n. 下级，属下；[印] 下标符号

infinite [ˈinfinət] adj. 无穷的，极大的 n. 无限，无穷大

influence [ˈɪnfluəns] n. 影响；势力 vt. 影响；感染；支配；对……起作用

infrastructure [ˈɪnfrəstrʌktʃə(r)] n. 基础设施；基础建设

ingredient [inˈgriːdiənt] n. 成分，因素，组成部分

injection moulds 注塑模具

innate [ɪˈneɪt] adj. 先天的，固有的，与生俱来的

in-processing gauging 在程检测

in-process gauging 在线检测

input signal 输入信号

insert [inˈsəːt] vt. 插入，嵌入 n. 插入物

inspection [inˈspekʃn] n. 检查，检验，视察，检阅

instantaneous [ˌinstənˈteinjəs] adj. 瞬间的；即时的；猝发的

instantaneously [ˌɪnstənˈteɪnɪəslɪ] adv. 即刻，突如其来地

instate [inˈsteit] vt. 任命

instruct [inˈstrʌkt] vt. 指导；通知；命令；教授

integrate [ˈintigreit] v. 使一体化，集成，成为一体 adj. 整体的，完整的

integration [ˌintiˈgreiʃən] n. 集成；综合；集中

intellectual [ˌɪntəˈlektʃʊəl] adj. 智力的，聪明的，理智的 n. 知识分子

interact [ˌɪntərˈækt] v. 相互作用，互相影响；互动

interacts with 与……相互影响或作用

interconnect [ˌɪntəkəˈnekt] v. 使互相连接，互相联系

interface [ˈintəfeis] n. 接口 v. 接合，连接 vi. 相互作用（或影响）；交流，交谈

interlock [ˌɪntəˈlɒk] n. （计）互锁设备；双螺纹针织品 v. 互锁；联锁

Internet of Things 物联网

interoperability [ˌɪntərˈɒpərəbɪlətɪ] n. 互通性，互操作性

interpolation [ɪnˌtɜːpəˈleɪʃn] n. 插补

interpolation algorithm 插补算法

interpret [inˈtəːprit] v. 解释，说明；口译 vi. 做解释；做口译

intricate [ˈɪntrɪkət] adj. 错综复杂的；难理解的；曲折；盘错

Invar steel 殷瓦钢

invest [ɪnˈvest] vt. 投资；授予 vi. 投资，入股

involvement [inˈvɒlvmənt] n. 参与；卷入

iron [ˈaiən] n. 铁，熨斗；烙铁 adj. 铁的；刚强的

isostatic [ˌaɪsəˈstætɪk] adj. 均衡的

issuance [ˈɪʃjuːəns] n. 发布，发行

jack [dʒæk] n. 千斤顶；插座；插孔 vt. 提醒；用千斤顶顶起

jeweler [ˈdʒuːələ] n. 珠宝商；宝石匠

jig [dʒig] n. 夹具；带锤子的钓钩 v. 抖动；用夹具或钻模等加工

junction [ˈdʒʌŋkʃn] n. 连接，会合处，交叉点

Kazan［kə'zaːn］n. 喀山（伏尔加河中游城市）
KCL（Kirchhoff current laws） 基尔霍夫电流定律
keep down 抑制；控制
keep in mind 记住，谨记
key［kiː］n. 键，电键，开关；楔，销；钥匙
kidney［'kɪdni］n. 肾，肾脏
kinematic［ˌkɪnɪ'mætɪk］adj. 运动学的，运动学上的
knee［niː］n. 升降台；膝盖 vt. 用膝盖碰
knob［nɒb］n. 疙瘩；球形把手；小块；旋钮
know－how［'nəuhau］n. 诀窍；实际知识；专门技能
KVL（Kirchhoff voltage laws） 基尔霍夫电压定律
ladder［'lædə（r）］n. 梯子，阶梯；梯状物；途径 vi. 使（袜子）发生抽丝现象
ladder logic 梯形图
lane［leɪn］n. 小路，小巷；规定的单向行车道；车道
lapping［'læpiŋ］n．研磨；抛光；搭接
latency［'leitənsi］n. 等待时间，延迟，潜伏期
lathe［leɪð］n. 车床，机床 vt. 用车床加工
laureate［'lɒrɪət；'lɔː－］adj. 戴桂冠的；荣誉的 n. 得奖者 vt. 使戴桂冠
layout［'leiˌaut］n. 布局，设计，规划
LCD screen（liquid crystal display screen）液晶屏幕
lead［liːd］n. 铅；导线；引线 vt. &vi. 导致；领导 adj. 带头的；最重要的
LED（Light Emitting Diode）发光二极管
legacy［'legəsi］n. 遗产；遗赠；传统
lengthwise travel 纵向运动
lettering［'letərɪŋ］n. 刻字 v. 用字母写；用印刷体写（letter的ing形式）
lever［'liːvə（r）］n. 杠杆，控制杆，操作杆
liable［'laiəbl］adj. 有责任的，有义务的，应受罚的，有……倾向的
likelihood［'laiklihuːd］n. 可能，可能性
Liquid Crystal Display（LCD） 液晶显示器
liver［'lɪvə(r)］n. 肝脏
LNG abbr. 液化天然气（Liquefied Natural Gas）
locksmith［'lɒksmɪθ］n. 钳工；修锁工，锁匠
logistic［lə'dʒɪstɪk］adj. 后勤学的
longitudinal［ˌlɔndʒi'tjuːdinəl］adj. 长度的，纵向的
loop［luːp］n. 环，圈，弯曲部分 vi. 打环；翻筋斗 vt. 以环连接
lot size 批量
M2M：machine to machine 机器到机器
machinability［məˌʃiːnə'biləti］n. 可加工性，可切削性，机械加工性
machine control unit（MCU）机床控制装置
machined［mə'ʃiːnd］adj. 机械加工的（machine的过去式与过去分词形式）
machinery［mə'ʃiːnəri］n.（总称）机器；组织；机器的运转部分；机械装置
machining［mə'ʃiːnɪŋ］n. 切削，制造，机械加工 v. 加工；开动机器
machining center 加工中心
machining sequence 加工工序
machinist［mə'ʃiːnɪst］n. 机械师
magnetic［mæg'netik］adj. 磁铁的，磁性的；有吸引力的
magnifying glass n. 放大镜
maintenance［'meintənəns］n. 维持；保养；维修；保管；维护
malleable［'mæliəbl］adj. 可锻的，有延展性的，易适应的，可塑的
manganese［'mæŋgəˌniːs，ˌmæŋgə'niːz］n.

［化］锰
manhole［ˈmænhəʊl］n. 检修孔，检查井
manifest［ˈmænifest］v. 证明，表明；显示，出现 n. 货单 adj. 显然的
manipulate［məˈnipjuleit］vt. 操纵；操作；巧妙地处理
manipulator［məˈnɪpjuleɪtə］n. 操纵者；操纵器；翻钢机
manual programming 手动编程
manufacture［ˌmænjuˈfæktʃə(r)］v. 制造；捏造
manufacturing line 生产线
Massachusetts Institute of Technology（MIT）麻省理工学院（MIT）
materialize［məˈtɪəriəlaɪz］vi. 具体化；实质化，实现 vt. 使成真实现
meaningful［ˈmiːniŋful］adj. 有意义的，意味深长的
means［miːnz］n. 方法，手段；收入
meanwhile［ˈmiːnwaɪl］adv. 同时，其间 n. 其间，其时
mechanical［məˈkænɪkl］adj. 机械的，机械学的；呆板的；体力的；手工操作的
mechanical means 机械设备（工具）
mechanism［ˈmekənɪzəm］n. 机制，原理；机械装置
mechatronics［ˌmekəˈtrɔniks］n. 机电一体化
melt flow front 熔融物料前沿
memorandum［meməˈrændəm］n. 备忘录；便笺
memory［ˈmeməri］n. 内存，存储器；记忆，记忆力
mesh［meʃ］n. 网眼；网丝；圈套 vi. 相啮合 vt.［机］啮合；以网捕捉
metallic［məˈtælik；me－］adj. 金属的，含金属的 n. 金属纤维
metallurgy［məˈtælədʒi］n. 冶金学，冶金术
metalworking［ˈmetəlˌwəːkiŋ］n. 金属加工，金属制造
meticulous［məˈtɪkjələs］adj. 一丝不苟的；小心翼翼的
microchip［ˈmaɪkrəutʃɪp］n. 晶片；微型集成电路片
microcomputer［ˈmaɪkrəukəmˌpjuːtə］n. 微电脑；［计］微型计算机
microcontroller［ˌmaikrəukənˈtrəulə］n. 微控制器，微控器，单片机
micropore［ˈmaɪkrəpɔː（r）］n. 微孔
microprocessor［ˌmaɪkrəʊˈprəusesə（r）］n.［计］微处理器
microprocessor－based 基于微处理器的
milliammeter［ˌmiliˈæmitə］n.［电］毫安计
milling［ˈmiliŋ］n. 铣削 v. 碾磨，磨成粉；滚（碾轧）金属
mind－blowing 令人兴奋的
miniaturization［ˌmɪnətʃəraɪˈzeɪʃn］n. 小型化，微型化
minimize［ˈminimaiz］vt. 把……减至最低数量（程度）
minimizing［ˈminimaiziŋ］n. 极小化；求最小参数值
mixture［ˈmɪkstʃə(r)］n. 混合；混合物
mnemonic［nɪˈmɒnɪk］adj. 记忆的；助记的；记忆术的
mnemonics［niːˈmɒniks］n. 记忆术
mobile communication network 移动通信网络
Model A Ford 福特A型汽车
moderate［ˈmɒdərət］adj. 有节制的，适度的，中等的
modifying［ˈmɒdifaiŋ］v. 修改，更改（modify的现在分词）；改变
modulation［ˌmɒdjʊˈleɪʃn］n.［电子］调制；调整
module［ˈmɒdjuːl］n. 模数；模块；组件
mould assembly 模具组件

mould - making 模具制作
mould release agent 脱模剂
molecular [məˈlekjələ(r)] adj. 分子的，由分子组成的
molten [ˈməultən] adj. 融化的，熔融的；铸造的，浇注的；灼热的
monitor [ˈmɒnɪtə(r)] n. 监视器 v. 监视
mounting point 安装位置点
multi - functional [ˌmʌltiˈfʌŋkʃənl] adj. 多功能的
Multi - Jet Modelling 多重喷射成型
multimeter [mʌltiˈmitə] n. [电] 万用表
multiple [ˈmʌltipl] adj. 多样的；许多的；多重的 n. 并联；倍数
navigation [ˌnævɪˈgeɪʃn] n. 航行（学）；航海（术）；海上交通
network [ˈnetwəːk] n. 网络；广播网；网状物 vt. & vi. 将……连接成网络；建立工作关系
neutral [ˈnjuːtrəl] n. 中立者 adj. 中性的
neutral rail 零线
nickname [ˈnɪkneɪm] n. 绰号；昵称 vt. 给……取绰号
nodal [ˈnəudəl] adj. 节的；结的；节似的
non - feedback system 非反馈系统
nonferrous metal 有色金属
nonmetallic [ˈnɒnmɪˌtælɪk] adj. 非金属的 n. 非金属物质
non - volatile [ˈnʌnˌvɒlətail] adj. 非易失的；不挥发的
nourish [ˈnʌrɪʃ] vt. 滋养，施肥于；抚养，教养；使健壮
nozzle [ˈnɔzl] n. 喷嘴；管口；鼻
nuclear island 核岛
numerical [njuːˈmerɪkl] adj. 数字的，用数字表示的，数值的
nursing [ˈnɜːsɪŋ] n. 护理，看护；养育
nut [nʌt] n. 螺母；难对付的人；难解的问题；坚果 vi. 采坚果
obstacle [ˈɒbstəkl] n. 障碍
offshore [ˌɒfˈʃɔː(r)] adj. 离开海岸的；国外的 adv. （指风）向海地，离岸地
open kinematic chain 开放运动链
open - loop [ˌəupənlˈuːp] adj. 开环的
open - loop system 开环系统
operand [ˈɒpərænd] n. 操作数；运算对象
Operational Technology 经营技术；操作工艺
optical [ˈɔptikəl] adj. 光学的，视觉的，有助视力的
optimization [ˌɒptɪmaɪˈzeɪʃən] n. 最佳化，最优化；优选法
optimum [ˈɒptɪməm] n. 最佳效果，最适合条件 adj. 最适宜的；最优化的
organic [ɔːˈgænik] adj. 有机的，组织的，器官的
orient [ˈɔːrient] v. 标定方向；使……向东方；以……为参照 vt. 使适应；确定方向
orientation [ˌɔːrienˈteiʃn] n. 方向，定位，取向
orientation mechanism 定向机构
ornament [ˈɔːnəmənt] n. 装饰；装饰物；教堂用品 vt. 装饰，修饰
orthographic [ˌɔːθəˈgræfik] adj. 正字法的；正交的；投影的
oscillate [ˈɒsɪleɪt] vt. 使动摇；使振荡 vi. 振荡；摆动
oscilloscope [əˈsiləskəup] n. [电子] 示波器；示波镜
outfitting 码头舾装；舾装设备
output response 输出响应
overflow [ˌəʊvəˈfləʊ] v. 泛滥，溢出，充满，洋溢 n. 泛滥，溢值，剩出
overhead guide rail 悬空导轨
oversimplify [ˌəuvəˈsɪmplɪʃaɪ] v. （使）过分地单纯化
overview [ˈəʊvəvjuː] n. 综述；概观

oxidation [ˌɒksɪˈdeɪʃn] n. 氧化
oxide [ˈɔksaid] n. 氧化物，氧化层，氧化合物
oxyacetylene [ˌɔksiəˈsetiliːn, - lin] adj. [化] 氧乙炔的
panel [ˈpænl] n. 控制板；嵌板 vt. 选定（陪审团）；把……分格；把……镶入框架内
parameter [pəˈræmitə] n. [数] 参数；<物><数>参量；限制因素；决定因素
paramount [ˈpærəmaʊnt] adj. 最重要的；至高无上的 n. 最高统治者
parentheses [pəˈrenθisiːz] n. 圆括号（parenthesis 的复数形式）
part drawing 零件图
part/tool transfer vehicles 工件/刀具传输装置
particular [pəˈtikjulə] adj. 特殊的，特别的
parting line/ parting surface 分型线/分型面
pastry [ˈpeɪstri] n. 糕点；油酥糕点；油酥面皮
patience [ˈpeɪʃ(ə)ns] n. 耐性，耐心；忍耐，容忍
pattern [ˈpæt(ə)n] n. 模式；图案；样品
pellet [ˈpelɪt] n. 小球，小子弹
perfection [pəˈfekʃ(ə)n] n. 完善；完美
performance [pəˈfɒːməns] n. 表演；表现
peripheral [pəˈrifərəl] adj. 外围的；外设的；次要的 n. 外围设备
permanent [ˈpəːmənənt] adj. 永久的，永恒的；不变的
persistence [pəˈsɪst(ə)ns] n. 持续；固执；存留；坚持不懈；毅力
phasor [ˈfeɪzə] n. 相位复（数）矢量；相图；彩色信息矢量；相量
photopolymer [ˌfəʊtəʊˈpɒlɪmə] n. 光聚合物，光敏聚合物；感光性树脂
physical [ˈfizikl] adj. 身体的；物质的，物理学的 n. 体格检查
physical properties 物理性质
pitch [pɪtʃ] n. 场地；最高点 vt. 扔，投
plague [pleɪg] n. 瘟疫；灾害；折磨 vt. 使染瘟疫；使痛苦，造成麻烦
plane [plein] n. 平面；飞机 vi. 刨 vt. 刨平，用刨子刨；掠过水面
planer [ˈpleɪnə(r)] n. 龙门刨床，刨床，刨工；刨路机；水榆；平刨
plant [plɑːnt] n. 被控对象
plaster [ˈplɑːstə(r)] n. 灰泥，涂墙泥；石膏；膏药
plastic grinder 塑料磨粉机
plug [plʌg] n. 栓；插头；塞子 vi. 塞住；用插头接电源
pneumatic [njuːˈmætɪk] adj. 气动的；有气胎的；充气的 n. 气胎
point - to - point 点对点
polymers [ˈpɔliməs] n. [高分子] 聚合物
polyphase [ˈpɒlɪfeɪz] adj. 多相的
pornography [pɒːˈnɒgrəːfɪ] n. 色情文学，色情描写；色情
porous [ˈpɔːrəs] adj. 多孔的
portion [ˈpɔːʃn] vt. 把……分成份额；分配 n. 一部分；一份遗产
position mechanism 定位机构
positive feedback 正反馈
positive phase sequence 正相序
powder [ˈpaʊdə] n. 粉；粉末 vt. 使成粉末；撒粉 vi. 搽粉；变成粉末
power [pauə] n. 功率；力量；能力；政权 vt. 激励 v. 快速前进 adj. 权力的；机械能的
power supply 电源
powerhouse [ˈpauəhaʊs] n. 精力旺盛的人；发电所，动力室；强国
precaution [prɪˈkɔːʃn] n. 预防，警惕；预防措施
precise [prɪˈsaɪs] adj. 精确的；正规的；

精密的
precision [prɪˈsɪʒ(ə)n] n. 精度，精密度；精确度
precision grinding 精磨
predominant [prɪˈdɒmɪnənt] adj. 占优势的，主要的
preferential [ˌprefəˈrenʃl] adj. 优先的；特惠的
presentation [ˌprezənˈteiʃən] n. 介绍，陈述
pressure [ˈpreʃə] n. 压力，压强 vi. 施加压力；迫使；使（机舱等）增压
pressurized [ˈpreʃəraɪzd] adj. 加压的，受压的
preventive [prɪˈventɪv] n. 预防，防止；预防措施；预防药 adj. 预防的，防止的
preventive maintenance 预防性维修
price – to – performance ratio 性价比
prior [ˈpraiə] adj. 在先的，在前的，优先的
prismatic [prɪzˈmætɪk] adj. 棱镜的
privacy [ˈprɪvəsi] n. 隐私，秘密，私事
proactively [ˌprəʊˈæktɪvlɪ] adv. 主动地
process [prəˈses; (for n.) ˈprəuses] vt. 处理；加工 n. 过程，进行；方法
process cycle 工艺周期
product [ˈprɒdʌkt] n. 产品；结果；乘积；作品
product tagged 加标签的产品
productivity [ˌprɒdʌkˈtivəti] n. 生产率，生产力
proficiency [prəˈfɪʃ(ə)nsɪ] n. 精通，熟练
profile [ˈprəufail] n. 侧面，轮廓，外形，剖面
profit [ˈprɒfit] n. 利润，收益，盈利
program coordinate 编程坐标
prohibited [prəˈhibitid] v. 禁止，阻止 adj. 被禁止的
prolonged [prəˈlɒŋd] adj. 延长的，拖延的，持续很久的
prominently [ˈprɔminəntli] adv. 显著地
propaganda [ˌprɒpəˈgændə] n. 宣传，传递；宣传运动
proportionate [prəˈpɔːʃənət] adj. 成比例的
proprietary [prəˈpraɪətri] n. 所有权；所有人 adj. 所有的；专利的
protective [prəˈtektiv] adj. 保护的，防护的
protocol [ˈprəʊtəkɒl] n. 礼仪；（数据传递的）协议 vt. 把……写入议定书
prototype [ˈprəʊtətaɪp] n. 原型；标准，模范
prototype work 原始模型工件
provenance [ˈprɒvənəns] n. 出处，起源
psychological [ˌsaɪkəˈlɒdʒɪkl] adj. 心理的；心理学的；精神上的
publish [ˈpʌblɪʃ] vt. 出版；发表；发行
pulley [ˈpuli] n. 滑轮，带轮 v. 用滑轮升起，用滑车推动；给……装滑车
pullout [pulaut] n. 拔；拉；撤退
pulse [pʌls] n. 脉冲 vt. 使跳动 vi. 跳动，脉跳
punch [pʌntʃ] v. 打孔；用拳猛击
purchase [ˈpɜːtʃəs] vt. 购买
quadricycle [ˈkwɒdrɪsaɪkl] adj. 四轮的 n. 脚踏四轮车
quantity [ˈkwɒntəti] n. 量，数量；大量；总量
quench [kwentʃ] vt. 解（渴）；终止；（用水）扑灭；将（热物体）放入水中急速冷却
quiescent [kwiˈesnt] adj. 不活动的，静态的，休眠的
quotation [kwə(ʊ)ˈteɪʃ(ə)n] n. [贸易] 报价单；引用语；引证
radian [ˈreidiən] n. 【数学】弧度
Radiation Sensor Board 辐射传感板
radio frequency identification 射频识别
radius [ˈreidjəs] n. 半径（距离），半径范

围；桡骨

raised［reizd］adj. 凸起的，阳文的，浮雕的；有凸起的花纹（或图案）

RAM（random access memory）随机存取存储器

Rapid Prototyping（RP）快速成型

ratio［ˈreiʃiəu，－ʃəu］n. 比率，比例 vt. 求出比值，除

ream［riːm］vt. 铰削；扩大 n. 令（纸张的计数单位）；大量

reaming［ˈriːmɪŋ］n. 铰孔；铰刀 v. 令（纸张的计数单位）

reciprocate［riˈsiprəkeit］vt. 互换；报答 vi. 互给；酬答；往复运动

recrystallization［riːˌkristəlaiˈzeiʃən］n. 再结晶

red－hot［redhɔt］adj. 赤热的；激烈的，恼怒的；近期的，新的

refer to as　称为，叫作；当作；参考作为；所说的；提到的

rehearse［rɪˈhɜːs］vt. 排练；预演 vi. 排练；演习

relative motion　相对运动

relatively［ˈrelətɪvlɪ］adv. 相关地

relay［ˈriːleɪ］n. 接替人员；接力赛；继电器 vt. 转播，传达；使接替；分程传递

reliability［rɪˌlaɪəˈbɪlətɪ］n. 可信度；可靠性，可靠度

relieve［riˈliːv］v. 解除，减轻，使不单调乏味

repetitive［rɪˈpetətɪv］adj. 重复的，啰嗦的

replicate［ˈreplɪkeɪt］vt. 复制，复写；［生］复制

represent［reprɪˈzent］v. 表现，体现，作为……的代表；描绘；回忆

resemblance［riˈzembləns］n. 相似；相似之处；相似物；肖像

reservoir［ˈrezəvwɑː（r）］n. 水库，蓄水池；液压油箱

resistive［rɪˈzɪstɪv］adj. 抗［耐、防］……的；电阻的；有抵抗力的

resistor［riˈzistə］n. 电阻器

restrict［rɪˈstrɪkt］vt. 限制，限定，约束

result in　导致，结果是

retrieve［rɪˈtriːv］v. 重新获得，找回 n. 取回；［计］检索

retrofit［ˈretrəufit］n. 改型（装，进）；（式样）翻新　vt. 给机器设备装配（新部件）

reversal［riˈvəːsəl］n. 逆转；［摄］反转；［法］撤销

revolution［ˌrevəˈluːʃn］n. 革命；旋转；运行

revolve［rɪˈvɒlv］vi. & n. 旋转；循环出现 vt. 使……旋转；使……循环；

RFID abbr. 无线射频识别（radio frequency identification devices）

rib［rɪb］n. 肋骨，排骨，肋状物 vt. 戏弄；装肋于

rigid［ˈrɪdʒɪd］adj. 严格的；僵硬的；（规则、方法等）死板的

rise［raɪz］vi.&n. 上升，增强，起立，高耸

rivet［ˈrivit］n. 铆钉　vt. 铆，铆接；把……固定住

robot［ˈrəubɒt］n. 机器人；遥控装置

roll［rəʊl］v. 滚动；卷 n. 滚；卷

rolls［rɒlz］n. 名册（roll 的名词复数）；滚翻 v.（使）转动；卷；把……卷成筒状

ROM（read only memory）只读存储器

rotate［rəʊˈteɪt］n. 使旋转或转动

rotating spindle　旋转主轴

rotation［rəuˈteiʃn］n. 旋转，转动；循环

rough casting　铸造毛坯

rung［rʌŋ］n. 阶梯，梯级；（地位上升）一级 v. 把……圈起来（ring 的过去分词）

sacrifice [ˈsækrifais] n. 牺牲，牺牲品，损失 v. 牺牲，献出
satisfactory [sætisˈfæktəri] adj. 令人满意的，满意的，符合要求的
scale [skeil] n. 刻度；比例；天平 vi. 攀登；衡量 vt. 测量；依比例改变
scaled [skeild] adj. 按比例缩放的
scenario [səˈnɑːrɪəʊ] n. （行动）方案；剧情概要；分镜头剧本
schematic [skiːˈmætɪk] adj. 纲要的；示意的；图解的；有章法的 n. 电路原理图
sci – fi blockbuster 科幻大片（sci – fi 是 Science – fiction 的缩写形式）
scrap [skræp] n. 废料
screw [skruː] vt. & vi. 转动；旋，拧；压榨，强迫 n. 螺钉；吝啬鬼
seal [siːl] n. [机] 密封垫；焊接；封蜡；封印
seaming [siːmɪŋ] n. 接缝缝合（卷边接合）
seamless [ˈsiːmləs] adj. 无缝的；不停顿的；无漏洞的
Seattle [siˈætl] n. 西雅图（美国一港市）
security [sɪˈkjuərətɪ] n. 安全；保证，担保
seep [siːp] v. 渗出，渗漏 n. 渗漏；小泉；水坑
selective [sɪˈlektɪv] adj. 精心选择的，不普遍的；淘汰的
Selective Laser Sintering 选择性激光烧结
semiautomatic [ˌsemiɔːtəˈmætik] adj. 半自动的
semiconductor [ˌsemɪkənˈdʌktə(r)] n. [电子] [物] 半导体
sensing system 传感系统
sensor [ˈsensə] n. 传感器，敏感元件
sensor network 传感网
sequence [ˈsiːkwəns] n. 次序，顺序，序列
Sequential Function Charts 顺序功能图
servo control system 伺服控制系统
servomotor [ˈsəːvəʊˌmautə] n. 伺服电动机
set [set] n. 设置
set – point [ˌsetpˈɔɪnt] n. 设定值
set – top boxes 机顶盒（set – top box 的名词复数）
setup [ˈsetʌp] n. 安装；设备；机构
setup sheet 设置表
shaft [ʃɑːft] n. 柱身；连杆；传动轴，旋转轴；轴 vt. 利用，在……上装杆
shaper [ˈʃeɪpə] n. 牛头刨床；造型者
shaping [ˈʃeɪpɪŋ] n. 成形，造型，塑造
shear [ʃɪə(r)] vt. 剪羊毛；切断；剪切；剥夺 n. 大剪刀；剪下的东西；剪切
show up 露面；露出；揭露
shrink [ʃrɪŋk] vi. 收缩，退缩，萎缩；畏惧，害怕 vt. 使收缩
shrinkage [ˈʃrɪŋkɪdʒ] n. 收缩；减少；损失，损耗
shuttle [ˈʃʌtl] n. 梭子；航天飞机 vt. & vi. 穿梭般来回移动
silicon [ˈsɪlɪk(ə)n] n. 硅，硅元素
simplified [ˈsimplifaid] v. 使简单（简明）
simplify [ˈsimplifai] vt. 简化；使单纯；使简易
simulation [ˌsimjuˈleiʃn] n. 仿真
single – point cutting tool 单刃刀具
single – spindle [ˌsɪŋglspˈɪndl] adj. 单主轴的
sintered material 烧结材料
sintering [ˈsɪntərɪŋ] n. 烧结 v. 烧结；使熔结
sketch [sketʃ] n. 素描；略图，梗概
slide [slaɪd] v. 滑落；下跌 n. 幻灯片
slope [sləup] n. 斜坡，坡度；山坡 vi. 倾斜；悄悄地走，溜
small – lot 小批量
socioeconomic [səusɪəuˌiːkəˈnɔmik] adj. 社会经济学的
soften [ˈsɑfn, ˈsɔːfn] v. 使软化，使柔和，

变软弱
solder [ˈsɔldə] vi. 焊接 vt. 焊接；使连接在一起 n. 焊料；接合物
soldering iron 焊铁，烙铁
solenoid [ˈsəulənɒɪd] n. 螺线管
solidify [səˈlɪdɪfaɪ] v. 变固体，凝固；变坚定；使团结
sophisticated [səˈfɪstɪkeɪtɪd] adj. 复杂的；精致的；久经世故的
spark [spɑːk] n. 火花；闪光 v. 发火花
special – purpose 专用的，特殊用途的
special – variety [ˌspeʃlvəˈraɪəti] adj. 特殊品种
specification [ˌspesɪfɪˈkeɪʃn] n. 规格；详述
spectral [ˈspektrəl] adj. [光] 光谱的
spectrum [ˈspektrəm] n. 范围；光谱
speed sensor 速度传感器
sphere [sfɪə] n. 球（体）；（兴趣或活动的）范围；势力范围
spindle [ˈspindl] n. 纺锤，纱锭；轴
spray [spreɪ] v. 喷雾，喷射，扫射 n. 喷雾，喷雾器，水沫
sprue [spruː] n. 主流道
squeeze [skwiːz] v. 挤（压），压榨，勒索，紧握
stack [stæk] vt. & vi. 堆成堆；堆起来或覆盖住 n. 堆，堆叠；存储栈，堆栈
stain [steɪn] vt. 玷污，给……着色 vi. 污染，玷污 n. 污点，瑕疵
stakeholder [ˈsteɪkhəuldə(r)] n. 股东；利益相关者
stamping [ˈstæmpiŋ] n. 冲压，冲击制品
standardization [ˌstændədaiˈzeiʃn] n. 标准化，标定，规范化
standardize [ˈstændədaɪz] vt. 标准化，统一标准；标定，校准
standardized product 【工程设计】标准化产品
stand – off 托脚
starch [stɑːtʃ] n. 淀粉
state – of – the – art 最先进的；已经发展的；达到最高水准的
stearin [ˈstɪərɪn] n. 硬脂，硬脂酸甘油酯；[化] 硬脂精
stem [stem] n. 干；茎；血统 vt. 阻止 vi. 阻止；起源于某事物；逆行
sterling [ˈstəːliŋ] adj. 英币的；纯正的；纯银制的 n. 英国货币；标准纯银
Steve Ballmer 史蒂夫·鲍尔默
stink [stɪŋk] vi. 散发出恶臭；招人厌恶；糟透 n. 恶臭；难闻的气味
strip [strɪp] n. 带，条状 vt. 剥夺，剥去
stripper ring 脱膜圈
structural [ˈstrʌktʃərəl] adj. 结构的，构造的；建筑的，建筑用的
stunning [ˈstʌnɪŋ] adj. 令人晕倒（吃惊）的；出色的；令人震惊的
stuff into 把……塞入
subassembly [ˌsʌbəˈsemblɪ] n. 部件，组件
submersible [səbˈmɜːsɪb(ə)l] adj. 能潜水的；能沉入水中的
subscriber – based 基于用户
subset [ˈsʌbset] n. 子集
substantial [səbˈstænʃl] adj. 大量的；坚固的；实质的；可观的 n. 本质；重要部分
substitute [ˈsʌbstɪtjuːt] vt. & vi. 代替，替换，代用 n. 代替者；替补
subsume [səbˈsjuːm] v. 包容，包含；归入
subtract [səbˈtrækt] v. 减
subtraction [səbˈtrækʃn] n. 减少；减法，差集
subtractive [səbˈtræktɪv] adj. 减去的；负的；有负号的
successive [səkˈsesɪv] adj. 连续的，继承的，依次的，接替的
sufficient [səˈfiʃnt] adj. 足够的，充足的

sulfide［ˈsʌlfaɪd］n. 硫化物，含硫系列，硫醚
superimpose［ˌsuːpərimˈpəuz］vt. 添加，附加
supervision［ˌsuːpəˈviʒn］n. 监督，管理
switch［swɪtʃ］n. 开关；转换，转换器 vt. & vi. 转变，改变；转换；关闭电流；鞭打
symbolic［simˈbɒlik］adj. 象征的；符号的
symbolically［simˈbɒlikəli］adv. 象征性地；用符号表示
table［ˈteibl］n. 工作台，台面 vt. 嵌合；搁置；制表
tag［tæg］n. 标签 v. 尾随；起浑名
tank［tæŋk］n.（盛液体、气体的）大容器，槽，罐，坦克
tap［tæp］vt. 轻敲，轻拍；攻螺纹 vi. 轻拍，轻击 n. 水龙头
taper［ˈteɪpə(r)］n. 蜡芯；尖锥形；渐弱 v. 逐渐变小，逐渐消失
tarnish［ˈtɑːnɪʃ］n. 污点；无光泽 vt. &vi. 玷污；使……失去光泽
team spirit 团队精神
tedious［ˈtiːdiəs］adj. 单调沉闷的；冗长乏味的；令人生厌的
telepresence［ˈteliprezns］n. 远程监控
telnet［ˈtelnet］远程登录
temper［ˈtempə(r)］n.（钢等）回火；脾气，性情；倾向 v. 调和，使回火
temperature uniformity 温度均匀性
tempering［ˈtempərɪŋ］n. 回火
tensile strength 抗拉强度
terminated［ˈtəːmineitid］v. 使终结；解雇（terminate 的过去式）
textured［ˈtekstʃəd］adj. 起纹理的，构造成的；使具有某种结构的
the best－seller lists 畅销书目录
the circuit element 电路元件
the Electronic Industries Association（EIA）（美国）电子工业协会
the electronic industry 电子工业
the MITS Altair MITS Altair 电脑
The Road Ahead 未来之路
the video equipment 视频设备
thermodynamic［ˌθɜːməʊdaɪˈnæmɪk］adj. 热力的；热力学的
thermoforming［θəməˈfɔːmɪŋ］n. 热成型；热压成型
thermoplastic［ˌθɜːməʊˈplæstɪk］adj. 热塑性的 n.［塑料］热塑性塑料
thermoset［ˈθɜːməset］n. 热凝，热固性；热变定法；热熔塑料
threaded shaft 丝杠
Three－Dimensional Printing 三维打印
tight［tait］adj. 牢固的；绷紧的；密集的，紧凑的 adv. 紧紧地；牢固地
timer［ˈtaɪmə(r)］n. 定时器；计时器；点火调节装置；跑表，时计
time－shifted 时移
tissue［ˈtɪʃuː］n. 组织；纸巾
to convert... into 把……转换成
toggle［ˈtɒg(ə)l］vt. 拴牢，系紧 n. 开关，触发器；拴扣；［船］套索钉
tolerance［ˈtɒlərəns］n. 宽容，容忍；忍受性；（配合）公差，容限
tonnage［ˈtʌnɪdʒ］n. 连接，肘节
tool list 刀具清单
tool magazine 刀库
tool monitoring system 刀具监测系统
tooling industry 刀具业
torch［tɔːtʃ］n. 火炬，火把；［机］气炬；吹管
torque［tɔːk］n. 转矩；力矩；扭矩
trace out 描绘出
traffic［ˈtræfɪk］n. 交通 v. 交易，用……做交换
trailer［ˈtreɪlə(r)］n. 拖车；追踪者；（电影或电视的）预告片

transducer [trænz'djuːsə] n. 传感器；变换器；换能器
transfer [træns'fəː(r)] vt. 使转移；使调动 vi. 转让；转学
transform [træns'fɔːm] vt. 改变，使变形；转换 vi. 变换，改变；转化
transformation [ˌtrænsfə'meɪʃn] n. 变化；<核>转换；<语>转换
transistor [træn'zɪstə] n. 晶体管，半导体管
transponder [træns'pɒndə] n. 应答器；转调器，变换器
transverse [trænz'vəːs] adj. 横向的，横断的
trendsetter ['trendsetə(r)] n. （在服装式样等方面）创新风的人
trial and error 反复试验
tricky ['trɪkɪ] adj. 狡猾的；复杂的；棘手的；难处理的；需要技巧的；（形势、工作等）复杂的；机警的；微妙的
trigger ['trɪgə(r)] vt. 引发，引起；触发 n. 扳机
trim [trɪm] v. 修剪，削减；装饰；调整帆以适应风向
troubleshoot ['trʌblʃuːt] vt. 检修，排除故障 vi. 充当故障检修员
tsunami [tsuː'naːmi] n. 海啸
tubing ['tjuːbiŋ] n. 管子；装管 v. 把……装管；使成管状
tunnel ['tʌnl] n. 隧道；坑道
turning ['tɜːnɪŋ] n. 车削；旋转；转向；转弯处
turret head 转塔头
twist [twɪst] n. 扭曲 vt. 捻，拧，扭伤 vi. 扭动，弯曲
twist drill 麻花钻
two separate steel tanks 分别两个钢制的储气瓶
ultra ['ʌltrə] adj. 极端的，过分的
ultra large - scale integration (ULSI) 超大规模集成电路
undercut [ˌʌndə'kʌt] vt. 从下部切开 n. 底切；砍口；切球凸雕；浮雕
undergo [ʌndə'gəu] vt. 经历，经受；忍受
unicast ['juːnɪkaːst] n. 单播
uniformity [ˌjuːnɪ'fɔːmətɪ] n. 均匀性，一致性
urban ['ɜːbən] adj. 都市的；具有城市或城市生活特点的；市内
urbanization [ˌɜːbənaɪ'zeɪʃn] n. 都市化；文雅化
vacuum ['vækjuəm] n. 真空
valve [vælv] n. 电子管，真空管，阀门 vt. 装阀于；以活门调节
variable ['veəriəbl] n. 变量
variable type programming 变量编程
variations [ˌveəriː'eɪʃənz] n. 变更；变化
vaseline ['væsəliːn] n. 凡士林 vt. 在……上涂凡士林
vector ['vektə] n. 矢量；向量；带菌者；航线 vt. 用无线电导航
vehicle ['viːəkl] n. 搬运装置，传送装置；车辆
velocity [və'lɒsəti] n. 速率
venting ['ventɪŋ] n. 通风，排气
verify ['verifai] v. 校验
version ['vəːʃən] n. 版本；译文；倒转术；
vertebral ['vəːtibrəl] adj. 椎骨的；脊椎的
vertical ['vəːtikəl] adj. 垂直的，直立的，竖式的 n. 垂直线；垂直面［建］竖杆；垂直位置
vertical milling machine 立式铣床
vibration [vaɪ'breɪʃn] n. 摆动；振动；感受；（偏离平衡位置的）一次性往复振动
video ['vɪdɪəʊ] n. ［电子］视频 adj. 录像的，视频的
video on demand (VOD) 视频点播（VOD）
viscous ['viskəs] adj. 黏性的

vision [ˈvɪʒn] n. 视力；眼力；想象力 vt. 想象

visual [ˈvɪʒʊəl] adj. 视觉的，视力的；栩栩如生的

visualization [ˌvɪzjʊəlaɪˈzeɪʃən] n. 形象化；清楚地呈现

vocational[və(ʊ)ˈkeɪʃ(ə)n(ə)l] adj. 职业的，行业的

voltage [ˈvəultidʒ] n. [电] 电压；伏特数

weld [weld] v. & n. 焊接，熔焊

weld line 熔接痕

welder [ˈweldə] n. 焊接工

well - earned rest 应得的休息

wheel [wiːl] n. 车轮；方向盘 vt. 转动；给……装轮子 vi. 旋转

whitish [ˈwaɪtɪʃ] adj. 带白色的；发白的

winding [ˈwaɪndɪŋ] n. 线圈；弯曲；缠绕物 adj. 弯曲的，蜿蜒的；卷绕的

wire guided vehicle 导线导向车

with the exception of 除……之外

work volume 加工区域

work - hardening range 淬火温度区

work - holding 工件夹持

World Skills Competition 世界技能竞赛

worldwide [ˈwɜːldwaɪd] adj. 全世界的，世界范围的 adv. 在世界各地

wrist [rɪst] n. 腕，手腕；腕关节

wrought [rɔːt] adj. 锻造的；加工的；精细的

zinc [zɪŋk] n. 锌 vt. 镀锌于……；在……上镀锌

References（参 考 文 献）

[1] 庄朝蓉．电子信息专业英语［M］．2 版．北京：北京邮电大学出版社，2013.
[2] 呼枫．机电专业英语［M］．北京：人民邮电出版社，2016.
[3] 朱一纶．电子技术专业英语［M］．4 版．北京：电子工业出版社，2015.
[4] 盛楠．机械英语综合教程［M］．北京：人民邮电出版社，2016.
[5] 王兆奇，刘向红．数控专业英语［M］．2 版．北京：机械工业出版社，2012.
[6] 刘小芹，刘骋．电子与通信技术专业英语［M］．4 版．北京：人民邮电出版社，2014.
[7] 徐存善．机电专业英语［M］．2 版．北京：机械工业出版社，2012.
[8] 徐存善．电子与通信工程专业英语［M］．2 版．北京：机械工业出版社，2017.
[9] 徐存善．自动化专业英语［M］．2 版．北京：机械工业出版社，2017
[10] 谷学静，王志良，郭宇承．物联网专业英语［M］．北京：机械工业出版社，2015.
[11] 胡成伟．通信技术专业英语［M］．2 版．北京：人民邮电出版社，2015.
[12] 杨梅，黄红辉．模具专业英语［M］．上海：上海科学技术出版社，2015.
[13] 赵中颖．职业英语——计算机类［M］．北京：机械工业出版社，2011.
[14] 赵杰．职业英语——机电类［M］．北京：机械工业出版社，2011.
[15] 黄星，王晓平．电气自动化专业英语［M］．2 版．北京：人民邮电出版社，2015.